Edited by Rolf Schäfer and Peter C. Schmidt

Methods in Physical Chemistry

Related Titles

Scott, R. A. (ed.)

**Applications of Physical
Methods to Inorganic and
Bioinorganic Chemistry**

2007
Hardcover
ISBN: 978-0-470-03217-6

Hammes, G. G.

**Physical Chemistry for the
Biological Sciences**

2007
Hardcover

Edited by Rolf Schäfer and Peter C. Schmidt

Methods in Physical Chemistry

Volume 2

SJS

WILEY-VCH Verlag GmbH & Co. KGaA

The Editors

Prof. Rolf Schäfer
Technische Universität Darmstadt
Eduard-Zintl-Institut für Anorganische
und Physikalische Chemie
Petersenstr. 20
64287 Darmstadt

Prof. Peter C. Schmidt
Technische Universität Darmstadt
Eduard-Zintl-Institut für Anorganische
und Physikalische Chemie
Petersenstr. 20
64287 Darmstadt

All books published by **Wiley-VCH** are
carefully produced. Nevertheless, authors,
editors, and publisher do not warrant the
information contained in these books,
including this book, to be free of errors.
Readers are advised to keep in mind that
statements, data, illustrations, procedural
details or other items may inadvertently be
inaccurate.

Library of Congress Card No.: applied for

**British Library Cataloguing-in-Publication
Data**
A catalogue record for this book is available
from the British Library.

**Bibliographic information published by the
Deutsche Nationalbibliothek**
The Deutsche Nationalbibliothek
lists this publication in the Deutsche
Nationalbibliografie; detailed bibliographic
data are available on the Internet at
<http://dnb.d-nb.de>.

© 2012 Wiley-VCH Verlag & Co. KGaA,
Boschstr. 12, 69469 Weinheim, Germany

Cover Design Adam-Design, Weinheim
Typesetting Laserwords Private Limited,
Chennai, India
Printing and Binding Strauss GmbH, Mörlenbach

Printed on acid-free paper

Print ISBN: 978-3-527-32745-4
ePDF ISBN: 978-3-527-63685-3
ePub ISBN: 978-3-527-63684-6
mobi ISBN: 978-3-527-63686-0

Foreword

"Real progress requires either a new idea or a new method." These were the words of my mentor Heinz Gerischer while I was a graduate student, and they characterize the enormous development of physical chemistry since its founding more than a hundred years ago. While at first mainly offering a bridge between physics and chemistry, this discipline has evolved into a cornerstone of all natural sciences, chiefly through the advent of novel methods. Several years ago, the Bunsen-Gesellschaft für Physikalische Chemie recognized that there was a need among graduate students and younger researchers (and others) for a bridge between traditional textbooks and review articles and, therefore, initiated a series of articles on modern methods of physical chemistry in its journal Bunsen-Magazin. This series was started by Peter C. Schmidt, then editor of the Bunsen-Magazin, and continued by his successor Rolf Schäfer. These contributions sometimes even became part of advanced lecture courses and hence the idea to publish the whole collection emerged. For this purpose the individual contributions were shaped into a uniform structure, leading to the present handbook. It is hoped that it will find widespread interest as a valuable guide to the methods of physical chemistry.

Gerhard Ertl

Contents to Volume 1

Contents to Volume 2

List of Contributors

Barbara Albert
Technische Universität
Darmstadt
Eduard-Zintl-Institute of
Inorganic and Physical Chemistry
Petersenstr. 18
64287 Darmstadt
Germany

Jessica Balbo
Heidelberg University
Cellnetworks Cluster &
Department of Physical
Chemistry
Im Neuenheimer Feld 267
69120 Heidelberg
Germany

Matthias Bauer
Technische Universität
Kaiserslautern
Fachbereich Chemie
Erwin-Schrödinger-Strasse 54
67663 Kaiserslautern
Germany

Klaus-Dieter Becker
Technische Universitaet
Braunschweig
Institute of Physical and
Theoretical Chemistry
Hans-Sommer-Strasse 10
38106 Braunschweig
Germany

Rüdiger Berger
Max-Planck-Institut für
Polymerforschung
Ackermannweg 10
55128 Mainz
Germany

Helmut Bertagnolli
Stuttgart University
Institut für Physikalische Chemie
Pfaffenwaldring 55
70569 Stuttgart
Germany

Gunther Brunklaus
University of Münster
Institute of Physical Chemistry
Corrensstrasse 30
48149 Münster
Germany

Roger A. De Souza
RWTH Aachen University
Institut für Physikalische Chemie
Landoltweg 2
52056 Aachen
Germany

Volker Deckert
Friedrich-Schiller
University of Jena
Institute for Physical Chemistry
Helmholtzweg 4
07743 Jena
Germany

and

IPHT-Institute for Photonic
Technology
Albert-Einstein-Str. 9
07745 Jena
Germany

Klaus-Peter Dinse
Free University Berlin
Physics Department
Arnimalle 15
14195 Berlin
Germany

Karl Doblhofer
Fritz-Haber-Institut der
Max-Planck-Gesellschaft
Faradayweg 4-6
14195 Berlin
Germany

Alfons Drochner
Technische Universität
Darmstadt
Ernst-Berl-Institut für Technische
und Makromolekulare Chemie
Petersenstr. 20
64287 Darmstadt
Germany

Hellmut Eckert
University of Münster
Institute of Physical Chemistry
Corrensstrasse 30
48149 Münster
Germany

Martin Engel
Merck KGaA
Frankfurter Str. 250
64293 Darmstadt
Germany

Frank Filsinger
Fritz-Haber-Institut der
Max-Planck-Gesellschaft
Department of Molecular Physics
Faradayweg 4-6
14195 Berlin
Germany

Klaus Funke
University of Muenster
Institute of Physical Chemistry
Corrensstrasse 30
48149 Muenster
Germany

Hans Georg Breunig
JenLab GmbH
Science Park 2
Campus D1.2
66123 Saarbrücken
Germany

Martin Gruebele
University of Illinois
Department of Chemistry
Department of Physics and
Center for Biophysics and
Computational Biology
Urbana, IL 61801
USA

Jochen Gutmann
Universität Duisburg-Essen
Fakultät für Chemie
Universitätsstraße 2
45141 Essen
Germany

Sebastian Günther
Technische Universität München
Chemie Department
Physikalische Chemie mit
Schwerpunkt Katalyse
Lichtenbergstr. 4
85748 Garching
Germany

Philipp Gütlich
Johannes-Gutenberg University
Mainz
Institute of Inorganic and
Analytical Chemistry
Staudingerweg 9
55128 Mainz
Germany

Martina Havenith-Newen
Ruhr University Bochum
Department of Physical
Chemistry II
Universitätsstraße 150
44801 Bochum
Germany

Georg Held
University of Reading
Department of Chemistry
Whiteknights
Reading RG6 6AD
United Kingdom

Dirk-Peter Herten
Heidelberg University
Cellnetworks Cluster &
Department of Physical
Chemistry
Im Neuenheimer Feld 267
69120 Heidelberg
Germany

Kathrin Hofmann
Technische Universitaet
Darmstadt
Eduard-Zintl-Institute of
Inorganic and Physical Chemistry
Petersenstrasse 18
64287 Darmstadt
Germany

Wolfram Jaegermann
Technische Universität
Darmstadt
Fachbereich Material-und
Geowissenschaften
Petersenstrasse 23
64287 Darmstadt
Germany

Jürgen Janek
Justus-Liebig-Universität Gießen
Institute of Physical Chemistry
Heinrich-Buff-Ring 58
35392 Gießen
Germany

Gunnar Jeschke
ETH Höngerberg
Department of Physical
Chemistry
Wolfgang-Pauli-Str. 10
8093 Zürich
Switzerland

Gregor Jung
Saarland University
Biophysical Chemistry
Gebäude B2 2
Postfach 15 11 50
66123 Saarbrücken
Germany

Hans-Joachim Kleebe
Technische Universität
Darmstadt
Institut für Angewandte
Geowissenschaften
Schnittspahnstr. 9
64287 Darmstadt
Germany

Andreas Klein
Technische Universität
Darmstadt
Fachbereich Material-und
Geowissenschaften
Petersenstrasse 23
64287 Darmstadt
Germany

Dieter M. Kolb[†]
Universität Ulm
Institut für Elektrochemie
Albert-Einstein-Allee 47
89069 Ulm
Germany

Jochen Küpper
Fritz-Haber-Institut der
Max-Planck-Gesellschaft
Department of Molecular Physics
Faradayweg 46
14195 Berlin
Germany

and

Center for Free-Electron Laser
Science
Deutsches Elektronen-
Synchrotron DESY
Notkestrasse 85
22607 Hamburg
Germany

and

University of Hamburg
Department of Physics
Luruper Chaussee 149
22761 Hamburg
Germany

Stefan Lauterbach
Technische Universität
Darmstadt
Institut für Angewandte
Geowissenschaften
Schnittspahnstr. 9
64287 Darmstadt
Germany

David M. Leitner
University of Nevada
Department of Chemistry
Reno, NV 89557
USA

Jörg Lindner
Rheinische
Friedrich-Wilhelms-Universität
Institut für Physikalische und
Theoretische Chemie
Wegelerstraße 12
53115 Bonn
Germany

[†] We communicate with great sorrow that
Professor Kolb passed away in October 2011.

Heiko Lueken
RWTH Aachen University
Institute of Inorganic Chemistry
Landoltweg 1
52074 Aachen
Germany

Bjoern Luerßen
Justus-Liebig-Universität Gießen
Institute of Physical Chemistry
Heinrich-Buff-Ring 58
35392 Gießen
Germany

Melissa M. Mariani
Mount Sinai School of Medicine,
Department of Developmental &
Regenerative Biology,
One Gustave L. Levy Place,
New York, NY 10029
USA

Manfred Martin
RWTH Aachen University
Institute of Physical Chemistry
Landoltweg 2
52056 Aachen
Germany

Thomas Mayer
Technische Universität
Darmstadt
Fachbereich Material-und
Geowissenschaften
Petersenstrasse 23
64287 Darmstadt
Germany

Gerard Meijer
Fritz-Haber-Institut der
Max-Planck-Gesellschaft
Department of Molecular Physics
Faradayweg 4-6
14195 Berlin
Germany

Mathis M. Müller
Technische Universität
Darmstadt
Institut für Angewandte
Geowissenschaften
Schnittspahnstr. 9
64287 Darmstadt
Germany

Eva Mutoro
Justus-Liebig-Universität Gießen
Institute of Physical Chemistry
Heinrich-Buff-Ring 58
35392 Gießen
Germany

Jens Röder
CERN
Physics Department
1211 Geneva 23
Switzerland

and

RWTH Aachen University
Institute of Physical Chemistry
Landoltweg 2
52056 Aachen
Germany

Arina Rybina
Heidelberg University
Cellnetworks Cluster &
Department of Physical
Chemistry
Im Neuenheimer Feld 267
69120 Heidelberg
Germany

Rolf Schäfer
Technische Universität
Darmstadt
Eduard-Zintl-Institute of
Inorganic and Physical Chemistry
Petersenstraße 20
64287 Darmstadt
Germany

Marie Anne Schneeweiss
Previously:
Universität Ulm
Institut für Elektrochemie
Albert Einstein-Allee 47
89081 Ulm
Germany

Christian Schröder
Universität Bayreuth und
Eberhard Karls Universität
Tübingen
Zentrum für Angewandte
Geowissenschaften
Umweltmineralogie &
Umweltchemie
Sigwartstr. 10
72076 Tübingen
Germany

Detlef Schröder
Institute of Organic Chemistry
and Biochemistry
Flemingovo nam. 2
Prague 6
16610
Czech Republic

Tinka Spehr
Technische Universität
Darmstadt
Institute for Condensed
Matter Physics
Hochschulstr. 8
64289 Darmstadt
Germany

Hans-Wolfgang Spiess
Max-Planck-Institut für
Polymerforschung
Ackermannweg 10
55128 Mainz
Germany

Henrik Stapelfeldt
Aarhus University
Department of Chemistry
Langelandsgade 140
8000 Aarhus C
Denmark

and

Aarhus University
Interdisciplinary Nanoscience
Center (iNANO)
Ny Munkegade 120
8000 Aarhus C
Denmark

Bernd Stühn
Technische Universität
Darmstadt
Institute for Condensed
Matter Physics
Hochschulstr. 8
64289 Darmstadt
Germany

Andreas Thissen
Technische Universität
Darmstadt
Fachbereich Material- und
Geowissenschaften
Petersenstrasse 23
64287 Darmstadt
Germany

G. Herbert Vogel
Technische Universität
Darmstadt
Ernst-Berl-Institute for Technical
Chemistry and Macromolecular
Science
Petersenstrasse 20
64287 Darmstadt
Germany

Peter Vöhringer
University of Bonn
Institute for Physical and
Theoretical Chemistry
Wegelerstrasse 12
53115 Bonn
Germany

Konrad G. Weil[†]
Technische Universität
Darmstadt
Institut für Physikalische Chemie
Petersenstraße 20
64287 Darmstadt
Germany

Karl-Michael Weitzel
Philipps-Universität Marburg
Chemistry Department - Division
of Physical Chemistry
Hans Meerweinstrasse
35032 Marburg
Germany

[†] We communicate with great sorrow that
Professor Weil passed away in May 2009.

Part III
Interfaces

Methods in Physical Chemistry, First Edition. Edited by Rolf Schäfer and Peter C. Schmidt.
© 2012 Wiley-VCH Verlag GmbH & Co. KGaA. Published 2012 by Wiley-VCH Verlag GmbH & Co. KGaA.

13

Raman Spectroscopy: Principles, Benefits, and Applications

Melissa M. Mariani and Volker Deckert

Method Summary

Acronyms

- Surface-enhanced Raman spectroscopy (SERS)
- Tip-enhanced Raman spectroscopy (TERS)
- Resonance Raman spectroscopy (RRS)
- Fourier-transform Raman spectroscopy (FT-RS)
- Coherent anti-stokes Raman spectroscopy (CARS)
- Principle component analysis (PCA)
- Multivariate curve resolution (MCR)

Benefits (Information Available)

- technique is non-invasive and is of non-destructive nature.
- no extensive sample preparation is necessary
- can be applied to gases, liquids, and solids
- allows speciation
- qualitative and quantitative analysis
- imaging and single molecule detection.

Limitations (Information Not Available)

- signal is comparatively weak
- sensitivity range is limited
- interpretation of spectra can be challenging.

Methods in Physical Chemistry, First Edition. Edited by Rolf Schäfer and Peter C. Schmidt.
© 2012 Wiley-VCH Verlag GmbH & Co. KGaA. Published 2012 by Wiley-VCH Verlag GmbH & Co. KGaA.

13.1
Introduction

Raman scattering was first discovered in 1928 [1], when it was described as the "molecular diffraction of light." This form of photon scattering was found to differ in energy and signal intensity from Rayleigh scattering, with Raman scattering occurring in only approximately every 1 in 10^6–10^8 photons. Without delay, Raman spectroscopy found a position in the chemical and materials analysis field, studying mixtures, pure substances, identifying compositions, and characterizing chemical structures. As little to no sample preparation is required, samples could be non-invasively studied, lending the technique to a wide variety of applications.

It was not until the advent of the laser in the 1960s that Raman spectroscopy advanced to the next level, providing constant and intense light sources under reproducible conditions. Following this development, detection electronics also became considerably more efficient and, hence, compatible with smaller sample volumes. From the wealth of multivariate information in a single Raman spectrum and the ability to better assess small sample components non-invasively and non-destructively, exploratory applications involving biological matter were initiated. As biological endeavors grew in prevalence, tailored instrumentation, laser wavelengths, and sample preparation became the focus of scrutiny.

However, biological samples provide numerous challenges. They are often present only in limited quantities, possess heterogeneous composition and compartmentalized targets, require microscopy, and produce high auto-fluorescence. This results in complex spectra and often difficult convolution of signals. These areas of concern were addressed through the enhancement of associated instrumentation and also novel techniques for spectra evaluation, such as multivariate data analysis, better suiting studies of this nature, and aiding the development of the micro-Raman spectroscopy widely used today.

This chapter will provide an outline encompassing electromagnetic waves and Raman scattering through to associated instrumentation. Following the background, early applications of Raman scattering will be discussed and then areas of innovation enabling technique development and some more recent fields of application. This chapter aims to provide a thorough background to the technique and associated instrumentation of Raman spectroscopy itself, as well as the diversity of applications and subsequent data processing methods.

13.2
Basic Principles: Raman Scattering

In contrast to the generally better known absorption spectroscopies, the excitation of energy levels in Raman proceeds via a "detour." In microwave, infrared (IR), or UV–vis (ultraviolet–visible) absorption spectroscopy the intensity loss at specific wavelengths is detected. This photon absorption happens when the energy

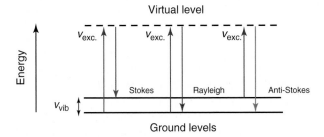

Figure 13.1 Induced excitation states of molecules via Stokes, Rayleigh, or anti-Stokes excitation patterns. Following laser excitation, molecules raise their excitation levels to short-lived but high-energy states and return to their ground state by photon emission. Emitted photons can be of higher excitation than the incident light (anti-Stokes), or of lower excitation than theincident light (Stokes), or of equal excitation to the incident light (Rayleigh scattering).

difference between two levels matches a wavelength of the source. This is valid for all levels, namely rotational, vibrational, and electronic and is illustrated in Figure 13.1 for a vibrational level that corresponds to IR absorption spectroscopy. In the same figure, the quantum state picture of Raman spectroscopy for the same vibrational levels is shown. In contrast to absorption spectroscopy, where either a broadband excitation is used or the wavelength is scanned over the entire range of interest, for Raman spectroscopy a monochromatic light source is required. If this monochromatic source is far from a resonance, most of the photons will not interact with the sample, which is evidently the case for transparent media. Nevertheless, a few photons excite the molecule in a so-called virtual level. From that level most photons are simply re-emitted at the same frequency (ω_0) giving rise to elastic or Rayleigh scattering, a radiation that can be detected orthogonal to the excitation direction. Even fewer photons are re-emitted at wavelengths shifted in energy with respect to the excitation. The amount of the difference in energy is exactly the difference between the initial state and the final energy level of the molecule. This inelastic scattering is called *Raman scattering*, resulting in the red-shifted Stokes lines ($\omega = \omega_0 - \omega_R$) and blue shifted anti-Stokes lines ($\omega = \omega_0 - \omega_R$). Most often this is used for vibrational spectroscopy, but the same arguments hold for rotational or electronic energy levels.

The quantum mechanical level description is straightforward, see Figure 13.1. To visualize the involved energy levels one can also use classical theory to understand the processes and the molecule properties that in particular influence the intensity of Raman lines. As outlined in Box 13.1, the intensity of the Stokes lines is given by

$$I_{Stokes} = \text{const} \cdot I_0 \cdot (\omega_0 - \omega_R)^4 \cdot \left(\frac{\partial \alpha}{\partial q}\right)^2$$

where I_0 and ω_0 are the intensity and frequency of the incident light and ω_R corresponds to the frequency of the inelastically scattered light; $\partial \alpha / \partial q$ is the change in the polarizability with respect to the change in the vibrational coordinate.

Box 13.1: Raman Scattering Intensities

An incoming electromagnetic field (E) acting on a target molecule is characterized for a specific time by a frequency (ω_0) and amplitude (E_0):

$$E = E_0 \cdot \cos \omega_0 t \tag{13.1}$$

Any electromagnetic field (E) induces an electrical dipole (μ)

$$\mu_{\text{ind}} = \alpha \cdot E \tag{13.2}$$

where α is the polarizability of the molecule.

If the oscillating electromagnetic field (Equation 13.1) is combined with the induced dipole (Equation 13.2), the dipole then oscillates with respect to the field

$$\mu_{\text{ind}} = \alpha \cdot E_0 \cdot \cos \omega_0 t \tag{13.3}$$

On top of the oscillating induced dipole the molecule also vibrates, either in its ground state or also in higher excited states. A vibrating molecule can then be expressed as:

$$q(t) = q_0 \cdot \cos \omega_R t \tag{13.4}$$

Where ω_R is the resonance frequency of the vibrating molecule and q_0 the normal coordinate. In turn, the polarizability can now be expanded around $q = 0$ using a Taylor expansion:

$$\alpha = \alpha(q) = \alpha_0 + \left(\frac{\partial \alpha}{\partial q} \right)_{q=0} \cdot q + \ldots \tag{13.5}$$

This expression for the polarizability can now be inserted into Equation 13.3 for the dipole oscillation, as follows:

$$\mu_{\text{ind}} = \left[\alpha_0 + \left(\frac{\partial \alpha}{\partial q} \right)_{q=q_0} \cdot q_0 \cdot \cos \omega_R t \right] \cdot E_0 \cos \omega_0 t \tag{13.6}$$

Expanding this term, it can be rewritten as:

$$\mu_{ind} = \alpha_0 E_0 \cos \omega_0 t + \frac{1}{2} \left(\frac{\partial \alpha}{\partial q} \right)_{q=0} q_0 E_0 \cos \left[(\omega_0 - \omega_R)t \right] + \frac{1}{2} \left(\frac{\partial \alpha}{\partial q} \right)_{q=0}$$
$$q_0 E_0 \cos \left[(\omega_0 + \omega_R)t \right] \tag{13.7}$$

In this equation, the first term corresponds to elastic scattering as it is not affected by the inherent influence of the vibration of the molecule. The second and third terms refer to inelastic scattering. In particular, the second term refers to the red-shifted Stokes Raman scattering and the third term to the blue-shifted anti-Stokes Raman scattering. Hence, Equation 13.7 readily explains the distinct Raman band positions of molecules.

The intensity of molecular polarizability can also be obtained classically using the emitted power of a hertzian dipole, since it is also valid for an induced dipole. The angular independent part of the intensity of a hertzian induced dipole is

$$I = \frac{\omega^4}{32\pi^2\varepsilon_0 c^3}\mu_{ind}^2 \qquad (13.8)$$

The induced dipole (Equation 13.2) can be combined with the above equation to obtain the intensity of molecular polarizability:

$$I = \frac{\omega^4}{32\pi^2\varepsilon_0 c^3}\alpha^2 E^2 \qquad (13.9)$$

For the Stokes portion of Raman scatting, $\omega = \omega_0 - \omega_R$ must be inserted to describe the induced oscillating dipole, and because the squared field is proportional to the intensity, $I \sim E^2$, one gets

$$I_{Stokes} = const\cdot I_0\cdot(\omega_0 - \omega_R)^4\cdot\alpha^2 \qquad (13.10)$$

where all the constants have been collected and I_0 indicates the intensity of the incoming field.

Using again the classical description of scattering intensities and following Equation 13.5, α must be replaced by $\partial\alpha/\partial q$

$$I_{Stokes} = const\cdot I_0\cdot(\omega_0 - \omega_R)^4\cdot\left(\frac{\partial\alpha}{\partial q}\right)^2 \qquad (13.11)$$

From equation (13.11) one can see that, apart from molecular properties influencing the polarizability of the molecule, the Raman intensity depends directly on the intensity of the excitation source and on the fourth power of the excitation frequency. The latter dependence has practical aspects as a change in excitation wavelength from, for example, 800 to 400 nm is accompanied by a 16-fold increase in scattering intensity. In practice this can be difficult to monitor directly as spectral dependences of the instrumentation can influence the observation. Other aspects, such as electronic resonances, must also be considered before simply blue shifting the Raman excitation to increase the intensity.

A major advantage of Raman spectroscopy is that samples can be either heterogeneous, homogeneous, or middle of the road. An example for liquid or liquid mixtures is provided in Figure 13.2. Here the vibrational Raman spectra of benzene, deuterated benzene, and a mixture of both is shown. This example visualizes not only the effect of isotope labeling which is, for instance, used to assign vibrational modes, but also how an assignment of different compounds can be made simply by comparison with known spectra.

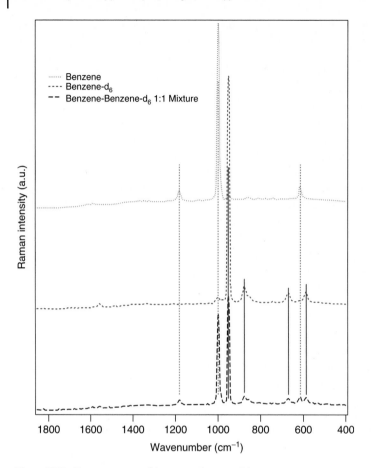

Figure 13.2 Raman spectra of benzene, deuterated benzene, and a mixture of both. Additional bands in the deuterated spectrum arise due to a Fermi resonance (shoulder at 876 cm^{-1}) and due to "impurities" of mixed H/D compounds with different selection rules, due to a different geometry.

13.3
Experimental Methods

13.3.1
General Aspects

13.3.1.1   Peak Intensity and Quantitative Measurements
The proportionality between Raman scattering and the number of illuminated molecules is the basis of quantitative Raman measurements [2]. In general, an increase in concentration is reflected in a proportional increase in the Raman signal.

This is strictly valid, however, only in diluted systems as molecules (e.g., solute and solvent) interact with one another, and this interaction influences the vibrational behavior and eventually the band intensity of the Raman scattering. Whenever molecules are in close contact, intermolecular forces will develop, influencing band positions and scattering cross sections [2, 3]. Considering such effects is crucial when performing quantitative Raman measurements. In comparison with its capabilities in structural assignment and qualitative analysis, the quantitative aspects of Raman spectroscopy are certainly less often used.

Raman scattering occurs in between 1 and 10 [4–6] emitted photons. This produces a fairly weak overall signal since most scattered light produces Rayleigh scattering, where the emitted photon is of equivalent excitation to the excitation light source [4]. In addition, the amount of inelastically scattered light ("Raman signal") varies from molecule to molecule and is referred to as the *Raman scattering cross section*. Interestingly, water has a small Raman scattering cross section and consequently does not interfere in Raman measurements like it does with IR, where the absorption coefficient is very high and often obscures the signals of the actual sample. As water is a common solvent, in particular in biology, Raman spectroscopy can be used in a much more straightforward way, and this enables the easy adaptation of Raman spectroscopy to a variety of sample types [5].

13.3.1.2 General Aspects of the Experimental Set-Up

In this section the practical aspects of Raman spectroscopy will be briefly addressed. As in most Raman experiments only the Stokes shifted light is detected we will focus our discussion on this, nevertheless the same instrumentation can also be used for the collection of the anti-Stokes region of the Raman spectrum. In general monochromatic, collimated light (typically near-UV, visible, or near-infrared (NIR)) from a laser is used to excite the sample. The resulting Stokes emission is commonly collected using either a 90° or 180° backscattering geometry (Figure 13.3).

Since the difference in energy between incident and excited photons corresponds to a transition from one molecular state to another, energy levels and resulting Raman spectra are unique for each molecule, as discussed in the previous section [7].

Additionally, when using polarized light, the symmetry of a vibrational mode can be explored using depolarization ratios, hence, providing additional sample information. The measurement scheme requires then a polarizer in front of the sample (excitation path) and one after the sample (detection path) in front of the entrance slit of the spectrometer. Comparing the Raman intensities with both polarizers oriented parallel or crossed it is possible to distinguish the totally symmetric vibrational modes as they will be much weaker in comparison with the other modes when the polarizers are crossed. Such a scheme works nicely in solution or in liquid samples, however, due to multiple scattering the polarized detection of polycrystalline samples (powders) is not reliable. When investigating single crystals (also liquid crystals) Raman can be used to determine even the orientation of the molecules in the crystal lattice.

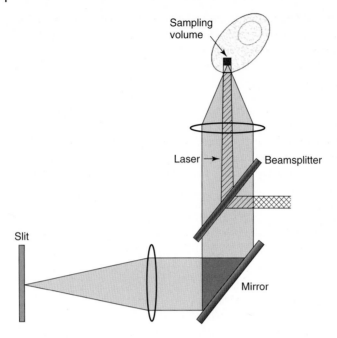

Figure 13.3 Example of a 180° collection geometry for Raman scattering. "Slit" denotes the entrance of the Raman spectrometer.

13.3.1.3 Presentation of Raman Spectra

Raman scattering is displayed in the form of a spectrum with, usually, the intensity as the y-axis and relative wavenumbers (cm^{-1}) as the x-axis. The relative wavenumbers represent the distinct vibrational states of a molecule. As it was shown in the previous section that the absolute spectral position of the scattering depends on the spectral position of the excitation source, it is practical to reference the Raman band positions to the spectral position of the source. This warrants that, independent of the excitation, the spectral positions in the spectra are identical. Because of this special scale very often the term "*Raman Shift*" can be found instead of the correct term "*wavenumber*."

13.3.2
Raman Instrumentation

Instrumentation has progressed dramatically since the 1960s, helping to optimize sample analysis and minimize the occurrence of fluorescence. Originally, Raman was better known for its susceptibility to fluorescence, weak signals, and cumbersome instrumentation. This has mostly changed during the past 20 years following dramatic enhancements in instrumentation. Specific areas of focus included laser sources (see Box 13.2), fiber optics, optical filters, spectrographs, general microscopes, and photon detectors. With the current abundance of commercially available equipment, addressing every component and their subsequent improvements would be too lengthy. However, an overview of the basic instrumentation components and their revisions will be given

Box 13.2: Lasers in Spectroscopy

Modernspectroscopy would not exist in its current state without the advent of lasers, as this is the most widely used method to achieve controlled sample excitation (with respect to time or intensity). Although Einstein theorized "stimulated emission" years prior to its actual manifestation, "stimulated emission" was eventually applied to the development of what is currently known as the *laser*. This development was particularly important for Raman spectroscopy; hence, it will be explained in somewhat more detail.

A solid ruby laser was the first to be developed in 1960 by Theodore Maiman [6]. Shortly after, a variety of other lasers were developed, including the more popular helium–neon (He–Ne) gas laser developed by Ali Javan in 1960, semiconductor lasers, liquid lasers, and chemical lasers [83]. By 1964, argon, xenon, and krypton ion lasers had also been developed [84].

Lasers were originally considered an answer without a problem, although this was a great underestimate, exemplified by their recent increase in applications. Aside from more commercial laser applications like compact disk production, surgical use, and barcode scanning, laser use has become of paramount importance in the spectroscopy industry. Research applications, despite the availability of "easier-to-handle" solid state laser systems, often utilize one of four laser types, namely argon, He–Ne, xenon, and krypton lasers, and are typically gas powered. He–Ne lasers are comprised of neon and helium gas in ratios 1:5 to 1:20 held within a closed container in which the gases are excited by an electrical discharge in the cavity, causing atoms to collide. These collisions produce the gain required for the laser [8].

Argon, krypton, and xenon are all used in ion lasers. Ion lasers differ by gain medium, where ion excitation is used to create the requisite energy level transitions for laser propagation. These lasers are most frequently applied to research-based spectroscopy because of their stable and high current.

Dependent on the gain medium, different wavelengths can be produced. Wavelength selection is critical for spectroscopic applications, especially for Raman spectroscopy. Laser wavelength selection must consider competing processes like fluorescence. Gas lasers like argon typically emit monochromatic visible light in the green, blue–green, and violet ranges and, if equipped with specific mirrors, also in the UV and deep-UV regions. Neodymium:YAG (Nd:YAG) can be used as a gain medium for the NIR region [83]. The selection of the excitation wavelength can be clearly understood from the spectra of biological analytes. Auto-fluorescence often develops in biological sample analysis, based on variables like pH and molar absorptivity following excitation in the green region, but can be overcome with a NIR laser. This change in wavelength prevents fluorescence emission (Figure 13.4) [85].

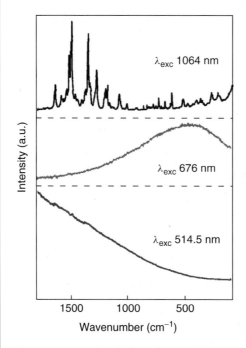

Figure 13.4 Wavelength dependencies. The same wavenumber range is shown for three Raman spectra of the same compound measured with different excitation wavelengths (see inset). The dependence of the fluorescent background is evident. Adapted from Ref. [86].

Great efforts are currently being made to replace the bulky and energy inefficient gas laser by all solid state lasers, mainly diode lasers, which are much easier to operate.

A basic Raman spectroscopic set-up begins with a beam of monochromatic light from a laser. This is focused onto the location of interest, *sample volume* V_s, enabling precise excitation. Scattered light from the sample is gathered by a collection optic at an angle Θ_c and focused by an imaging optic onto the entrance slit of the wavelength selection system (spectrograph or interferometer). Wavelength selection is crucial in order to obtain good quality Raman scatter and to suppress unwanted wavelengths like the Rayleigh scatter, allowing only a small portion of the collected light to be converted into a spectral image following transmission through the detection system (Figure 13.5).

More powerful and stable continuous wave lasers are now commercially available, simplifying and shortening analysis time, thus extending the range from the

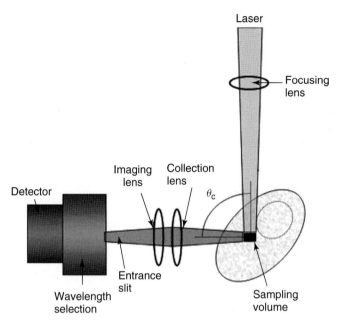

Figure 13.5 Example of an instrumentation set-up for Raman excitation of a sample. (Adapted from Laserna [8].) A sample is excited by a focused laser at an area of interest (sampling volume). When an electromagnetic (EM) wave is incident on a molecular structure, the molecule is excited and the atoms oscillate. These oscillations release their own EM wave, the Raman scatter. The scattered light is collected by a lens and imaged on the spectrometer system where the wavelength selection is performed and finally transmitted to a detector. This is generally coupled to a computer system for digital analysis.

NIR to the UV and broadening research applications [9]. Longer wavelength NIR compared to the shorter wavelength (visible) excitation offers the significant advantage of greater freedom from potentially interfering fluorescence of biological material. The ability to detect many wavelengths at once with so-called multichannel detectors has since rendered single-filter instruments outdated. In turn, holographic notch or efficient dielectric filters that reject elastically scattered light more effectively also maximize the photon throughput. Implementation of efficient charge-coupled device (CCD) detectors has also contributed to dramatic increases in spectrophotometric sensitivity and selectivity at multiple wavelengths [10].

By focusing the laser beam into a smaller area of the sample, the sample volume V_s is decreased but the intensity at the sample I_{in} will increase. This is an effective way to increase detection signals collected from the sample, although optical limitations remain and warrant consideration. For instance, incongruities in sample homogeneity will be amplified as V_s is decreased, resulting in significant inter-sample variation. Furthermore, by decreasing the V_s by focusing the laser beam more tightly, there is a simultaneous increase in the laser power at V_s and potential for sample photodegradation [8].

Because Raman scattering occurs at such a low intensity and sample pho-todegradation is likely, laser power intensity and aperture optimization are crucial. Ultimately, the selected laser power must provide spectra with low background, while at an optimized intensity to maintain sample integrity.

13.3.2.1 Filter Systems

Filter systems select wavelength from the back-scattered light before transmission to a detector. The resulting spectra will possess lower overall background noise, no Rayleigh scattering, and can be selected to fit the needs of the sample (e.g., scattering intensity, reflectivity of sample). Maintaining low spectral noise levels is critical for downstream analysis as the higher the noise level, the more they contribute to overall spectral features or the more data pre-treatment is needed to decrease the noise. This increase in filter specificity decreases spectral intensity variations and disruptive noise, providing high quality spectra.

Holographic notch or dielectric edge filters are currently the most prevalent, replacing spectrographs and interferometers as filter stages. In combination with a single monochromator, the collection efficiency increases considerably and, in combination with a microscope, allows efficient spectroscopic mapping. Certain applications require selected wavelengths rather than the whole spectrum and band-pass filter systems have been developed to enable the transmission of single or even multiple selected wavelengths. These systems are ideally suited for quantitative Raman imaging when complete wavenumber (cm^{-1}) information is not needed.

Optical bandpass filters (OBFs) can be used to obtain a whole image at a specific wavenumber position, hence replacing spectrographs and interferometers. OBFs transmit one spectral band or wavelength at a given time, but have recently been applied to transmitting more wavelengths concurrently, hence generating images at different wavelengths during such a scan. These filters are often used for Raman imaging or with single-element detectors for generic, high-throughput instruments. OBFs are available in a few different forms, with dielectric interference filters (DIFs), acoustic-optic tunable filters (AOTFs), and birefringent filters (BFFs) being the most frequently used [2].

DIFs are widely available and can be modified to a variety of specifications. Filter sizes, permissible wavelengths, transmission bandwidths, and peak transmittance can all be modified to the specifications of the lab. However, should a more powerful filter be needed, AOTFs are more suitable. Such systems are well suited for bio-medical studies [11]. These filters possess no moving parts as the band-pass selection is electrically tunable and the transmitted wavelength can be modified in microseconds to accommodate the requirements of the sample. AOTFs can also be used with larger acceptance angles for incoming light, compared with dielectric filters. The overall diffraction efficiency reaches more than 80%, making AOTFs ideally suited to fast and consistent quantitative Raman imaging.

Another type of popular filter is the BFF. These can be used to produce tunable bandpass filters from the wavelength-dependent polarization properties of the birefringent crystal [12, 13]. This is achieved by placing a retarder between two polarizers. The retarder alters the polarization of the incident light by the amount necessary to enable selection of only the desired wavelength. In a simple BFF model, the retarder is situated at a 45° angle to the first polarizer. The resulting signal transmission is sinusoidal and these filters can be combined to produce a bandpass filter capable of transmitting spectra from individual simple BFF and combining them into the spectra of a multi-stage BFF. As the light propagates through each of these filters, there is a twofold increase in light retardation compared with the previous stage [12, 13]. This results in the bandpass being determined by the highest retarded stage or last BFF in the series.

While all the direct imaging systems provide an intuitive way of obtaining spectra, often very quickly, intrinsically a large part of the information content is not used because of the nature of such a set-up. Consequently, if the whole spectral information is needed at every sample position, a mapping system that detects the entire spectral information is more advantageous.

13.3.2.2 **Detectors**

Once filter systems separate the signal from undesired wavelengths, the remaining collected light is dispersed to the detector. Photomultiplier tubes (PMTs) used to be the commonly used detector, but have mainly been replaced by CCDs [11]. CCD detectors are analog shift registers composed of a series of square detector elements (photosensitive capacitors) commonly referred to as *pixels*. Each pixel collects analog signals from electrical charges in the absorbed light with very low noise levels. The amount of stored photons in each pixel is relative to the number of photons that hit a particular pixel. The number of pixels in a CCD detector is typically in the range of thousands to millions and each pixel can range in size from 5 to 30 μm [11]. Collected charge signals are moved through a series of photosensitive capacitors, controlled by a timed release, to a charge-sensing amplifier where the output is digitized and sent to a computer for use.

Each pixel has an input threshold, or full-well potential. Typically, the full-well potential is in the range of 10^5 photoelectrons until "spilling" into adjacent pixels in either the identical row or column occurs. This overflowing of pixels can affect the resulting spectrum, distorting its shape, and can be prevented by optimizing spectral acquisition times [14].

As noted, the contents of each pixel are read out through capacitors, amplifiers, and finally a computer. This process of moving photoelectrons from each full pixel to the computer must be carried out sequentially to prevent image distortion. A standard CCD contains, for example, 256 rows, containing 1024 pixels each, forming a total of 262 144 pixels. This is a large amount of data and, therefore, automation is crucial. Additionally, the amount of data obtained can be moderately controlled, while decreasing the potential for inaccurate readout prior to digitizing, by binning [15]. Binning combines a series of pixels to form "super-pixels," generally along the height axis, reducing the number of total pixels for readout and concordantly, the overall noise level.

CCDs are an area of constant exploration to improve signal acquisition, minimize noise levels (see Section 13.4) and reduce the effect of cosmic rays. In turn, innovative methods to produce better signal acquisition are in constant development, expanding reproducibility, ease of use, and overall applications.

13.3.3
Coupling Raman with Microscopy

The microscope has long been an ideal platform for sample analysis, allowing the region of interest to be targeted. The incorporation of confocal microscopy with Raman spectroscopy revolutionized the scope of adaptations and, in 1990, was first applied to the study of single cells and chromosomes [16]. This pairing simplified biological studies and enabled a three-dimensional image translation and magnification of the sample relative to the microscope objective. Moreover, the confocal pinhole implements a geometry that ensures only selected photons are collected from the back-scatter and reach the detector. This precision enhances the ability to quantify spectra by providing spatio-temporal information of their origin, useful in identifying and quantifying specific localitions.

By guiding a laser through an objective lens, a near-diffraction-limited spatial resolution and increased collection efficacy can be achieved [11]. The focal spot can be made much tighter, increasing the intensity per area, and can provide a spatial resolution below 1 μm, enabling biologically important structures, including individual nuclei, mitochondria, cilia, and regions of cell-interaction, to be analyzed [17–20]. The minimally invasive micro-Raman process maintains sample integrity, and direct sampling in either air or aqueous environments can be carried out, maintaining viability. Micro-Raman spectroscopy can also provide confocal resolution with appropriate wavelength selection, permitting measurements on different planes below the sample from a reduced absorption of the laser by the sample [21].

Most micro-Raman spectroscopic measurements implement a confocal set-up, exciting and collecting the backscatter from the same location using the 180° collection geometry (Figure 13.3). High numerical apertures provide increased Raman performance but can be hindered by spherical aberrations [11]. This can cause optical rays in the center of the lens, causing the center of the focal plane to focus on a different position. This reduces the efficacy of the numerical aperture and the overall image contract.

Micro-Raman spectroscopy has enabled the collection of detailed information pertaining to analyte molecular structure and composition on multiple focal plains. It continues to be a growing area of research, providing a spatial resolution surpassing that of IR spectroscopy and providing quantitative chemical insight. Its versatility can be best observed by the broad range of applications. Despite its relatively recent development, it has been applied to a variety of different problems. Examples include the characterization of molecular structure in biological samples (e.g., cells, tissues) and distinguishing differences between cells as a result of growth cycles, physiological behavior, or even physiological states (e.g., activation states) [22].

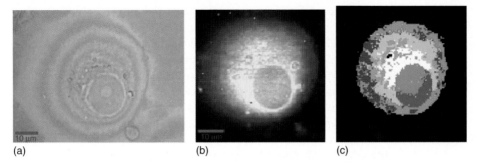

(a)   (b)   (c)

Figure 13.6 Micro-Raman imaging of a single MCF-7 cell and subsequent qualitative analysis [23]. An image is mapped by an automated stage system. Each pixel consists of an individual spectrum and the total mapped spectra are compared by the signal intensity of a selected band. Intensities are assigned a color in relation to intensity (bright orange = high intensity, black = low intensity) and the image is converted into a spectral map. (a,b) Visible image of an MCF-7 cell and chemical imaging specific for the intensity of CH-stretching throughout the cell (80×80 pixels and a 0.5 μm spot size). (c) Hierarchical clustering of the spectral image specific for the spectral "fingerprint region" (1800–675 cm^{-1}) of each individual pixel allows a further distinction and displays the localitions of sub-cellular components.

Through technological advancements, fully automated xy-maps and line scans can be taken of selected sample areas, upholding the spatial resolution of the analyte requisite for both single cell and full tissue analysis. A sampling region is broken into a series of pixels that are color coded based on the intensity of a chosen peak (see, for example, Figure 13.6). Spectroscopic imaging of cells is still a rather new and exciting area of research, although it shows great promise for providing novel ways of interpreting samples. As a result, both micro-Raman mapping spectroscopy and micro-Raman imaging possess great potential for use in diverse applications.

13.3.4
Benefits of Raman Spectroscopy

Interest in Raman spectroscopy has grown considerably. Attractive features include its non-invasive and non-destructive nature. In addition, no extensive sample preparation is necessary. Raman spectroscopy can also be easily combined with multivariate sample analysis, ideal for complex biological systems. The combination of these advantages illustrates the high impact of Raman spectroscopy.

Foremost, the high spatial resolution employed by Raman spectroscopy (1–100 μm) permits the study of individual species and their structures in complex samples. This is an attribute specifically advantageous for single cell and imaging studies, something more difficult to achieve with related techniques like IR.

Advantages also include the ability to work using aqueous solutions in conventional glassware and the opportunity to monitor reactions. For pharmaceutical studies on drug activity, this is tremendously beneficial. Final product conformations can also be investigated and their activity can be studied in a natural environment, shedding light on *in vivo* compound reactions. Moreover, as most drug active molecules contain a degree of unsaturation, producing a strong signal in the Raman "fingerprint region" between 400 and 2000 cm^{-1}, detecting symmetrical stretching provides a significant advantage over IR spectroscopy.

13.4
Applications of Raman Spectroscopy

From the wealth of information in a Raman spectrum, and its versatility, an expansive range of applications have followed. In particular, biological and bio-medical applications have become vast, encompassing the detection of sub-cellular reactions and components (including protein and nucleic acids) [24], single cell studies [25–27], sample imaging, and tissue characterization [28–32]. Moreover, chemical studies focusing on drug permeability and conformations [33] are still carried out alongside environmental [34], chemical [35, 36], industrial [37], pharmaceutical [38, 39], polymer [40, 41], and food science [42, 43] industry involvement.

13.4.1
Chemical Applications

Raman instrumentation provides many associated benefits for chemical analysis. By the use of fiber optics, Raman measurements can be conducted from a distance and can even penetrate casings like plastic, acetone, or glass, alleviating the need for direct sample contact. This is highly advantageous for identifying caustic or other dangerous samples, particularly in field research. Spectral acquisition times are also rapid (seconds to minutes) enabling quick sample turnaround times, specifically in the case of inorganic samples, unsaturated compounds, aqueous solutions, and odd-shaped samples [44]. This is particularly of interest for security uses, including airport and police security procedures, or for maintaining quality control by screening samples prior to sale [45–47].

Micro-Raman spectroscopy has all the aforementioned benefits while also contributing micro-scale resolution for the measurement of nano-structures and other nanoscale samples. This derivative of Raman can also identify modifications in chemical compounds, delineating where structural changes have occurred and identifying active components, as used in pharmaceuticals [33, 48]. The structure and form of polymers in relation to areas as specific as crystallinity, crystallite size, and polymorph screening can also be probed using this technique. Polymorph screening is particularly important for pharmaceutical screenings since polymorphs are linked to chemical solubility [44, 49, 50]. This enables reactions to be monitored through Raman spectra, providing real-time detection capabilities [51].

Furthermore, an often overlooked application in the chemical industry includes the ability to measure gases. This is most often not carried out under ambient conditions but is possible using closed environments. Raman has been particularly applied to the study of gases of industrial importance, such as homonuclear diatomic molecules [52].

From such a broad range of uses in industrial, pharmaceutical, and environmental disciplines, there is no doubt that Raman spectroscopy will continue to push the standards for commercial sample analysis.

13.4.2
Biological Applications

Recently, a myriad of papers have been published applying micro-Raman imaging to spatially explore chemical variations in tissue samples for diagnostic and research purposes [30, 31, 53–55]. Some examples of subject areas include prostate, lung, and bladder cancers and these have been extensively reviewed [56]. These applications are a direct result of the unparalleled sensitivity and the relatively simple, label-free analyte preparation associated with the technique. From the development of micro-Raman spectroscopy, detailed information can be obtained of analyte molecular structure and composition using a high spatial resolution with minimal invasion and direct sample analysis. This unmatched sensitivity can supply both quantitative and qualitative results by measuring emission or adsorption patterns, distinct for all functional groups and organic compounds, and is critical considering the increasingly quantitative nature of the biological field.

The incorporation of confocal microscopy with Raman spectroscopy enabled biological studies and imaging by providing a three-dimensional image translation and sample magnification. Moreover, the confocal pinhole implements a geometry that ensures only selected photons are collected from the back-scatter and reach the detector. This precision enhanced the ability to compare spectra by providing spatio-temporal information of their origin, useful in identifying specific locations.

Micro-Raman analysis can be carried out using either point scans or full sample imaging. Imaging is typically the preferred method as the microscope pairing enables a quick selection of specific locations, while automated stages allow for precise *xy* sample mapping or line scanning and are generally more flexible. From the high lateral resolution (below 1 μm), the resulting spectral images can provide direct insight into the composition of cellular components or full tissue organization patterns with molecular resolution. As a result, this information-rich method of analysis can provide a truly objective view of pathology or cell functionality, displaying its bio-medical advantages [57]. Cellular components have been studied using micro-Raman imaging in a variety of samples. Using single spectrum analysis, specific proteins, organic pigments, and nucleic acids have been analyzed [58]. Individual cells [26, 30] and nuclear components have also been explored using this method [59].

Biological sample images are formed by examining the intensity of individual peaks for the spectra of each pixel, like the commonly used Amide I or CH-stretching modes. Similar regions are often color-coded based on similarities and the colors are plotted for each pixel (Figure 13.6) [60]. Variations in spectral groups related to spatial and temporal classifications can then be identified, illustrating the differences between the grouped regions.

In concurrence with the effects of water on Raman spectra, many buffers and medias used in biological sample growth and preparation all possess low Raman signals, reducing interference in the resulting data [61]. This flexibility for cell sampling can permit cells to be fixed, dried, or even measured *in vivo* [62–65]. Furthermore, flexible sampling permits subsequent analysis by other means, including IR spectroscopy, traditional histological methods, quantitative gene expression analysis, and a variety of other techniques.

Micro-Raman spectroscopy has also received great interest for bacterial, yeast, and viral studies. Resulting from bacterial and other microorganism outbreaks in hospitals and contamination within food facilities and pharmaceutical packaging sites, the detection and identification of the bacteria and microorganisms present is critical in upholding quality control standards and ensuring that facilities are free from potential infections. By applying micro-Raman spectroscopy and collecting a single spectrum from each analyte, individual strain variations can be detected and important phylogenetic information can be obtained and quantitatively assessed [66, 67]. Using a single organism, quick, reliable, and non-destructive analysis can provide spectral data pertinent to treatment or cleaning conditions [67].

13.4.3
Variations of Raman Spectroscopy

Raman spectroscopy as an individual technique has expanded in popularity since its onset during the 1960s. Attributed to surrounding interest in the technique, the versatility of Raman spectroscopy can be truly understood when looking at the series of spin-off enhanced techniques. These associated techniques include, but are not limited to, surface-enhanced Raman spectroscopy (SERS), tip-enhanced Raman spectroscopy (TERS), resonance Raman spectroscopy (RRS), Fourier-transform Raman spectroscopy (FT-RS), and coherent anti-stokes Raman spectroscopy (CARS). Each technique is a more specialized version of Raman spectroscopy, while providing added benefits like increased sensitivity, speed, and imaging capabilities.

13.4.3.1 Surface-Enhanced Raman Spectroscopy
The most notable off-shoot of Raman spectroscopy is SERS. SERS provides enhanced spectral results through the use of specific substrates. To achieve this spectral enhancement, samples are placed onto metal nanoparticles or colloid covered surfaces. Primarily gold, silver, or copper are used as enhancers. Figure 13.7 demonstrates a typical example of the sensitivity of the technique. Enhancement in the obtained spectra has been reported to reach $10^{14}-10^{15}$, permitting single molecule detection. This technique was first pioneered in 1974 by Martin Fleischman from Southampton University using pyridine on an electrochemically roughened silver coated surface [68]. Initially, the detected signal enhancement was not associated with the metal substrate and was thought to result from the concentration of excited molecules. However, the enhancement was later determined

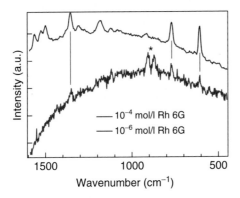

Figure 13.7 Surface enhanced Raman spectra of Rhodamine 6G excited at 514 nm, acquisition time 5 s each spectrum. 1 μl of sample was dropped onto a silver island film of 20 nm thickness and covered approx 1 cm² which corresponds to roughly 10 000 molecules/μm². Spectra were detected in a special near-field optical geometry corresponding to 100 molecules in the sample volume in the 10^{-6} M case. The star denotes spurious signal due to computer screen illumination. (Adapted from Ref. [87]).

by two groups to be a direct result of the metal substrate. This led to two theories being proposed to explain the enhancement, the electromagnetic theory and the chemical theory [69, 70].

The electromagnetic theory was originally proposed by Jeanmaire and Van Duyne in 1977. The methodology of this theory was based on the idea that an enhancement in the electric field of the analyte occurs by the metal surface interaction with the excitation light, creating the signal enhancement [69]. Resulting from surface plasmon excitation, the field enhancement intensifies the incident light and also the Raman signal, which is also enhanced by the same principle [14]. This theory can be applied to most samples through the use of gold, silver, or copper metals from their plasmon resonance frequencies residing in the wavelength ranges of visible to NIR radiation. However, the enhancement factor has been found to be inconsistent and wavelength and metal types are not compatible with all sample types, being particularly tricky with biological samples [71].

The chemical theory can be applied to any molecule study but does not fully explain the breadth of spectral enhancement since it is based on a charge transfer between the chemisorbed species and the metal surface. From its inability to explain the spectral enhancement seen in SERS spectra, it is speculated that this theory works in concert with the electromagnetic theory [72].

Limitations to the use of SERS are present but minimal. These include the difficulty in carrying out quantitative measurements as there is preferential adsorption of some materials onto the substrates [2]. The analyte is also required to contact the metal substrate which may not be ideal for all sample types or *in vivo* measurements, and undermines the advantages of non-invasive and non-contact traditional Raman analysis. Moreover, surface reproducibility has proven to be a challenge.

Aside from the theory behind SERS, this technique has received acclaim alongside the rise of traditional Raman spectroscopy, particularly in biological applications, for its ability to keep background fluorescence to a minimum. SERS has also been coupled with imaging and has been applied to studies of chemical composition in live cells, focusing on contents like DNA and phenylalanine [73]. Additionally, drug discovery, enzyme activity, DNA labeling, and glucose detection are amongst the diverse range of SERS applications [74–76].

13.4.4
Data Analysis

Large amounts of information are collected in a Raman spectrum, making analysis both overwhelming and arduous. Once data is obtained, it must be meticulously and reliably assessed, extrapolating all of the content. The actual information that is extracted is, in many cases, a mere contrast or a yes/no decision. In other words in many cases it is necessary to extract just the important information and visualize this information clearly.

Depending on the application, spectral peaks will need to be labeled for their relation to the sample and, potentially, compared for cluster analysis. Novel algorithms can be developed specific for the analysis needs of a sample, a very important aspect for biological cell studies [26, 67, 77]. These algorithms can be used to determine sample homogeneity during processing and to classify all spectra into groups based on spectral similarities. Single spectrum analysis can also be assessed relative to the sample, identifying differences or similarities by using readily available methods like principle component analysis (PCA) or multivariate curve resolution (MCR). Both of these analysis methods assume that the spectra from a mixture of chemicals can be viewed as a linear combination of the mixture's component spectra [7]. For PCA, singular-value decomposition calculates a basis spectrum while MCR extracts basis spectra similar to the original Raman spectra of the chemicals present [78]. These techniques are both useful when little is known prior to analysis, since the extraction of embedded chemical information can still be carried out [7]. These methods also allow a comparison of data relating to the composition of the sampled population and can be applied using commercially available software for rapid data pretreatment and analysis.

Data acquired from imaging studies can be assessed using multiple methods. In one study, oocytes molecular architecture was investigated and unsupervised hierarchical cluster analysis was applied to the resulting spectral images, assessing sample heterogeneity [79]. Micro-Raman imaging creates spectral images by examining the intensity of selected individual peaks for each pixel's spectrum. These spectral images can be compared using a variety of analysis methods, including K-means [80] and Ward's algorithm [81], as frequently used examples to achieve hierarchical clustering. This form of analysis can assess an image by clustering image spectra based on similarities. False-colors are used to identify different regions in a heterogeneous sample, simplifying the analysis of otherwise complex and detailed spectra. The number of pixels forming each cluster can be compared with the total number of pixels in the image to provide a quantitative view of analyte distribution in the spectral image (Figure 13.6c). The molecular composition of each cluster can also measure individual specific peak intensities between clusters within a sample. Using cluster analysis methods, spectral clustering can be carried out specific to the spectral region(s) of interest, such as the "fingerprint region" between 400 and 2000 cm^{-1} for general RNA, DNA, and protein content, the CH-stretching region between 2800 and 3050 cm^{-1} for lipid content, or the ester carbonyl band from 1720 to 1750 cm^{-1} specific for types of lipid content.

13.5
Conclusion

Since its discovery, Raman scattering has become a valuable means of sample analysis, propelling the area of spectroscopy to new heights. Coupled with the development of commercially available lasers, modern day Raman spectroscopy has grown into an innovative and highly attractive technique. Attributes like its intrinsic multivariate nature, little to no sample preparation, and non-invasive analysis make this technique appealing for studies ranging from chemical compound analysis to tissue architecture [8, 82]. This flexibility permits the analysis of samples in various states, including solids, gases, aerosols, vapors, and liquids, in addition to *in situ* measurement.

Also contributing to the new found interest in Raman spectroscopy and its derivatives are the successive improvements of experimental techniques and instrumentation [8]. Improvements to the requisite instrumentation, such as filters, detectors, and fiber optics have all helped shape the industry of spectroscopy and the capabilities of Raman spectroscopy. With the capacity to select specific wavelengths more easily, detect the scattering signal and display results by means of digital spectra using a computer system, the speed and ease of use have increased tremendously.

In parallel to these advancements, the pairing of Raman spectroscopy with microscopy has enabled the breadth of Raman applications to expand considerably in the direction of the bio-medical sciences. This combination has lead to the non-invasive analysis of specific locations or the imaging of whole samples, providing a resolution below the micrometer level while maintaining spatio-temporal organization.

As industry and academia begin to work together more closely, the demands for individual scientists have followed suit. Bridging the gap between biology and chemistry has become of paramount importance and, with this change, a basic knowledge of the various techniques and their functionalities has become requisite at the very minimum. Consequently, Raman spectroscopy, its applications and its associated techniques have begun branching into a multitude of disciplines, as illustrated by its current success. Therefore, Raman spectroscopy has become a mainstay technique, not only for traditional chemistry but increasingly in the realms of biology, and shows signs of continuing to be a technique of choice in many laboratories.

References

1. Raman, C. and Krishnan, K. (1928) *Nature*, **121**, 501.
2. Pelletier, M. (1999) *Introduction to Applied Raman Spectroscopy*, Blackwell Science, Oxford.
3. Lewis, I. and Edwards, H.G.M. (2001) *Handbook of Raman Spectroscopy*, Marcel Dekker, New York.
4. Pistorius, A.M.A. (1995) *Spectrosc. Eur.*, **7**, 8–15.
5. Wood, B.R., Heraud, P., Stojkovic, S., Morrison, D., Beardall, J., and McNaughton, D. (2005) *Anal. Chem.*, **77**, 4955–4961.
6. Maiman, T. (1960) *Nature*, **187**, 493–494.
7. Shafer-Peltier, K.E., Haka, A.S., Motz, J.T., Fitzmaurice, M., Dasari, R.R., and Feld, M.S. (2002) *J. Cell Biochem. Suppl.*, **39**, 125–137.
8. Laserna, J. (1996) *Signal Expressions in Raman Spectroscopy*, John Wiley & Sons, Inc., New York.
9. Tobin, R.P., Peard, K.A., Bode, G.H., Rozsa, K., Donko, Z., and Szalai, L. (1995) *IEEE J. Sel. Top. Quantum Electron.*, **1**, 805–810.
10. Shim, M.W. and Wilson, B.C. (1997) *J. Raman Spectrosc.*, **28**, 131–142.
11. Pelletier, M. (1999) *Analytical Applications of Raman Spectroscopy*, Blackwell Publishers, Oxford.
12. Gaenswein, P.U. and Winkler, H. (1973) *Zeiss Inf.*, **21**, 44–46.
13. Miller, P. (1990) *SPIE Instrum. Astron. VII*, **1235**, 466–473.
14. Smith, E. and Dent, G. (2005) *Modern Raman Spectroscopy: A Practical Approach*, John Wiley & Sons, Inc., New York.
15. Epperson, P. and Denton, M.B. (1989) *Anal. Chem.*, **61**, 1513–1519.
16. Puppels, G.J., de Mul, F.F., Otto, C., Greve, J., Robert-Nicoud, M., Arndt-Jovin, D.J., and Jovin, T.M. (1990) *Nature*, **347**, 301–303.
17. Timlin, J.A., Carden, A., and Morris, M.D. (1999) *Appl. Spectrosc.*, **53**, 1429–1435.
18. Paschalis, E., DiCarlo, E., Betts, F., Sherman, P., Mendelsohn, R., and Boskey, A. (1996) *Calcif. Tissue Int.*, **59**, 480–487.
19. Kazanci, M., Roschger, P., Paschalis, E., Klaushofer, K., and Fratzl, P. (2006) *J. Struct. Biol.*, **156**, 489–496.
20. Huang, Y., Karashima, T., Yamamoto, M., and Hamaguchi, H. (2005) *Biochemistry*, **44**, 10009–10019.
21. Krafft, C., Beleites, M., and Schackert, G.S. (2007) *Anal. Bioanal. Chem.*, **389**, 1133–1142.
22. Diem, M., Romeo, M., Boydston-White, S., Miljkovic, M., and Matthaus, C. (2004) *Analyst*, **129**, 880–885.
23. Mariani, M., Day, P., and Deckert, V. (2010) *Integr. Biol.*, **2**, 94–101.
24. Tuma, R. (2005) *J. Raman Spectrosc.*, **36**, 307–319.
25. Jess, P.R., Smith, D.D., Mazilu, M., Dholakia, K., Riches, A.C., and Herrington, C.S. (2007) *Int. J. Cancer*, **121**, 2723–2728.
26. Crow, P., Barrass, B., Kendall, C., Hart-Prieto, M., Wright, M., Persad, R., and Stone, N. (2005) *Br. J. Cancer*, **92**, 2166–2170.
27. Swain, R.J. and Stevens, M.M. (2007) *Biochem. Soc. Trans.*, **35**, 544–549.
28. Krafft, C., Kirsch, M., Beleites, C., Schackert, G., and Salzer, R. (2007) *Anal. Bioanal. Chem.*, **389**, 1133–1142.
29. Crow, P., Molckovsky, A., Stone, N., Uff, J., Wilson, B., and WongKeeSong, L.M. (2005) *Urology*, **65**, 1126–1130.
30. Haka, A.S., Shafer-Peltier, K.E., Fitzmaurice, M., Crowe, J., Dasari, R.R., and Feld, M.S. (2005) *Proc. Natl. Acad. Sci. U.S.A.*, **102**, 12371–12376.
31. Stone, N., Kendall, C., Smith, J., Crow, P., and Barr, H. (2004) *Faraday Discuss.*, **126**, 141–157; discussion 169–183.

32. Zhang, G., Moore, D.J., Flach, C.R., and Mendelsohn, R. (2007) *Anal. Bioanal. Chem.*, **387**, 1591–1599.

33. Frank, C. (1999) *Review of Pharmaceutical Applications of Raman Spectroscopy*, Blackwell Science, Oxford.

34. Tripathi, A., Jabbour, R.E., Treado, P.J., Neiss, J.H., Nelson, M.P., Jensen, J.L., and Snyder, A.P. (2008) *Appl. Spectrosc.*, **62**, 1–9.

35. Brennan, B.A., Cummings, J.G., Chase, D.B., Turner, I.M., and Nelson, M.J. Jr. (1996) *Biochemistry*, **35**, 10068–10077.

36. Rothschild, K.J., Andrew, J.R., De Grip, W.J., and Stanley, H.E. (1976) *Science*, **191**, 1176–1178.

37. Andrew, J.J., Browne, M.A. Clark, I.E., Hancewicz, T.M., and Millichope, A.J. (1998) *Appl. Spectrosc.*, **52**, 790–796.

38. Doub, W.H., Adams, W.P., Spencer, J.A., Buhse, L.F., Nelson, M.P., and Treado, P.J. (2007) *Pharm. Res.*, **24**, 934–945.

39. Clarke, F.C., Jamieson, M.J., Clark, D.A., Hammond, S.V., Jee, R.D., and Moffat, A.C. (2001) *Anal. Chem.*, **73**, 2213–2220.

40. Schaeberle, M.D. and Treado, P.J. (2001) *Appl. Spectrosc.*, **55**, 257–266.

41. Appel, R., Zerda, T.W., and Waddell, W.H. (2000) *Appl. Spectrosc.*, **54**, 1559–1566.

42. Lepock, J.R., Arnold, L.D., Torrie, B.H., Andrews, B., and Kruuv, J. (1985) *Arch. Biochem. Biophys.*, **241**, 243–251.

43. Archibald, D.D., Kays, S.E., Himmelsbach, D.S., and Barton, F.E. (1998) *Appl. Spectrosc.*, **52**, 22–31.

44. Lipp, E.D. and Leugers, M.A. (1999) in *Analytical Applications of Raman Spectroscopy* (ed. M. Pelletier), Blackwell Science, Oxford.

45. Eliasson, C., Macleod, N.A., and Matousek, P. (2008) *Anal. Chim. Acta*, **607**, 50–53.

46. Eliasson, C., Macleod, N.A., Jayes, L.C., Clarke, F.C., Hammond, S.V., Smith, M.R., and Matousek, P. (2008) *J. Pharm. Biomed. Anal.*, **47**, 221–229.

47. Matousek, P. and Parker, A.W. (2006) *Appl. Spectrosc.*, **60**, 1353–1357.

48. Pivonka, D.E., Chalmers, J.M., and Griffiths, P.R. (2007) *Applications of Vibrational Spectroscopy in Pharmaceutical Research and Development*, John Wiley & Sons, Inc., New York.

49. Lu, J., Wang, X.J., Yang, X., and Ching, C.B. (2007) *J. Pharm. Sci.*, **96**, 2457–2468.

50. Ali, H.R., Edwards, H.G., Hargreaves, M.D., Munshi, T., Scowen, I.J., and Telford, R.J. (2008) *Anal. Chim. Acta*, **620**, 103–112.

51. Nelson, E. and Scranton, A.B. (1995) *Polym. Mater. Sci. Eng.*, **72**, 413–414.

52. Murphy, W. (1991) *Chemical Applications of Gas-Phase Raman Spectroscopy*, Chapter 9, John Wiley & Sons, Inc., New York.

53. Crow, P., Uff, J., Farmer, J., Wright, M., and Stone, N. (2004) *BJU Int.*, **93**, 1232–1236.

54. de Jong, B.W., Schut, T.C., Maquelin, K., van der Kwast, T., Bangma, C.H., Kok, D.J., and Puppels, G.J. (2006) *Anal. Chem.*, **78**, 7761–7769.

55. Harvey, T.J., Faria, E.C., Henderson, A., Gazi, E., Ward, A.D., Clarke, N.W., Brown, M.D., Snook, R.D., and Gardner, P. (2008) *J. Biomed. Opt.*, **13**, 064004.

56. Krafft, C., Steiner, G., Beleites, C., and Salzer, R. (2009) *J. Biophotonics*, **2**, 13–28.

57. Crow, P., Uff, J.S., Farmer, J.A., Wright, M.P., and Stone, N. (2004) *BJU Int.*, **93**, 1232–1236.

58. Bhosale, P., Ermakov, I.V., Ermakova, M.R., Gellermann, W., and Bernstein, P.S. (2003) *Biotechnol. Lett.*, **25**, 1007–1011.

59. Chan, J.W., Taylor, D.S., Lane, S.M., Zwerdling, T., Tuscano, J., and Huser, T. (2008) *Anal. Chem.*, **80**, 2180–2187.

60. McCreery, R. (2000) *Raman Spectroscopy for Chemical Analysis*, Wiley Interscience, New York.

61. Swain, R.J. and Stevens, M.M. (2007) *Biochem. Soc. Trans.*, **35**, 544–549.

62. Krafft, C. (2004) *Anal. Bioanal. Chem.*, **378**, 60–62.

63. Krafft, C., Codrich, D., Pelizzo, G., and Sergo, V. (2008) *Analyst*, **133**, 361–371.

64. Uzunbajakava, N., Lenferink, A., Kraan, Y., Willekens, B., Vrensen, G., Greve, J., and Otto, C. (2003) *Biopolymers*, **72**, 1–9.

65. Mariani, M., Lampen, P., Popp, J., Wood, B., and Deckert, V. (2009) *Analyst*, **134**, 1154–1161.

66. Keller, G. (2000) *Anal. Chem.*, **72**, 732A–733A.

67. Rösch, P., Harz, M., Schmitt, M., Peschke, K.D., Ronneberger, O., Burkhardt, H., Motzkus, H.W., Lankers, M., Hofer, S., Thiele, H., and Popp, J. (2005) *Appl. Environ. Microbiol.*, **71**, 1626–1637.

68. Fleishman, M.,Hendra,P.J., andMcQuilan,A.J. (1974) *Chem. Phys. Lett.*, **26**, 163–166.

69. van Duyne, R. and Jeanmaire, D.P. (1977) *J. Electroanal. Chem.*, **84**, 1–20.

70. Albrecht, M.G. and Creighton, J.A. (1977) *J. Am. Chem. Soc.*, **99**, 5215–5219.

71. Moskovits, M. (2006) *Surface-Enhanced Raman Spectroscopy: A Brief Perspective*, Springer.

72. Lombardi, J.R.B., Lu, R.L., and Xu, J. (1986) *J. Chem. Phys.*, **84**, 4174–4180.

73. Kneipp, K., Haka, A.S., Kneipp, A.S., Badizadegan, H., Yoshizawa, K., Boone, N., Shafer-Peltier, C., Motz, K.E., Dasari, J.T., and Feld, R.R. (2002) *Appl. Spectrosc.*, **56**, 150–154.

74. Faulds, K., Fruk, L., Robson, D.C., Thompson, D.G., Enright, A., Smith, W.E. and Graham, D. (2006) *Faraday Disc.*, **132**, 261–268.

75. Shafer-Peltier, K.E., Haynes, C.L., Glucksberg, M.R., and Van Duyne, R.P. (2003) *J. Am. Chem. Soc.*, **125**, 588–593.

76. Moore, B.D., Stevenson, L., Watt, A., Flitsch, S., Turner, N.J., Cassidy, C., and Graham, D. (2004) *Nat. Biotechnol.*, **22** (9) 1133–1138.

77. Nicolaou, N. and Goodacre, R. (2008) *Analyst*, **133**, 1424–1431.

78. Wold, S., Esbensen, K., and Geladi, P. (1987) *Chemom. Intell. Lab. Syst.*, **2**, 37–52.

79. Wood, B.R., Chernenko, T., Matthaus, C., Diem, M., Chong, C., Bernhard, U., Jene, C., Brandli, A.A., McNaughton, D., Tobin, M.J., Trounson, A., and Lacham-Kaplan, O. (2008) *Anal. Chem.*, **80**, 9065–9072.

80. Steinhaus, H. (1956) *Bull. Acad. Polon. Sci.*, **C1. 3**, 801–804.

81. Ward, J. (1963) *J. Am. Stat. Assoc.*, **58**, 236–244.

82. Krafft, C., Knetschke, T., Funk, T., and Salzer, R.H.W. (2006) *Anal. Chem.*, **78**, 4424–4429.

83. Hecht, E. (1987) *Optics*, 2nd edn, Addison-Wesley Publishing Company, Reading, MA.

84. Gordon, E.I., Labuda, E.F., and Bridges, W.B. (1964) *Appl. Phys. Lett.*, **4**, 178–180.

85. Tsien, R. and Waggoner, A. (1995) *Fluorophores for Confocal Microscopy*, Springer.

86. Engert, C., Michelis, T., and Kiefer, W. (1991) *Appl. Spectrosc.*, **45** (8) 1333–1339.

87. Zeisel, D., Deckert, V., Zenobi, R., and Vo-Dinh, T. (1998) *Chem. Phys. Lett.*, **283**, 381–385.

14

Diffuse Reflectance Infrared Fourier Transform Spectroscopy: an *In situ* Method for the Study of the Nature and Dynamics of Surface Intermediates

Alfons Drochner and G. Herbert Vogel

Acronyms, Synonyms

- Attenuated total reflection (ATR)
- (Discrete) Fourier transformation ((D)FT)
- Diffuse reflectance infrared Fourier transform spectroscopy (DRIFTS)
- Diffuse reflectance spectroscopy (DRS)
- Emission spectroscopy (ES)
- Infrared (IR)
- Infrared emission spectroscopy (IRES)
- Nuclear magnetic resonance (NMR)
- Photo acoustic spectroscopy (PAS)
- Temperature programed desorption (TPD)
- Ultraviolet (UV)
- Visible (VIS).

Benefits (Information Available)

- IR technique especially suited to application on solid powdered samples
- Measurements without additional dilutants like KBr are possible
- IR measurement on "dark" probes such as carbon are possible

Methods in Physical Chemistry, First Edition. Edited by Rolf Schäfer and Peter C. Schmidt.
© 2012 Wiley-VCH Verlag GmbH & Co. KGaA. Published 2012 by Wiley-VCH Verlag GmbH & Co. KGaA.

- Suitable for *in situ* experiments
- Reactions on solid surfaces can be examined
- Information about the nature and kinetics of surface intermediates can be obtained
- Qualitative and quantitative analysis.

Limitations (Information Not Available)

- Only applicable to solid scattering probes (e.g., powders or rough materials)
- Experiments under elevated pressures are limited due to the resistance of the cell windows
- Maximum temperature is limited due to the infrared emission at high temperatures.

14.1
Introduction

Many of the experimental methods that are applied in the study of heterogeneously catalyzed reactions are based on the interaction of particle beams, such as electrons or ions, with the catalyst [1, 2]. From the shift of the particles' energy and momentum one can draw conclusions about the structure of the solid. These methods normally work under *ex situ* conditions. To investigate catalysts under *in situ* conditions electromagnetic radiation (NMR, IR, VIS, UV, X-ray) is used, as the radiation can penetrate into the reactive atmosphere. In this context infrared spectroscopy is a well established method.

14.1.1
IR Spectroscopy of Solids

There are different ways of applying infrared spectroscopy to solids. The most common and probably best-known technique is the transmission measurement. Here the solid sample is mostly diluted with an IR transparent material (e.g., potassium bromide) and then formed into a pressed tablet. For the investigation of reactions on solid surfaces, for example, in heterogeneous catalysis, disadvantages arise from this type of sample preparation. The pressing process itself can cause structural modifications and thus lead to changes in the IR spectrum. When diluents are used, interactions or even reactions between the sample and the diluent can occur [3, 4]. An alternative to this method is the so-called self-supporting wafer, obtained by compaction of the pure, undiluted powdered materials. The applied pressure, however, can also lead to phase transitions, and thus to misinterpretation. Moreover, in the preparation of a self-supporting wafer one must ensure that the layer is not too thick, otherwise the entire IR radiation will be completely absorbed by the sample material. Samples with high absorption coefficients or reflectivity (e.g., carbon or metals) cannot, therefore, be investigated with this preparation

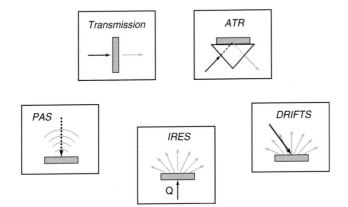

Figure 14.1 Different kinds of infrared spectroscopy (ATR: attenuated total reflection, PAS: photoacustic spectroscopy, IRES: infrared emission spectroscopy, DRIFTS: diffuse reflectance infrared Fourier transform spectroscopy). Transmission and ATR spectroscopies usually involve a compaction of the samples. PAS, IRES, and DRIFTS can be directly applied on samples in powdered form.

method. With both techniques, dilution matrix and self-supporting wafer, there is a further disadvantage for the *in situ* investigation of heterogeneously catalyzed reactions, namely gases and reactants can no longer flow freely through the catalyst. The use of compressed wafers leads to a "preparation-induced" inhibited mass transport, that is, the flow of the gaseous reactants to the active centers is reduced, and thus the rate of reaction is limited. In the extreme case, the diffusion of the reactants into and out of the pressed catalyst wafer is rate determining. This leads to artifacts, especially in kinetic experiments, and thus is an argument against the application of both of these techniques. Sometimes, however, techniques that allow the direct measurement of a powdered sample are necessary. This is the case for the study of adsorption processes in heterogeneously catalyzed reactions.

Various IR techniques exist that allow direct measurement of powders, for example, the measurement of a diffuse reflectance (DR), a photo acoustic (PA), or an emission spectrum (ES) [5, 6], see Figure 14.1. However, the measurable radiation intensities are in these three cases substantially lower than in transmission or attenuated total reflection (ATR) spectroscopy, that is, they were analytically uninteresting for a long time. In this context the introduction of FT-technology (FT: Fourier transformation) led to a major success. With it, the analysis time was drastically reduced and spectra could be acquired in a reasonable time, contributing to a sufficiently enhanced signal-to-noise ratio. The FT-technique, in combination with DR spectroscopy, then led to the development of DRIFTS (diffuse reflectance infrared Fourier transform spectroscopy) [7].

DRIFTS provides the possibility to investigate heterogeneously catalyzed gas phase reactions under working conditions (*in situ*). Thus, information on the interaction of the reactants with the surface of the catalyst and the kinetics of surface intermediates can be obtained, contributing to a deeper understanding of the catalysis mechanism.

14.2
Basic Principles

In DRIFTS the IR beam is directed onto a powder material, which gives rise to scattering due to the rough surface of the sample. Depending on the nature of the sample a superposition of reflection, diffraction, refraction, transmission, and absorption processes occurs. The scattered radiation distributes itself diffusely over the whole hemisphere above the sample (Figure 14.2).

The reflection on rough surfaces or powdered samples is sometimes subdivided into three different categories [8]. The first is the regular, directed reflection of the incident radiation on the boundary layers parallel to the "macroscopic" surface of the sample or the powder. Analogous to a smooth surface, the angle of incidence and the angle of reflection are equal (specular reflection). This type of isotropic reflection represents the regular proportion of the so-called Fresnel reflection [9].

Reflection can also occur at the boundary layers that are not oriented parallel to the "macroscopic" surfaces. Single or multiple reflections thereby lead to a diffuse component of Fresnel reflection. Fresnel reflection can affect the resolution and can lead to a shift or distortion and even affectation of IR bands. From the analytical point of view, however, the proportion of diffuse reflected radiation is interesting, where alongside the reflection processes the absorption of radiation also takes place. The IR radiation penetrates the powder, interacts via various processes (e.g., transmission and reflection) with the sample material and finally escapes as diffuse radiation (Kubelka–Munk path in Figure 14.3). The information on the incidence angle is thereby lost.

In DRIFTS experiments the main concern is aimed at the non-Fresnel reflection. The regular portion of the Fresnel reflection is omitted via a suitable geometry of the measuring set-up. Concerning the diffuse portion of Fresnel reflection, the most efficient reduction is achieved by diluting the analyte with a powdered absorption-poor matrix [10]. However, this is not always advantageous in practice due to the reasons given above.

The set-up for carrying out DRIFTS analyses is limited to the collection of the diffuse scattered radiation inside one part of the entire half space. In order to direct a sufficiently high proportion of the diffuse reflected radiation to the detector, appropriate optical systems are required [8, 10–15].

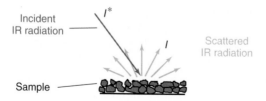

Figure 14.2 In DRIFTS the IR beam is focused on a powdered sample. One part of the radiation is diffuse reflected to the hemisphere above the sample.

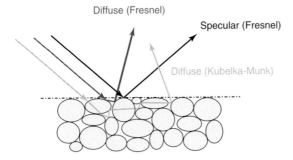

Figure 14.3 Scheme with different types of reflection. Specular reflection occurs directly at the global surface of the powder (Fresnel reflection). Single or multiple reflections can lead to a diffuse component of the Fresnel reflection. A combination of reflection and transmission equates to the diffuse reflection according to the Kubelka–Munk model [8].

The information from a DRIFT spectrum is practically identical to that from a transmission measurement. The intensities of the single bands can, however, be quite different. Since the penetration depth of the IR radiation depends on the absorption behavior of the solid, IR bands with small extinction coefficients from DRIFTS measurements are more pronounced than would be expected from a comparable transmission experiment. This is an advantage if the aim of IR experiments is the observation of adsorbate bands with low intensities.

DRIFTS is occasionally also suitable for quantitative analyses. Diverse models exist for the quantitative description of DR [16–22]. The most common and probably the most used is the Kubelka–Munk (KM) model, which relates the reflectance of a supposed infinitely thick layer R_∞ to the ratio of the absorption coefficient (absorption module) K and the scattering coefficient (scattering module) S (see Box 14.1).

$$R_\infty = 1 + \frac{K}{S} - \sqrt{\frac{K}{S} \cdot \left(2 + \frac{K}{S}\right)} \tag{14.1}$$

The equation for K/S is known as the *Kubelka–Munk transformation*:

$$F(R_\infty) \equiv \frac{(1 - R_\infty)^2}{2 \cdot R_\infty} = \frac{K}{S} \tag{14.2}$$

For most cases the scattering module S will be taken as constant and the absorption module K is proportional to the concentration c and the molar extinction coefficient ε, similar to the absorbance E (with $E = \varepsilon \cdot d \cdot c$) according to the Lambert–Beer (LB) equation (for details see Box 14.1):

$$K = 2 \cdot \ln(10) \cdot \varepsilon \cdot c \tag{14.3}$$

The diffuse scattering leads to differences in the usual relation between the transmission and absorption. This is illustrated in Figure 14.4, where the transmission (T), according to the LB law, and the reflectance of an infinitely thick

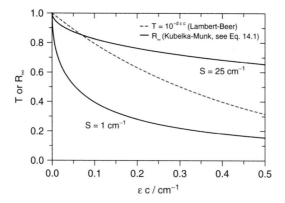

Figure 14.4 Lambert–Beer transmission (*T*) and Kubelka–Munk reflectance (R_∞) as a function of the absorption module *K* (divided by 2 · ln(10) according to Eq. 14.3) for different values of the scattering module *S*.

layer, according to the KM equation, are displayed. The layer thickness (*d*) for the transmission is set arbitrarily to 1 cm.

For small absorption modules ($\varepsilon \cdot c$ is small) the reflectance is smaller than the transmittance. Therefore, one expects a higher intensity of the respective bands obtained by DRIFTS compared to the transmission experiment. This behavior is more pronounced for smaller values of the scattering module.

For strong absorption, however, this situation is reversed. Considering only the intensity one would conclude that a transmission experiment would be advantageous. Note that this phenomenon is related to the penetration depth of the IR radiation into the sample. If only weak absorption appears, the penetration depth is correspondingly large, whereas with strong absorption only a small proportion of the sample material is involved.

Box 14.1: Kubelka-Munk Model

Reflection and absorption of a sample correlate with one another. This fact was described by Kubelka and Munk with a two-parameter approach [23, 24].

It considers a sample that is extended to infinity in a direction perpendicular to the plane of its surface. Only the radiation flux along this perpendicular axis is taken into account. By the interaction of the radiation with the sample, a weakening, and partially also an inversion of the direction, of the radiation flux results. This is described by an absorption module (*K*) and a scattering module (*S*). Both are wavenumber dependent and have the dimension of a reciprocal length.

The KM function is based on a layered structure of the solid with infinitesimally small layer thickness (*dx*). Every layer is irradiated from two sides; from the direction of the irradiated outer surface and from the reflected radiation of the layers underneath. Both radiation fluxes run opposite to one another and their intensities are modified due to absorption and scattering. An amplification of the radiation flux results from the scattering of the opposing radiation flux. At any point, the intensities of the radiation going down and the radiation going up, by reflection or scattering, amount to i and j, respectively. If one looks at a differential layer of thickness dx, then a constant proportion is absorbed and scattered by the radiation passing through the layer:

$$S \cdot dx + K \cdot dx = (S + K) \cdot dx \tag{14.4}$$

On passing through the layer the flux i is weakened by:

$$(S + K) \cdot i \cdot dx \tag{14.5}$$

Since the already scattered upward moving radiation (j) is in turn scattered downwards, its weakening can be described in the same way:

$$(S + K) \cdot j \cdot dx \tag{14.6}$$

However, the portion of the intensity, which equals the part of the radiation moving downwards decreased by scattering (Figure 14.5), increases the intensity of the upward-moving radiation and *vice versa*. This leads to the

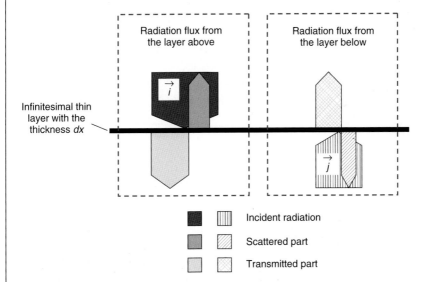

Figure 14.5 Transmission and scattering on an infinitesimal thin layer. Both the intensity of the layer above as well as the intensity of the layer below (i and j) are changed due to absorption or scattering (cf. [10]).

following two simultaneous differential equations:

$$-di = -(S + K) \cdot i \cdot dx + S \cdot j \cdot dx \tag{14.7}$$

$$dj = -(S + K) \cdot j \cdot dx + S \cdot i \cdot dx \tag{14.8}$$

Here, the direction of the spatial coordinate x was chosen in such a way that it runs from the bottom to the top. By the division of both differential equations by i or j, and subsequent addition, a single differential equation is obtained:

$$\frac{dj}{j} - \frac{di}{i} = dln\frac{j}{i} = -2 \cdot (S + K) \cdot dx + \left(\frac{i}{j} + \frac{j}{i}\right) \cdot S \cdot dx \tag{14.9}$$

The ratio between j and i corresponds to the reflectance (r) on the observed differential layer:

$$r = \frac{j}{i} \tag{14.10}$$

After its introduction and the separation of the variables, the following equation arises:

$$\int_{R'}^{R} \frac{dr}{r^2 - 2 \cdot a \cdot r + 1} = S \cdot \int_0^H dx \quad \text{with} \quad a = \frac{S + K}{S} \tag{14.11}$$

R and R' are defined as the reflectances within the boundaries of the spatial coordinate $(r(x = 0) = R'$ and $r(x = H) = R)$. They are dimensionless and have values between zero and one. By expansion into partial fractions Equation 14.11 can be integrated:

$$\frac{1}{2 \cdot \sqrt{a^2 - 1}} \cdot \left(\int_{R'}^{R} \frac{dr}{r - a + \sqrt{a^2 - 1}} \right.$$
$$\left. - \int_{R'}^{R} \frac{dr}{r - a - \sqrt{a^2 - 1}} \right) = S \cdot \int_0^H dx \tag{14.12}$$

$$\ln \frac{\left(R - a - \sqrt{a^2 - 1}\right) \cdot \left(R' - a + \sqrt{a^2 - 1}\right)}{\left(R' - a - \sqrt{a^2 - 1}\right) \cdot \left(R - a + \sqrt{a^2 - 1}\right)}$$
$$= 2 \cdot S \cdot H \cdot \sqrt{a^2 - 1} \tag{14.13}$$

For an infinitely thick layer, the reflectance R' would be equal to zero. In this case the left side of Equation 14.13 goes to infinity or $R - a + (a^2 - 1)^{0.5}$ goes to zero, since $0 \le R \le 1$. This results in the definition of the reflectance of an infinitely thick layer (R_∞):

$$R_\infty = a - \sqrt{a^2 - 1} \quad \text{resp.}$$

$$R_\infty = 1 + \frac{K}{S} - \sqrt{\frac{K}{S} \cdot \left(2 + \frac{K}{S}\right)} \tag{14.14}$$

The formation of the inverted function is known as the *Kubelka–Munk transformation*. This transformation gives the ratio of the absorption and scattering modules in dependence on the reflectance of an infinitely thick layer:

$$F(R_\infty) \equiv \frac{(1 - R_\infty)^2}{2 \cdot R_\infty} = \frac{K}{S} \tag{14.15}$$

For most cases the scattering module S will be taken as constant. The absorption module K is, however, according to the LB law proportional to the concentration c and the molar extinction coefficient ε:

$$K = 2 \cdot \ln(10) \cdot \varepsilon \cdot c \tag{14.16}$$

In the above derivation the angular distribution of the radiation is not considered and merely a parallel and perpendicular incidence beam has been assumed. The factor of 2 in Equation 14.16 takes into account that the mean path of a diffuse radiation is twice as large as the actual thickness of the considered layer. In this context a parallel incident radiation at an angle of 60° leads to the same result as an ideal diffuse irradiation of the surface, since the inverse cosine of 60° is equal to 2. Kubelka addressed these issues in a later work [24].

Considering Equations 14.15 and 14.16, a proportionality between the inverted function $F(R_\infty)$ and the concentration c is established. This relationship is suitable for quantitative analysis.

The KM function is only valid if both the sample and the reference material fulfill the criterion of an infinitely thick layer, that means the IR radiation should not penetrate to the bottom of the sample holder. The penetration depth of the IR radiation depends on the scattering and absorption properties of the sample [25]. In the case of IR transparent powders it can be several millimeters [13].

$$\frac{R_\infty}{R_{\infty 0}} \quad \rightarrow \quad R_{\text{relative}} = \frac{I}{I_0} \tag{14.17}$$

Since absolute values of R_∞ are normally unavailable, in practice the intensities I and I_0 of reflected radiation are obtained for a sample and a reference probe (see Figure 14.6). An absolute value for the reflectance would only be

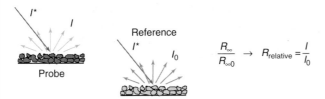

Figure 14.6 Scheme for the determination of the relative grade of reflectance.

obtained if the reference itself would have a reflectance of one over the entire wavenumber range.

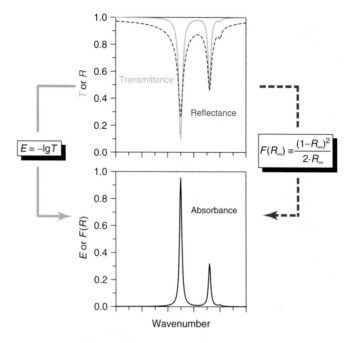

Figure 14.7 illustrates the intensity differences between a transmittance and a reflectance spectrum. A transmittance spectrum is converted into an absorbance spectrum via the LB law, a reflectance spectrum can be converted, for example, via a KM transformation in a similar way. On adoption of the strict validity of both laws (LB and KM), both cases would result in the same absorbance. However, it should be noticed that often neither LB nor KM are exactly fulfilled in practice.

Figure 14.7 Solid path: formation of an absorption spectrum (absorbance) from a transmission spectrum using Lambert–Beer's law ($E = \varepsilon \cdot c \cdot d$ and $E = -\lg T$); dashed path: formation of an absorption spectrum from a reflectance spectrum using the Kubelka–Munk transformation.

14.3
Experimental Set-Up

In addition to a commercially available FTIR-spectrometer, a special DRIFT accessory is required, as described in the following section. It can be placed either in the internal sample chamber of the spectrometer or in an additional external box.

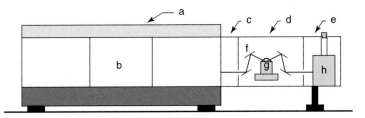

Figure 14.8 Scheme of a spectrometer with external accessories: a, FTIR spectrometer; b, internal sample chamber; c, mirror chamber; d, external sample chamber; e, detector chamber; f, mirror system; g, DRIFTS measuring cell; h, MCT-detector.

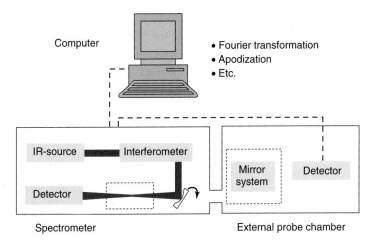

Figure 14.9 Simplified scheme of the interior of a FTIR spectrometer with additional, externally arranged accessories. By controlling a mirror in the spectrometer, the IR radiation can be directed to either the internal or the external sample chamber.

The interior of a FTIR spectrometer is quite simply laid out (see Figures 14.8 and 14.9). The main components are an infrared source, an interferometer, and a detector. With an external DRIFTS arrangement the detector would also be placed externally. For DRIFTS, photoelectric (and photovoltaic) detectors are normally used. They contain semiconductor materials which change their conductivity through interaction with the IR photons. Almost as standard, so-called MCT (mercury cadmium telluride) detectors are used, which are indirectly cooled with liquid nitrogen (77 K).

For a light source in the mid-IR (MIR) range a Globar is typically used, which is a conductive silicon carbide ceramic heated to temperatures of about 1500 K. The source is a thermal radiator, whose behavior approximates to that of a black radiator [26].

The IR radiation with its total bandwidth, that is, all of its wavelengths, is fed into the interferometer as a parallel bundle. For details concerning FT spectroscopy see Box 14.2.

Behind the interferometer the IR beam is directed to the sample which is placed in the DRIFTS accessory.

Box 14.2: Interferometry and Fourier Transformation[1]

The working principle of interferometry is exemplified by means of a Michelson interferometer (Figure 14.10).

The incident IR radiation is split by a semi-transparent beam splitter (e.g., germanium-coated potassium bromide). One part is diverted to a mirror (M1) and the other part passes the beam splitter and reaches the second mirror (M2). The returning radiation from both mirrors is partially reunited at the beam splitter. This leads to interference, that is, to the superposition of the single waves of both partial beams. Now one of the two mirrors is moved back and forth over a defined distance at a constant speed. This causes a difference in runtime or path length (δ) between both returning partial beams. Since the radiation passes the distance between the beam splitter and the mirror twice, the phase difference δ is twice as high as the deflection of the mirror (Δx).

$$\delta = 2 \cdot \Delta x \tag{14.18}$$

The consequences of this can be demonstrated by the example of monochromatic radiation. The intensity of both interfering waves would reach a maximum when the phase difference is an integer multiple of the wavelength (constructive interference). On the other hand, the intensities cancel each other out when the phases differ by one half wavelength

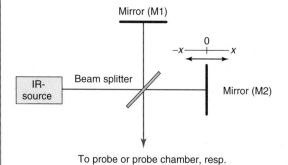

To probe or probe chamber, resp.

Figure 14.10 Schematic set-up of a Michelson interferometer.

---

[1] See also Section 6.2.3, page 202

(destructive interference) (see Figure 14.11):

$$\delta_{constructive} = 2 \cdot \Delta x_{constructive} = N \cdot \lambda \quad \text{with} \quad N = 0, 1, 2, \ldots$$
$$(14.19a)$$

$$\delta_{destructive} = 2 \cdot \Delta x_{destructive} = (N + 0.5) \cdot \lambda \qquad (14.19b)$$

In the monochromatic case, both waves interfere to a "new" wave whose frequency and wavelength remain unchanged. The amplitude is dependent merely on to what extent the phases are shifted relative to each other.

In the polychromatic case, the frequencies and the maximum amplitudes change. Thus, the amplitude pattern of the interfering waves can appear complex, which is shown for two interfering waves in Figure 14.12.

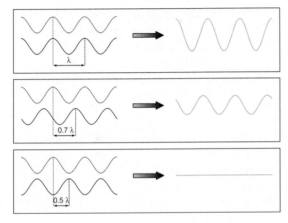

Figure 14.11 Constructive and destructive interference for the monochromatic case. A superposition in the same phase leads to a wave with the same wavelength but with twice the amplitude. If the phases of the interfering waves are shifted against each other the amplitude decreases. Total extinction results when the phase difference is equal to half the wavelength.

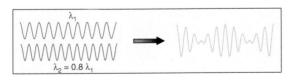

Figure 14.12 Two interfering cosines waves whose wavelengths differ by 20% are shown as a simple example for the polychromatic case. The interference leads in this case to a periodic beat.

An interferometer thus delivers the intensity of the polychromatic radiation brought to interference in dependence on the mirror deflection (Δx). In order to ascertain the information on the therein-contained single frequencies, a mathematical process is required, which is integrated in the software of the spectrometer. The mathematical process is the Fourier transformation:

$$S(\tilde{v}) = \int_{-\infty}^{\infty} I(\delta) \cdot \cos(2 \cdot \pi \cdot \tilde{v} \cdot \delta) \cdot d\delta \qquad (14.20)$$

This is a transformation of the spatial domain ($I(\delta)$ corresponding to the interferogram) to the frequency domain ($S(\tilde{v})$ corresponding to the spectrum). Figure 14.13a shows the FT of a single cosine function. The result is a Dirac delta-function, that is, an infinitesimally narrow, infinitely high pulse at the corresponding frequency. If several cosine functions overlap, the FT provides the respective frequencies of each individual function (see Figure 14.13b).

A real spectrum, however, does not consist of infinitesimally narrow lines, but of bands whose intensities are ultimately distributed over a certain wavenumber range. The intensity distribution leads to interferograms whose envelopes also give a distribution that tends to zero with increasing mirror deflection, which is shown for a Gaussian distribution in Figure 14.14.

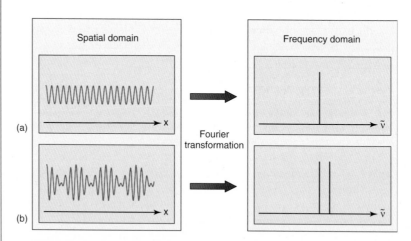

Figure 14.13 (a) Fourier transformation of a cosine function produces an infinitesimally narrow pulse. (b) Shows the overlapping of two cosine functions; after the Fourier transformation two pulse functions appear.

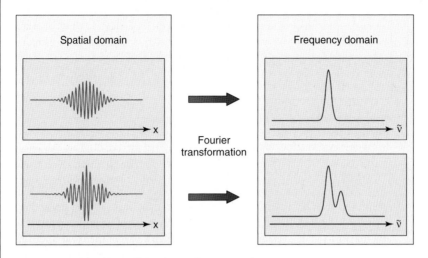

Figure 14.14 Interrelation between interferogram and spectrum or spatial domain and frequency domain, respectively (here: a single Gaussian distribution in comparison to a superposition of two Gaussian distributions).

Inevitably, the interferogram of the polychromatic IR radiation has its highest intensity when the distances from each mirror (M1 and M2 in Figure 14.10) to the beam splitter are equal ($x = 0$). In this case a constructive interference over the whole bandwidth results. This effect or condition is known as *center burst*. The center burst is exploited to metrologically determine the zero position of the mirror deflection.

To ascertain the mirror position (x or Δx), again the interferometer principle is applied, however, this time using a monochromatic light source. Typically a He–Ne-laser ($\lambda = 633$ nm or $\nu = 4.74 \cdot 10^{14}$ Hz) whose radiation is fed into the interferometer simultaneously with the IR primary beam so that the position of the mirror (M2) is recorded.

For the position measurement the zero-crossings of the laser signal are drawn upon. The measurement of the mirror position is inevitably digitalized. Therefore, the intensities at the detector are determined as a function of discrete values of the mirror deflection. This discretization in finite differences, however, also influences the result of the FT. Certain transformations and correction strategies (e.g., discrete Fourier transformation (DFT) in combination with a so-called zero-filling, by which additional "interim data" by means of interpolation are, in principle, generated) are used and the recording rates are adjusted accordingly.

The resolution of an interferogram and therefore the spectrum depends on the maximum mirror deflection. The best resolution is obtained with

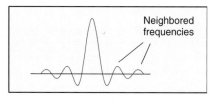

Figure 14.15 After Fourier transformation undesirable neighbor frequencies ("feet") arise from the finite mirror deflection of the interferometer. This can be suppressed by mathematical application of a so-called apodization function (prior to the Fourier transformation).

an infinite path of the mirror. In reality, the maximum mirror deflection, however, is limited. In calculating the spectra, a range of neighbored minima and maxima ("feet") result for a band, on which a part of the band intensity is distributed (see Figure 14.15). The mathematical procedure with which this is corrected is called *apodization* ("*removing the foot*").

Nowadays, all of these transformations and correction strategies for the avoidance of spectral artifacts are integrated in the spectrometer software so that the user remains unaffected by them [27].

14.3.1
DRIFTS Accessories

Carrying out DRIFT spectroscopy requires that a special mirror arrangement is placed in the sample chamber of the spectrometer. The mirror arrangement allows the IR radiation to focus on the sample and to collect a part of the emerging diffuse reflected radiation. Such systems are offered as DRIFTS accessories by various suppliers. The mirror system presented here contains two ellipsoid mirrors whose arrangement reminds one of the arm position of a mantis and thus led to the name "Praying Mantis™ system" [15].

The radiation is directed into the mirror arrangement in three dimensions. The IR beam directed into the sample chamber is focused on the sample via three mirrors. A part of the diffuse reflected radiation is again directed by three mirrors to the exit of the sample chamber, which is connected to the detector (Figure 14.16).

The solid sample for investigation is poured into an open sample holder and thus exposed to the prevailing conditions in the sample chamber. Primarily, this accessory for recording powder spectra is intended to reduce the time required for sample preparation for transmission measurements of solids. For the investigation of heterogeneous catalysts under *in situ* reaction conditions this arrangement is unsuitable since, in this case, all components of the external attachments would be exposed to the reactive gas phase. The common sample holder must be substituted with a hermetically sealed reactor chamber.

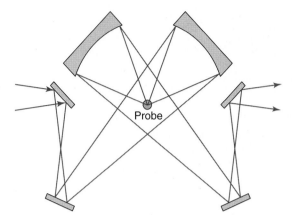

Figure 14.16 Principle of the beam path in a Praying Mantis™ system.

14.3.2
DRIFTS Reactor Cells for Heterogeneous Catalysis

In a reactor cell (see Figure 14.17) catalyst powders can be introduced without any prior mechanical manipulation, and especially without the addition of a diluent like potassium bromide. Mostly, it is possible to direct reaction gas mixtures through the loosely-poured and smooth coated catalyst powder. Combined with electrical heating of the sample holder, IR investigations can be carried out at elevated temperatures and in the presence of a reaction gas, that is, *in situ*. Heatable and evacuable cells to record DRIFT spectra under reaction conditions are described in the literature [15, 28–31].

Needless to say, the measuring cells are equipped with windows through which the incident and diffuse reflected IR radiation are directed. Typical IR transparent materials are used as windows, for example, potassium bromide, calcium fluoride, zinc selenide, and so on. For investigations under high pressure, DRIFTS cells are only partially suitable. The pressure resistance depends mainly on the dimensions and the window material.

The constructive arrangement of the windows (or window; there are cells with one as well as with two windows) varies. Ultimately the geometry of the beam path should remain untouched and the loss of intensity by reflection on the windows should be avoided as far as possible. The window size and the angle of the window to the beam path must be adjusted accordingly.

Conventional reactor cells possess only a single sample holder. In principle, this should be sufficient, since DRIFTS is concerned with a pure single beam technique and thus only one sample can be measured at a time. Sometimes, when carrying out long-term studies (over hours or days) or *in situ* experiments, a single sample holder proves to be disadvantageous. Thus, changes outside the sample can cause perturbing absorption contributions which are reflected in the spectra.

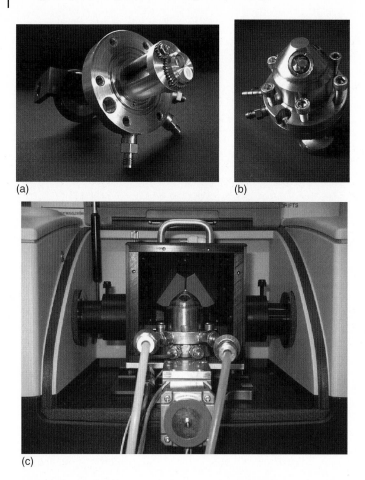

(a)

(b)

(c)

Figure 14.17 DRIFTS reactor cell with two sample holders (self-construction by TU Darmstadt). (a) Bottom part of the cell with a heatable sample stamp and a sample holder for two different probes which can be turned from the outside; (b) complete cell with its heatable dome; (c) an installed reactor cell in a Praying Mantis™ system.

In reactor cells with more than one sample holder a catalyst sample and a reference substance can be placed side by side. Both substances are stored in porous sample cups on a rotatable holder inside the cell, and can be positioned in the spectrometer beam path alternately (quasi-two-beam technique) [32, 33]. Since now both samples are exposed to the same reaction gas, IR absorptions that are caused by gas phase molecules in the reactor cell can be eliminated, as well as changes that occur due to deposits on the windows of the cells. With such DRIFTS reactor cells it is possible to detect adsorbates or surface reactions under *in situ* conditions.

14.4
Application Examples

14.4.1
NOx Reduction

The following example has been chosen from the field of automotive exhaust catalysis. It describes the application of DRIFTS in the context of a research area called *NSR-technology* (NSR: NOx storage reduction) [34]. The catalysts used consist of several components, which have independent functions in the global catalyst cycle, but only the whole unit works as a "catalytic converter" of pollutants into harmless gases. To better understand the NSR-technology it will be briefly explained. Thereafter, the potential of DRIFT spectroscopy to obtain mechanistic and kinetic information about the reactions taking place on the catalyst components will be demonstrated.

14.4.1.1 NOx Reduction via NSR-Technology

The reason for the installation of catalytic converters in new vehicles is to reduce the emission of carbon monoxide, unburned hydrocarbons (HC) and nitrogen oxides by transforming them into the health-safe compounds: water, carbon dioxide, and nitrogen. The oxidation of CO and HC while simultaneously reducing NOx, however, only succeeds for a defined stoichiometric oxygen to fuel ratio. In the case of modern catalytic converters, the partial pressure of the exhaust oxygen or its activity is measured directly over a lambda oxygen sensor and the air supply for optimal exhaust conversion is controlled according to the changing manner of driving.

In the context of the reduction of greenhouse gases, such as CO_2, the automotive sector now offers alternatives to the common Otto-engine: diesel and gasoline lean burn engines, as well as a combination of both technologies, known as "*Diesotto*"-engines, have a considerably higher fuel efficiency due to their operation with excess oxygen and, as a result, lower CO_2 emissions. The expensive NOx after-treatment under the lean conditions is a disadvantage. One technical solution is the selective catalytic reduction (SCR) in which NOx reacts with ammonia or HC, among others, to form nitrogen. Another solution is the NSR-technology (Figure 14.18).

Under lean conditions NOx is adsorbed on the catalyst components by the formation of nitrites and or nitrates. Then, during a short, fuel rich period with excess CO and HC the stored nitrites/nitrates are converted to N_2, CO_2, and H_2O on a redox catalyst (Figure 14.18). After regeneration of the whole catalyst, NOx is stored again and the cycle repeats. Precious metals, such as platinum or rhodium, are suitable redox catalysts [35] and the main alkali or alkaline earth metal oxides, such as BaO, are used as NOx storage components [36].

Cerium oxide can operate to store oxygen due to its redox change (Ce^{III}/Ce^{IV}) [44]. In addition, it appears that CeO_2 is also a NOx storage component [37] and especially at low temperatures, which is advantageous for the cold start phase.

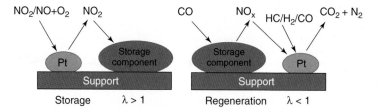

Figure 14.18 Composition and operation mode of a NSR catalyst. In addition to a supporting material (e.g., alumina) the catalyst contains a redox component (e.g., platinum) and a NOx storage component (e.g., barium oxide).

Moreover, CeO_2 is stable over a large temperature range, making it also a good support material. For this reason cerium oxide-based NSR catalysts are interesting and a subject of intense research [38]. In order to be able to understand the complex processes under real operating conditions it is recommended as a first step to investigate the interaction between single model exhaust gas components and individual catalyst components. This separation approach involves a multitude of measurements and, in the initial stage, does not represent the real system exactly. It has, however, the advantage that the understanding grows gradually and the contributions of individual components can be studied in a rational way.

14.4.1.2 Solid Spectra with Different Pre-Treatment

Results are presented below for selected systems that give insight into DRIFT spectroscopic research of NOx storage on NSR catalyst components.

For reasons of reproducibility it should be noted that before the actual experiment is performed the sample undergoes a standard pre-treatment. The effect of a pre-treatment of CeO_2 can be observed based on the DRIFT spectra shown in Figure 14.19. Each spectrum was recorded with KBr as reference sample.

In the top spectrum are large, wide bands caused by OH vibrations which disappear after thermal treatment under nitrogen. The OH vibrations originate from adsorbed water. A subsequent treatment with oxygen leads to narrow bands coming from cerium-oxygen vibrations. Reaching the final temperature stops the pre-treatment procedure and subsequent NOx adsorption experiments start. Overall, it is clear that a standardized pre-treatment is indispensible for the comparison of spectra.

14.4.1.3 Differential Spectra for Adsorbate Identification

The bottom spectrum in Figure 14.19 shows a solid-state spectrum after a treatment with NO (here: 1000 vol ppm). Contrasting the spectra before and after NO adsorption alone is confusing for the purpose of analysis and interpretation of the resulting changes in CeO_2. The generation of a differential spectrum remedies this (see Figure 14.20).

In the differential spectrum positive band contributions starting from the baseline can be recognized that, in this case, result from the formation of various bound

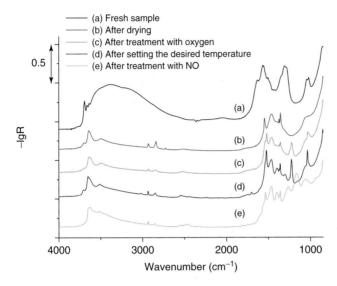

Figure 14.19 DRIFT spectra of CeO$_2$ in different pre-treatment states.

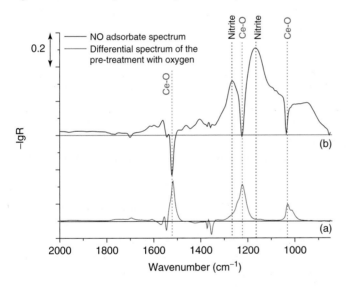

Figure 14.20 Differential spectra gained from the spectra after the pre-treatment with oxygen (a) and after subsequent NO adsorption (b).

nitrite species [37]. Thereby, monodentate, bidentate, or chelating nitrites, as well as cis- and trans-isomers, and also dimers of nitrites are found. These can be bound on the cerium oxide surface via both the nitrogen and the oxygen atom [39]. An exact identification of the species present is only possible to a limited extent without "band deconvolution" due to the overlapping of individual bands.

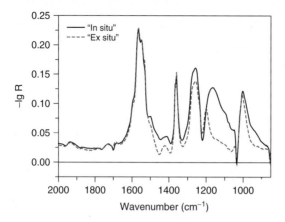

Figure 14.21 Differential spectrum of the adsorbates after treatment with NOx at 200 °C (solid curve) which partly changes on subsequent purging with pure nitrogen (dashed curve).

Furthermore, a band with negative intensity on the adsorbate spectrum is found. This indicates the consumption of a species or at least an IR-active component. Considering the differential spectrum of the oxidative pre-treatment in the lower part of Figure 14.20, it becomes clear that these "negative bands" can be assigned to cerium-oxygen bands. From this, mechanistic information can be deduced, that is, a part of the NO molecule is bound to the oxygen at the cerium oxide lattice, or has reacted with cerium oxide.

14.4.1.4 Advantages of *In Situ* Measurements

Not all adsorbates prove to be temperature stable and during the NOx adsorption and at particularly elevated temperatures the dynamics between ad- and desorption becomes apparent. If the NOx-treated sample is, for example, removed from the reactor measuring cell and exchanged with the KBr reference (as occurs with a cell with only one sample holder) and/or is not exposed to NOx for a short time, this can lead to the desorption of temperature-unstable species. The effect is shown based on the adsorbate spectra in Figure 14.21.

Moreover, this demonstration has not yet considered that further changes due to interactions with atmospheric air can occur. It remains to be emphasized that the "true" situation and changes on the catalyst are only actually accessible under *in situ* conditions.

14.4.1.5 Kinetics and Temperature Dependence of NOx Storage

If the spectra are recorded repeatedly during the NO treatment then the kinetics of the storage process can also be tracked by means of DRIFTS. The temperature dependence of the NO adsorption on CeO_2 is shown in Figure 14.22.

Within the presented temperature range (100–300 °C) it can be recognized in the spectra of the adsorbates that the maximum band intensity decreases with

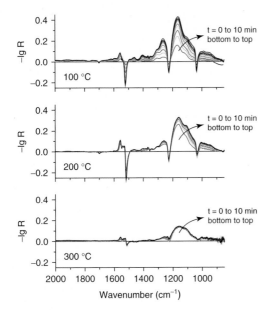

Figure 14.22 Differential spectra during the adsorption of NO on CeO$_2$ at different temperatures.

increasing temperature. Since the band intensities correlate with the amount of stored NO, it can be concluded that less NO is stored at high temperatures than at low temperatures. The reason for this is the low stability of the NO adsorbates on CeO$_2$ at high temperatures [39]. Parallel to that, thermogravimetric measurements on a microbalance and NOx storage capacity (NSC) measurements in a tubular reactor also confirm that the NSC decreases with increasing temperature.

Furthermore, it can be observed that the bands rise more quickly with increasing temperature, that is, the rate of adsorption increases. From a chemical viewpoint this is not surprising since the chemisorption, that is, the reaction of NOx with cerium oxide, is a thermally activated process. Based on the DRIFT spectra the kinetic progress of the bands can be observed and analyzed.

As an example, the kinetics of the most intense nitrite band at 1163 cm^{-1}, a chelating nitrite [37], is presented in Figure 14.23. The differences with regard to the temperature dependence of the formation of the nitrite species can clearly be seen. If a chemical model underlies the investigated reaction, this representation of the DRIFT spectra can be used for a later mathematical modeling and the evaluation of rate coefficients.

14.4.1.6   Influence of Oxygen on NO Adsorption

In a next step the investigated model system (adsorption of NO on CeO$_2$) has been extended by feeding additional oxygen. The results of the adsorption measurements at different temperatures are presented in Figure 14.24.

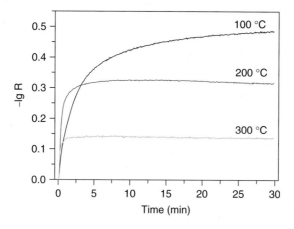

Figure 14.23 Kinetics of the nitrite band at $1163\,\mathrm{cm}^{-1}$.

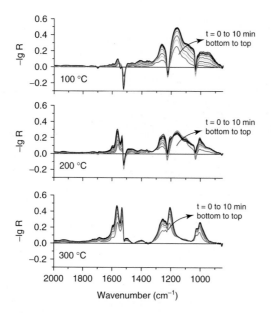

Figure 14.24 Differential spectra during the adsorption of NO/O_2 on CeO_2 at different temperatures.

The presence of oxygen (here: 10 vol%) has a large influence on the adsorbate spectra during the NO adsorption. Comparing the spectra of NO adsorption at $100\,^\circ$C in the absence and presence of oxygen (Figures 14.22 vs. 14.24), it becomes apparent that the intensities of the bands have increased. Otherwise, the band positions are almost identical; no other bands have appeared or disappeared. This indicates that the same adsorbate species are formed during NO storage at $100\,^\circ$C, however, in greater amounts.

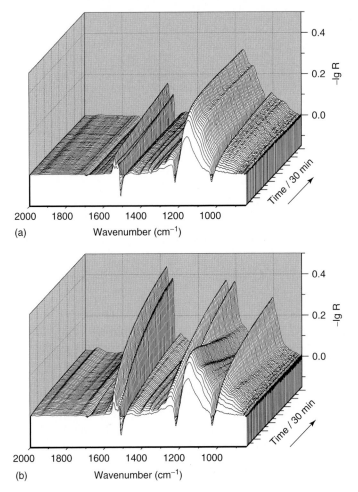

Figure 14.25 Evolution of the adsorbate spectra during the first 30 min at 200 °C: (a) without oxygen and (b) with oxygen.

On increasing the temperature to 200 °C, the kinetic picture changes. Remarkably, the nitrite band at 1163 cm^{-1} exhibits a maximum after about 3 min – in the remainder of the process the intensity decreases (see Figures 14.24 and 14.25).

The decrease in band intensity occurs because nitrite is consumed. Since the measurements without oxygen have shown that the bands display no loss in intensity up to 300 °C inside 30 min, the consumption of the nitrite species results from the addition of oxygen to the reaction gas. Furthermore, it can be observed that nitrate bands are formed in the region of 1560 and 1520 cm^{-1} [37] which are only partly visible in the spectra at 100 °C. Thus, it can be concluded from the information of the DRIFT spectra that a consecutive reaction takes place in which NO is stored in the form of cerium nitrite in the first step, and this intermediate is

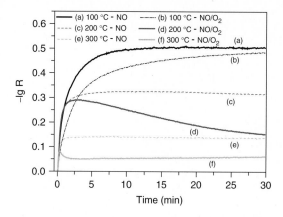

Figure 14.26 Kinetics of the nitrite band at 1163 cm^{-1} during the adsorption of NO without and with additional O$_2$ in the gas phase.

then oxidized to cerium nitrate by oxygen. Comparable results can be observed in the Pt-BaO/Al$_2$O$_3$ system [40].

At 300 °C nitrate bands (1256, 1205, and 1000 cm^{-1}) that were either invisible or too weak to see at 200 °C develop already in the first minutes of the storage. Similarly, the rate of nitrate formation is also higher. The consecutive reaction from nitrite to nitrate at 300 °C is so fast that the nitrite bands can no longer be seen after a minute. However, NO is stored in the form of chelating nitrites (among others) which can be determined from the kinetics shown in Figure 14.26.

The maximum of the band at 1163 cm^{-1} is reached during the first minute and the band intensity has subsequently stagnated at a level of about 0.05 $-$lg R, which is caused by an overlaying of the nitrate band at 1205 cm^{-1}. From the kinetic findings it is obvious that the addition of oxygen increases the rate of NO storage, especially at low temperatures.

14.4.2
Quantification of Oxygen Surface Groups

In this section DRIFTS experiments on a commercial activated carbon (AC) which was oxidized with ozone to form surface oxygen groups are considered. Oxidation was carried out in a stainless steel fixed bed reactor. For oxidation the reactor was filled with 200 mg of AC. After drying at 100 °C for half an hour, the AC sample was treated with 4000 vol ppm ozone in a N$_2$/O$_2$ (1.0 : 1.3) stream. Ozone was received from an ozonizer [41].

Depending on the reaction period, AC was obtained in different degrees of oxidation. Figure 14.27 shows the *ex situ* DRIFT spectra of the untreated and oxidized activated carbon (AC-ox) with KBr as reference.

The focus will be on the peaks in the range between 1500 and 1950 cm^{-1}. There is a band maximum at 1602 cm^{-1} resulting from polyaromatic C=C stretching

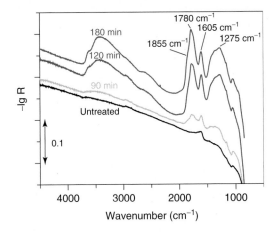

Figure 14.27 *Ex situ* DRIFT spectra of untreated activated carbon and activated carbon oxidized with 4000 ppm ozone at 60 °C with KBr as a reference [41].

vibrations. It is known that carbonyl groups conjugated with the aromatic system enhance these vibrational modes. Therefore, the peak at $1602 \, cm^{-1}$ is considered as a measure of highly conjugated carbonyl/quinone groups [42]. Furthermore, there is one peak at $1780 \, cm^{-1}$ showing a shoulder at $1855 \, cm^{-1}$. These wavenumbers are associated with C=O stretching modes, for instance in anhydrides and carboxylic acids [42, 43]. As expected, all of the mentioned peaks increase as oxidation proceeds.

Additionally, temperature programed desorption (TPD) experiments were carried out with the oxidized sample (AC-ox) that was divided into 10 portions. The samples were heated up to nine different end temperatures between 100 and 500 °C in a N_2 flow at a heating rate of $10 \, K \, min^{-1}$. Half an hour after the end temperature was reached, the reactor was cooled down to ambient temperature and an *ex situ* DRIFT spectrum of the resulting sample was collected. The unoxidized material (AC) was used as reference to receive a straight baseline for subsequent simulation of the spectra. Subsequent elemental analysis for N, C, and H was carried out and O was calculated as the difference from 100%.

During TPD the sample evolves CO and CO_2. Around 150 °C both CO and CO_2 release show a peak, resulting most likely from decomposing carboxylic acid and anhydrides [43].

The *ex situ* DRIFT spectra of the oxidized surface (AC-ox) and the nine samples received by TPD experiments are depicted in Figure 14.28 with AC as reference. The peaks show different decomposition kinetics while heating up.

To get a quantitative analysis the spectra were simulated using Gauss functions. In the range between 1500 and $2000 \, cm^{-1}$ it is expected that anhydrides appear in IR spectra as a doublet with peaks at $1750–1790 \, cm^{-1}$ and $1830–1880 \, cm^{-1}$ [42]. Lactones are supposed to show a single peak around $1750 \, cm^{-1}$. Therefore, simulations are based on five Gaussian, namely at $1603 \, cm^{-1}$ representing carbonyl,

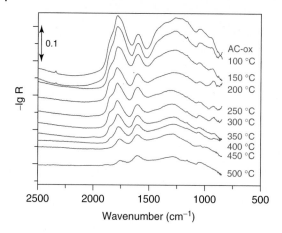

Figure 14.28 *Ex situ* DRIFT spectra of AC-ox and the nine samples received from TPD experiments of AC-ox with the unoxidized activated carbon (AC) as a reference [41].

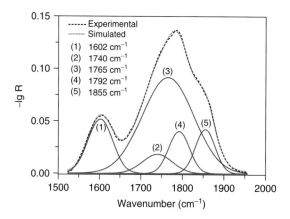

Figure 14.29 Example for simulation of a DRIFT spectrum with a set of five Gauss functions located at 1602, 1740, 1765, 1792, and 1855 cm^{-1} [41].

at 1744 cm^{-1} for carboxylic acid, lactone at 1740 cm^{-1} and two Gaussian at 1792 and 1855 cm^{-1} for anhydride. One example for the fitting is shown in Figure 14.29.

The integrals over the Gaussian for the simulation of the curves in Figure 14.28 give some insight into the formation and decomposition of the species at the carbon surface. In Figure 14.30 the change in the integrals as a function of temperature is shown for all five Gaussian.

The integral associated with the peak at 1602 cm^{-1} decreases slightly from 100 °C. The integral of the band at 1765 cm^{-1} decreases rapidly from 100 °C. At temperatures above 200 °C the decrease is extenuated and at 350 °C the integral is nearly zero. The integral of the 1740 cm^{-1} peak is constant up to 350 °C and

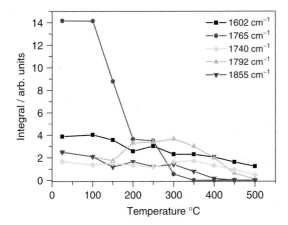

Figure 14.30 Integral values of the five peaks received from simulation of the DRIFT spectra in Figure 14.29 with five single Gauss functions [41].

then decreases, whereas the peak at 1855 cm^{-1} already does so from 300 °C. The peak at 1792 cm^{-1} increases from 150 to 300 °C and then decreases. Based on the development of the integral of each separated IR band (Figure 14.29) and the specific decomposition temperatures known from the literature, the peaks are assigned as follows. The peak at 1602 cm^{-1} belongs to the carbonyl groups which decompose to CO. Carboxylic acid groups appear in the DRIFT spectrum at 1765 cm^{-1} and decompose already from 100 °C to give CO_2. Another possible explanation for the decrease is that two neighboring carboxylic acid groups condense to give an anhydride group. According to the literature, the band at 1740 cm^{-1} would represent the lactones which evolve CO_2 at temperatures higher than 350 °C. The double band of anhydrides is located at 1792 and 1855 cm^{-1}. The increase in the 1792 cm^{-1} peak from 150 to 300 °C can be correlated with the formation of anhydride groups by condensation of two carboxylic acid groups. At temperatures above 300 °C, the anhydrides decompose to CO and CO_2. The weaker anhydride band at 1855 cm^{-1} does not show the same obvious increase at 150 °C but also reduces above 300 °C.

Finally, it shall be pointed out that the DRIFT spectra can give the basis for the measurement of the concentration of the different species on the carbon surface by assuming that the values of the integrals are proportional to the concentration and taking the quantities of CO and CO_2 evolved during TPD into account.

14.5
Summary

Nowadays DRIFT spectroscopy is an established infrared technique for the investigation of powdered materials. Hereby, a qualitative as well as a quantitative analysis is possible. Advantageously, no time-consuming sample preparation by compaction of the probe and no additional diluent is necessary in contrast to the well known IR transmission spectroscopy. Moreover, with the help of reactor cells, DRIFTS gives the opportunity to study reactions between gaseous compounds and solids. This is in great demand in some research fields, like heterogeneous catalysis, when *in situ* measurements are required. Here, changes on the solid catalysts and the nature as well as the kinetics of surface intermediates are observable.

References

1. Niemantsverdriet, J.W. (1993) *Spectroscopy in Catalysis*, Wiley-VCH Verlag GmbH, Weinheim.
2. Haw, J.F. (2002) *In Situ Spectroscopy in Heterogeneous Catalysis*, Wiley-VCH Verlag GmbH, Weinheim.
3. Krauß, K. (2000) Entwicklung und Einsatz einer DRIFTS-Meßzelle zur In situ-Spektroskopie heterogen katalysierter Gasphasenoxidationen. Dissertation. TU Darmstadt.
4. Drochner, A., Fehlings, M., Krauß, K., and Vogel, H. (1999) *GIT Laborfachzeitschrift*, **43**, 476–479.
5. Griffith, P.R. and Fuller, M.P. (1982) in *Advances in Infrared and Raman Spectroscopy*, Vol. 9, Chap. 2 (Eds R. Clark *et al.*), Heyden, London, pp. 63–129.
6. Woerkom, P.C.M. (1988) in *Advances in Applied Infrared Fourier Transform Infrared Spectroscopy* (Ed. M.W. Mackenzie), John Wiley & Sons, Inc., New York, p. 323ff.
7. Holmes, P.D., McDougall, G.S., Wilcock, I.C., and Waugh, K.C. (1991) *Catal. Today*, **9**, 15–22.
8. Mitchell, M.B. (1993) in *Structure–Property Relations in Polymers*, Advances in Chemistry Series, Vol. 236, Chap. 13 (Eds M.W. Urban and C.D. Craver), American Chemical Society, pp. 351–375.
9. Brimmer, P.J. and Griffith, P.R. (1988) *Appl. Spectrosc.*, **42**, 242–247.
10. Korte, E.H. (1990) *Anal. Taschenbuch*, **9**, 91–123.
11. Clark, R. *et al.* (1982) *Advances in Infrared and Raman Spectroscopy*, vol. **9**, Hexden, London, pp. 63–129.
12. Fraser, D.J.J. and Griffith, P.R. (1990) *Appl. Spectrosc.*, **44**, 193–199.
13. Fuller, M.P. and Griffith, P.R. (1978) *Anal. Chem.*, **50**, 1906–1910.
14. Kortüm, G. and Delfs, H. (1964) *Spectrochim. Acta*, **20**, 405–413.
15. Harrick Scientific Product, Inc. manual, Pleasantville, NY.
16. Mandelis, A., Boroumand, F., and Van den Bergh, H. (1991) *Spectrochim. Acta A*, **47** (7), 943–971.
17. Gade, R., Kaden, U., and Fassler, D. (1987) *J. Chem. Soc., Faraday Trans. 2*, **83**, 2201–2210.
18. Hecht, H.G. (1976) *J. Res. Nat. Bur. Stand. A. Phys. Chem.*, **80** (4), 567–583.
19. Kortüm, G. (1969) *Reflexionsspektroskopie*, Spinger-Verlag, Berlin, Heidelberg.
20. Melamed, N.T. (1963) *J. Appl. Phys.*, **34**, 560–570.
21. Simmons, E.L. (1975) *Appl. Opt.*, **14**, 1380–1386.
22. Dahm, D.J. and Dahm, K.D. (2007) *Interpreting Diffuse Reflectance and Transmittance*, NIR Publications, IM Publications, Chichester.
23. Kubelka, P. and Munk, F. (1931) *Z. Techn. Phys.*, **12**, 593–601.
24. Kubelka, P. (1948) *J. Opt. Soc. Am.*, **38**, 448–457. (errata); (1948) *J. Opt. Soc. Am.*, **38**, 1067.
25. Krivacsy, Z. and Hlavay, J. (1994) *Talanta*, **41** (7), 1143–1149.
26. Günzler, H. and Gremlich, H.-U. (2003) *IR-Spektroskopie*, Wiley-VCH Verlag GmbH, Weinheim.
27. Wartewig, S. (2003) *IR and Raman Spectroscopy*, Wiley-VCH Verlag GmbH, Weinheim.
28. Hamadeh, I.M., King, D., and Griffith, P.R. (1984) *J. Catal.*, **88** (2), 264–272.
29. Kubelkova, L., Hoser, H., Riva, A., and Trifiro, F. (1983) *Zeolites*, **3**, 244–248.

30. Sordelli, L., Psaro, R., Dossi, C., and Fusi, A. (1992) in *Catalysis and Surface Characterization* (Eds T.J. Dines *et al.*), The Royal Society of Chemistry, London pp. 127–135.

31. Bulushev, D.A., Paukshtis, E.A., Nogin, Y.N., and Balžzhinimaev, B.S. (1995) *Appl. Catal. A Gen.*, **123**, 301–322.

32. Fehlings, M., Drochner, A., Krauß, K., Vogel, H., Süttinger, R., and Hibst, H. (1999) Meßzelle, spektrometer sowie spektroskopieverfahren zur untersuchung von proben mittels elektromagnetischer strahlung, O. Z. 0050/49800 (03.09.1999). DE 19910291 A1.

33. Drochner, A., Fehlings, M., Krauß, K., and Vogel, H. (2000) *Chem. Eng. Technol.*, **23** (4), 319–322.

34. Motohisa, S., Tadashi, S., Naoto, M., Satachi, I., Kaichi, K., Toshiaki, T., and Syuji, T. (1994) EP0613714.

35. Hauptmann, W., Drochner, A., Vogel, H., Votsmeier, M., and Gieshoff, J. (2007) *Top. Catal.*, **42/43**, 157–160.

36. Fridell, E.,Skoglundh, Westerberg, B., Johansson, S., and Smedler, G. (1999) *J. Catal.*, **183**, 196–209.

37. Philipp, S., Drochner, A., Kunert, J., Vogel, H., Theis, J., and Lox, E.S. (2004) *Top. Catal.*, **30/31**, 235–238.

38. Symalla, M.O., Drochner, A., Vogel, H., Philipp, S., Göbel, U., and Müller, W. (2007) *Top. Catal.*, **42/43**, 199–202.

39. Philipp, S. (2007) Untersuchungen zur NOx-Einspeicherung an Ceroxid mittels IR-Spektroskopie in diffuser Reflexion. Disseration. TU Darmstadt.

40. Nova, I., Castoldi, L., Prinetto, F., Dal Santo, V., Lietti, L., Tronconi, E., Forzatti, P., Ghiotti, G., Psaro, R., and Recchia, S. (2004) *Top. Catal.*, **30/31**, 181–186.

41. Kohl, S., Drochner, A., and Vogel, H. (2010) *Catal. Today*, **150**, 67–70.

42. Zhuang, Q.-L., Kyotani, T., and Tomita, A. (1994) *Energy Fuels*, **8**, 714.

43. Wiederhold, H. (2006) Modifizierung von Carbon Black mit Ozon: Struktur und Kinetik der Oberflächenintermediate. Dissertation. TU Darmstadt.

44. Matsumoto, S. (2004) *Catal. Today*, **90**, 183–190.

15

Photoelectron Spectroscopy in Materials Science and Physical Chemistry: Analysis of Composition, Chemical Bonding, and Electronic Structure of Surfaces and Interfaces

Andreas Klein, Thomas Mayer, Andreas Thissen, and Wolfram Jaegermann

■ Method Summary

Acronyms, Synonyms
- X-ray photoelectron spectroscopy (XPS)
- Electron spectroscopy for chemical analysis (ESCA)
- Synchrotron induced X-ray photoelectron spectroscopy (SXPS)
- Ultraviolet photoelectron spectroscopy (UPS).

Benefits (Information Available)
- technique is surface sensitive with information depth 1–10 nm depending on electron kinetic energy (XPS, UPS)
- solids, liquids, and gases accessible (XPS, UPS)
- technique can determine chemical composition and oxidation states of elements (XPS, UPS)
- sensitive to chemical reactions at surfaces and interfaces (XPS, UPS)
- depth profiling by angle-resolved measurements or ion etching (XPS)
- morphological properties including growth modes of thin films and adsorbates (XPS)
- binding energies of conducting samples can be referenced to the Fermi energy providing (i) information on surface electrostatic potentials of semiconducting materials which are affected by doping and surface band bending, (ii) energy band alignment including Schottky barrier heights, and (iii) surface photovoltage effects (XPS, UPS)
- sensitive to orbital contributions to the density of states (SXPS)
- absolute determination of work functions possible (XPS, UPS)
- determination of momentum-resolved electronic structure of materials (UPS, SXPS).

Methods in Physical Chemistry, First Edition. Edited by Rolf Schäfer and Peter C. Schmidt.
© 2012 Wiley-VCH Verlag GmbH & Co. KGaA. Published 2012 by Wiley-VCH Verlag GmbH & Co. KGaA.

Limitations (Information Not Available)
- technique cannot identify H and He in solids
- lateral resolution limited to approximately $1\,\mu$m
- assessment of buried interfaces due to high surface sensitivity
- no atmospheric changes of surfaces due to high-vacuum environment
- no absolute binding energies for electrically insulating samples
- no reliable depth profiling for rough surfaces.

15.1
Introduction

Fundamental questions addressed in physical chemistry and materials science are the thermodynamics and kinetics of chemical interface reactions involving solids, liquids, and gases, and also electronic properties depending on bulk defect species and interface states. A wealth of important technological applications crucially depends on these processes including, for example, sensors, catalysis, ion batteries, fuel cells, and thin film solar cells. Many of the involved properties can be addressed by photoelectron spectroscopy. Among them are the composition and electronic structure of surfaces, the effects of doping, chemical reactions at solid/gas, solid/liquid, and solid/solid interfaces, growth modes of thin films, the chemical state of adsorbates, barrier heights at contacts, and even charge transport mechanisms and diffusion processes at interfaces.

As the photoemission technique is introduced in Chapter 16 we will only briefly summarize its basic principles. In addition, a large number of books and articles are available for more detailed information [1–6]. A short but rather comprehensive introduction into practical aspects of using X-ray photoelectron spectroscopy (XPS) is given in Ref. [7]. Some information on state-of-the-art equipment is provided in the information boxes at the end of this chapter.

The emphasis of our chapter is the experimental approach and the type of information accessible by photoemission in its application to modern technology areas where the functionality of devices may depend on complex surface or interface interaction involving submonolayer concentration of species. Different approaches for controlled sample preparation, which are crucial for obtaining reliable information on physico-chemical processes by photoelectron spectroscopy, will be described and illustrated by a number of examples. An overview of the information which can be extracted from X-ray photoelectron spectra will be given rather than an overview of the different mechanisms which can be studied. The different contributions that can affect photoemission spectra will be presented. We concentrate on the most important features from our viewpoint as materials scientists in order to provide insight into the prospects and limitations of the technique. A comprehensive review on the applicability of photoemission is not intended. Finally, some examples for the use of ultraviolet photoelectron spectroscopy (UPS) will also be given.

15.2
Experimental Procedure[1]

15.2.1
Basic Set-Up and Operation Principle

Photoelectron spectroscopy or photoemission is based on the external photoelectric effect, where illumination of a sample with photons of defined energy larger than the ionization energy causes electrons to be emitted from the sample [8]. A typical photoelectron spectrometer is shown in Figure 15.1. Various photon sources are available for lab-based photoemission set-ups. Among them are X-ray tubes, which typically utilize Mg or Al anodes, providing photons of 1253.6 and 1486.6 eV energy with energy widths of 0.7 and 0.85 eV, respectively. With single crystal X-ray monochromators coupled to Al anodes, the energy resolution of the X-ray source can be drastically reduced to <300 meV. The monochromatized X-ray source has the additional advantage of avoiding the typical satellite X-ray emissions [7]. This becomes particularly useful for recording valence band spectra, which are important for the analysis of the electronic structure of surfaces and interfaces (see also Section 16.3).

Photoemission spectra of valence bands, which are formed by the filled electronic (chemically bonding) states of materials with binding energies of up to ~15 eV below the Fermi energy, can also be excited with vacuum ultraviolet radiation. The difference between XPS and UPS is related to the different excitation energies. With UPS only valence electron states can be accessed. Direct information on the chemical bonding to the substrate can be obtained from UPS of molecular adsorbates, and band structure measurements of single crystalline solids further provide a

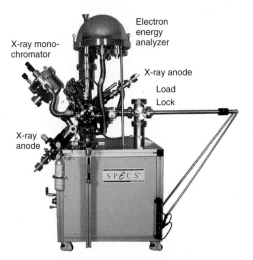

Figure 15.1 A typical photoelectron spectrometer.

[1] See also Section 16.3, page 518

detailed description of the electronic structure of materials [1–4]. In addition to the valence band region, XPS can access deeply bound core-level states, which are characteristic for the respective chemical element and do not contribute to the chemical bonding in a compound. The binding energy of the core levels, however, depends sensitively on the charge at the atomic site and, therefore, varies with the chemical bonds of the atom [6]. This sensitivity is the most exploited feature of XPS and the origin of the frequently used expression electron spectroscopy for chemical analysis (ESCA).

Synchrotron radiation is extremely useful for photoemission experiments. It extends the capabilities of photoemission by providing intense monochromatic light with variable photon energies from the infrared up to several keV. This enables, for example, studies with variable information depths, band structure, and resonant photoemission measurements. Describing the benefits of synchrotron radiation for photoemission experiments, although extensively used by the authors, is beyond the scope of this chapter and the reader is referred to other articles in this series and to literature (Chapter 16; [9]).

In photoelectron spectroscopy the information is contained in the energetic and spatial distribution of the emitted electrons. In vacuum the mean free path of the emitted primary electron is large enough to travel through an energy analyzer to a detector without scattering. Today, different methods are available for electron energy analysis and detection (see Sections 15.6.1–15.6.3). In recent years, the restriction of high vacuum environments has also been partially overcome. The development of high-pressure photoemission currently allows photoemission of samples stored at a pressure of up to 1000 Pa. The experimental set-ups used in such experiments have recently been reviewed [10]. It is, for example, possible to place the sample a few millimeters from the entrance aperture of a differentially pumped electrostatic lens system.

Photoelectron spectroscopy is a highly surface-sensitive technique with an information depth of only a few nanometers. The surface sensitivity is related to the inelastic mean free path of the primary electrons in solids, which obeys a rather general behaviour, as sketched in Figure 15.2 [11, 12]. Only the primary electrons

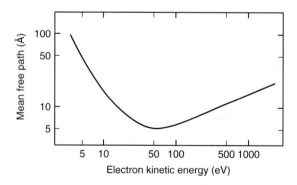

Figure 15.2 Inelastic mean free path of electrons in solids.

contribute to the important chemical and electronic information. Inelastically scattered electrons contribute to the background intensity or also to satellite emissions. The most surface-sensitive information is reached at the minimum path length of 0.5 nm for kinetic energies of the primary photoelectrons of around 50 eV. The minimum is mainly caused by excitation of plasmons. At higher kinetic energies, the mean free path λ follows roughly a square root dependence on energy. There are basically two means to vary the surface sensitivity: changing the electron emission angle or changing the kinetic energy when variable energy photon sources (synchrotrons) are available, or when several core levels of atoms of one element (e.g., Ga 2p with binding energy ~1120 eV and Ga 3d with binding energy ~20 eV) can be excited.

The energy resolution in standard photoemission experiments is limited by the lifetime of the excited state, the energy width of the photon source, and the resolution of the analyzer. In XPS, the light source and lifetime are most important. In UPS, or with synchrotron radiation, the energy resolution of the analyzer may also limit the resolution due the narrow energy width of the photon line, which is only ~2 meV for He I, and of similar magnitude for synchrotron radiation (~2 meV) or even better for UV laser sources (<0.5 meV).

The kinetic energy of the photoelectrons E_{kin} is given by

$$E_{kin} = h\nu - E_B - \phi_{SP} \tag{15.1}$$

where $h\nu$ is the photon energy, E_B the binding energy with respect to the Fermi energy of the sample holder, and ϕ_{SP} the work function of the analyzer. ϕ_{SP} is an instrumental parameter, which is adjusted by setting the Fermi edge emission of a clean metallic sample to zero binding energy. For conducting samples the Fermi energies of the samples and the spectrometer are aligned when electrical contact is established, thereby keeping the energy calibration. The measurement of binding energies with respect to the Fermi energy is a highly useful feature of photoelectron spectroscopy when variations of the chemical potentials of electrons are involved.

As an example, a typical XPS spectrum recorded from a $SrTiO_3$ single crystal surface is displayed in Figure 15.3. The survey spectrum shows emissions from all core levels at characteristic binding energies for the respective elements, which can be found in reference books [7] or online databases [13]. Due to spin−orbit interactions, the electrons emitted from shells with an orbital momentum (p, d, f) show a typical doublet with a L + 1/2 and L − 1/2 component. The intensity ratio of the two components is given by the degeneracy of the levels: (L + 1)/L, which is 2 : 1, 3 : 2, and 4 : 3 for p-, d-, and f-orbitals, respectively.

Photoelectron lines originate from a transition of a bound electron to vacuum. Their kinetic energy is given by Equation 15.1. In addition to photoelectron lines, Auger transitions also occur in a photoelectron spectrum. Two of them, the Ti LMM and O KLL transitions are indicated in the survey spectrum in Figure 15.3. Auger transitions are induced by the recombination of an electron from a lower bound energy level with the core hole created by the original photoemission process.

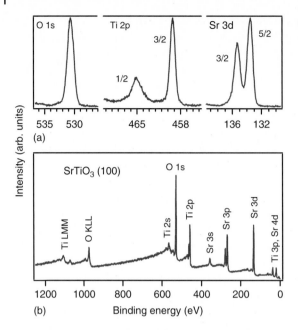

Figure 15.3 XPS survey (b) and detail (a) spectra of a SrTiO₃(100) single crystal recorded with monochromatized Al K radiation. The Ti 2p and Sr 3d levels show splitting into spin–orbit doublets. The sample has been cleaned by heating at 600 °C in 1 Pa oxygen to remove the typical hydrocarbon contaminations from ambient conditions.

The energy from the recombination is then transferred to another electron, which can also be emitted if it has sufficient energy. Auger transitions are indicated as "XYZ"-transitions where the electron shell of the initial core hole is indicated as "X," the shell of the electron which recombines with the core hole as "Y" and the shell of the electron which adopts the recombination energy as "Z." In our example, the O KLL emission in Figure 15.3, for example, is caused by a photo-excitation of the K-shell (1s), a recombination from the L-shell (L_1: 2s, $L_{2,3}$: 2p) and transfer of the energy difference to the L-shell. Photoemission and Auger transitions are compared in Figure 15.4.

According to Equation 15.1, the kinetic energy of a photoelectron depends on the excitation energy. In contrast, the kinetic energy of an Auger electron is fixed by the energy of the recombination process. Auger lines can thus be easily distinguished from photoelectron lines if the photon energy can be varied. The Auger process is an intra-atomic process and leaves the samples with two empty states on a single atom. The doubly charged final state of the Auger process, compared to the singly charged final state of the photoemission process, leads to considerable differences in the energy shifts of photoelectron and Auger lines, which are sometimes indispensable for the analysis of the chemical state [5–7, 14].

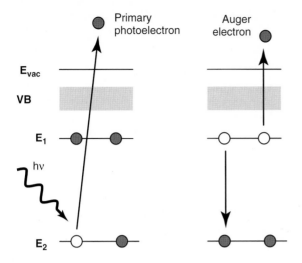

Figure 15.4 Primary photoemission and Auger process.

15.2.2
Sample Preparation

15.2.2.1 Vacuum Conditions

Photoemission experiments are typically performed in ultra-high vacuum, that is, at pressures below 10^{-7} Pa. The vacuum conditions are necessary to keep the inelastic mean free path of the electrons long enough in order to reach the detector without energy loss. In addition, vacuum is required to maintain clean surface conditions during the experiment. Once a surface is completely clean, it will take up a monolayer of adsorbates in 1 s in a vacuum of 10^{-6} Torr if the sticking coefficient of the adsorbates is 1 [1]. The gas dose is quantified in Langmuir units, where $1 L = 10^{-6}$ Torr $\cdot$ 1s $= 1.3 \times 10^{-4}$ Pa $\cdot$ s. Hence, a gas dose of 1 L corresponds roughly to one monolayer for a unity sticking coefficient. In a vacuum of 10^{-7} Pa, a monolayer may adsorb in 1000 s, which is less than a typical experiment time. Fortunately, the room temperature sticking coefficient of adsorbates from the residual gas in the vacuum chamber (N_2, H_2, H_2O, CO, CO_2) is usually much smaller than one.

The sticking coefficient depends crucially on the gas species, the substrate material, and temperature. In particular, for transition metal surfaces, it is more difficult to maintain clean surfaces, while chalcogenides are much less reactive. With oxide surfaces it can also be difficult to avoid hydroxide formation, as water is typically the dominant species in the residual gas. Hydroxide formation can be very difficult to identify on oxide surfaces as the oxygen and valence band emissions are dominated by the substrate lattice oxygen [15, 16]. In addition, peroxide species, which are also likely on oxide surfaces, can lead to very similar photoemission features [17].

15.2.2.2 Preparation of Solid Surfaces for Photoemission Experiments

In photoemission, one can measure samples prepared either *ex situ* or *in situ*, that is, outside or inside the vacuum system of the spectrometer unit. *In situ* experiments are possible either directly inside the spectrometer chamber or in vacuum chambers, which are connected by an ultra-high vacuum transfer system to the spectrometer. The layout of such a system, which is used in the surface science group at TU Darmstadt, is shown in Figure 15.5.

In situ sample deposition is a straightforward way to prepare clean surfaces of various materials. It is almost indispensable for the study of interface properties, which can be performed using stepwise deposition of a contact material onto a substrate. A large number of deposition methods can be used for such studies. With the system shown in Figure 15.5, deposition of films using magnetron sputtering, metal-organic chemical vapor deposition, and molecular beam epitaxy is possible. However, the preparation of thin films with any of these techniques will usually involve extensive developments, so that *in situ* deposition is not always the first choice.

Samples prepared outside the vacuum are always contaminated by adsorbed hydrocarbons and water leading to typical oxygen and carbon emissions at binding energies near 530 and 285 eV (see Figure 15.6). For oxide surfaces, these can be

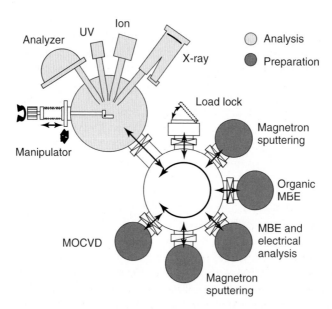

Figure 15.5 Layout of an integrated ultra-high vacuum system combining a multi-technique surface analysis system including a photoelectron spectrometer for XPS and UPS and several chambers for surface preparation and thin film deposition such as molecular beam epitaxy (MBE) or metal-organic chemical vapor deposition (MOCVD).

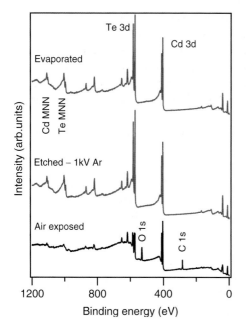

Figure 15.6 XPS-survey spectra of air exposed, Ar ion etched, and *in situ* evaporated CdTe thin film samples. The air exposed sample shows additional emissions of oxygen and carbon and also a different Cd-Te intensity ratio.

removed by heating to moderate temperatures of 400–600 °C. The presence of oxygen during heating can considerably promote the removal of hydrocarbons [18]. A widely used alternative technique for surface cleaning is argon ion sputtering (ion beam (IB)), or argon ion sputtering combined with sample annealing (IBA). Ion sputtering can effectively remove surface contaminations. However, sputtering leads to rupture of chemical bonds and therefore destroys the crystallinity and leads to the formation of defects. Due to preferential sputtering effects, the composition of a compound surface can also be severely modified. Sputtering of TiO_2 surfaces, for example, leads to a reduction of the surface by preferential removal of oxygen [19]. While the crystalline state can often be restored by subsequent annealing, this is usually not possible for the changed compositions. Care must, therefore, be taken in the interpretation of results of sputtered samples.

Argon ion sputtering can also be used to perform depth profiling with XPS. In such cases it is preferable to use low ion energies (<1 kV) to avoid the problems mentioned above. For the CdS/CdTe interface, the validity of the depth profile approach has been verified explicitly by a comparison with an *in situ* evaporation experiment [20]. Not only the stoichiometry of Ar ion etched and evaporated CdTe films are the same, as evident from Figure 15.6, but also the electronic properties are not affected. In this case, depth profiling offers additional opportunities as it could be used to study the effect of post-deposition treatments on interface properties [21], which is not possible in an evaporation experiment.

Alternatives for preparing clean surfaces of *ex situ* prepared materials are cleaving, fracturing, or scratching in vacuum. Cleaving can be particularly useful for preparing selected single crystal surfaces as, for example, the (110) surfaces of

III–V semiconductors like GaAs and InP. An elegant way for surface preparation is offered by the de-capping procedure, where a cap layer is deposited *in situ* onto a freshly prepared sample. When the sample is exposed to air, the cap layer oxidizes and becomes contaminated with adsorbates. When the sample is heated in a vacuum system, the cap layer evaporates and a clean substrate surface is exposed. The technique has been successfully used with an As cap layer on epitaxial GaAs [22] and with Se cap layers on polycrystalline Cu(In,Ga)Se$_2$ films [23].

15.2.2.3 Analysis of Solid/Solid Interfaces Using Photoemission

Interfaces between solids are particularly important for a large number of devices. Due to the high surface sensitivity of photoemission, the analysis of the chemical and electronic properties of interfaces in a completed structure is usually not possible as they are buried under subsequently deposited layers. To overcome this restriction, fundamental properties of interfaces can be studied during their formation in an experiment which repeats photoemission and film deposition in a number of steps, as sketched in Figure 15.7.

An experiment starts with a clean substrate surface. Then, the contact material is deposited in several steps until a complete attenuation of the substrate emission is obtained. For a layer-by-layer growth of the deposited material, this corresponds to

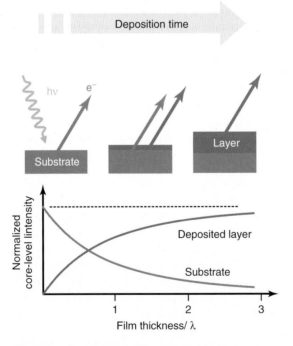

Figure 15.7 Schematic procedure of a stepwise deposition experiment for the analysis of chemical and electronic properties of interfaces between solids.

a film thickness a few times the inelastic mean free path, that is, typically 5–10 nm. Several deposition steps starting from sub-monolayer coverage are typically required in order to derive a complete set of interface properties. With such experiments it is possible to derive growth modes of the deposited films, chemical interactions at the interface, and electronic properties such as Schottky barrier heights, band discontinuities, changes in Fermi level positions, and work function. Examples will be given below. With suitable experimental set-ups, different deposition techniques and thus a wide range of materials have become accessible for such experiments. We have recently described, for example, interface experiments involving the controlled deposition of sub-monolayer $(Ba,Sr)TiO_3$ thin films by magnetron sputtering at substrate temperatures above 600 °C [24]. Systematic studies of the interface formation of magnetron sputtered ZnO films have also been performed [17].

15.2.2.4 Study of Solid/Liquid Interfaces Using Photoemission

Wet chemical processing is widely applied in chemistry, materials science, and device fabrication. A fundamental understanding of the solid/liquid interface on an atomic level is therefore strongly needed. In order to apply photoelectron spectroscopy to study electrochemical interfaces, we have developed the experimental station SoLiAS especially designed as a solid/liquid analysis system, which is run at the synchrotron light source BESSY [25, 26]. SoLiAS allows the SXPS analysis of electrochemical interfaces via three complementary experimental routes:

1) emersion of substrates (working electrodes) from liquid electrolytes and transfer to UHV under controlled, clean, and inert atmosphere [27]
2) freezing thin layers of the electrolyte by cooling emersed samples with liquid nitrogen during pumping to UHV to avoid loss of volatile species [28]
3) synthesis of frozen models of electrolytes *in vacuo* by coadsorption of solvent and redox active species onto liquid nitrogen cooled electrode surfaces [29].

Detailed insight into final surface composition in dependence on applied potential, intermediate reaction steps, and reaction products of electrochemical processes such as etching [27, 30], chemical bath deposition [31], and pulse plating [32] have been accomplished for a number of semiconductor/electrolyte systems. The interaction of solvent molecules with different crystallographic surfaces of semiconductor electrodes [33] and with adsorbed organic sensitizing dye molecules [34] has also been studied. In addition, results on fundamental but yet unsolved questions of electrochemistry have been achieved [35, 36], for example, concerning the redox potential of an electrolyte, the density of electronic states distribution in a redox electrolyte, and solvation of redox active species. The analysis of the electronic structure of the Helmholtz layer and of the bulk electrolyte are still ongoing challenges, as simple one electron redox pairs to be studied in UHV are rare.

We expect that the development of high pressure electron analyzers [10] will open up a route to overcome the restriction to use frozen models of the electrolyte in the near future. The liquid/vacuum interface has also been analyzed using nonvolatile electrolytes [37, 38]. The surface composition of volatile electrolytes has also been analyzed recently using a liquid microjet within a UHV chamber [39].

15.3
Case Studies: XPS

In XPS, the absolute values of intensities and binding energies contain important information. The extraction of information from peak intensities and binding energies is, at least partially, independent, and will thus be outlined separately in the following sections. For consistent experiments both sets of data have to be considered together. It often happens, for example, that the observed binding energies do not allow a unique assignment of the emissions to a particular compound at the surface. In this case, the relative intensities are also helpful to support or exclude possible explanations. The detailed analysis typically requires taking emissions from all elements into account.

15.3.1
Analysis of Peak Intensities

15.3.1.1   Elemental Analysis
The first step in evaluating XPS data is the analysis of the elements present at the surface of a sample. This can provide already very useful information as often unexpected species are observed, in particular for *ex situ* prepared samples. Elemental analysis is accomplished by identification of all (!) peaks in a survey spectrum, which spans the maximum possible energy range defined by the excitation source. Examples are provided in Figures 15.3 and 15.6. Reference spectra for the different elements, as displayed, for example, in the *Handbook of X-ray Photoelectron Spectroscopy* [7], are indispensable tools for elemental analysis. Many elements exhibit more than one emission with a characteristic intensity pattern. Noticeable deviation from the intensity pattern typically indicates the presence of additional elements. Excitation of levels with other photon energies (ghost lines) may be induced by a contaminated X-ray anode (to be inferred from calibration measurements using clean samples).

15.3.1.2   Composition Analysis
XPS is the most important technique for analyzing the chemical composition of surfaces. Although a quantitative analysis is possible with XPS, a large number of effects can contribute to the peak intensities and an absolute determination with an accuracy of 10% or better can be reached in only a few cases. The relative composition, however, is highly reproducible and the technique can thus provide valuable information on compositional changes dependent on sample treatment.

The first step in the analysis of a surface composition is the extraction of peak intensities, which starts with a background subtraction. While fitting of the inelastic background by linear or polynomial functions is only a mathematical operation, the Shirley [40] and the Tougaard [41] methods for background subtraction are based on physical approaches. Shirley assumed that the number of inelastic losses of primary photoelectrons is proportional to the integral number of photoelectrons excited. Therefore, the inelastic background function is proportional to the integral of the complete photoelectron spectrum starting from highest kinetic energies at the Fermi energy. For valence band spectra excited with low photon energies (UPS) this approximation is useful. For wide energy scans in XP spectra including core level lines the background is overestimated, because the energy losses are not simply cumulative but show a distribution of energy losses with a maximum around the typical plasmon energies. This generalized loss curve is specific for the material used. Tougaard facilitated a method to derive the inelastic background function based on this loss distribution curve.

Curve fitting can be required if the intensities of different chemical species of an element are to be evaluated. For non-metallic species, the lines are generally symmetric and reproduced by a convolution of lorentzian and gaussian curves. Emissions from metallic species exhibit an asymmetry toward higher binding energy due to interactions of the conduction electrons with the core hole [42]. The intensity of a photoelectron line of a particular atom depends on the density of atoms n in the detected area, their photoionization cross section σ, the photon flux, the angle between the photon source and the detector, the detection efficiency, the mean free path of the photoelectrons, and, for single crystalline samples, also on the orientation of the sample with respect to the detector and the polarization plane of the photons [2, 3]. A simplified analysis of sample composition is based on tabulated sensitivity factors S for individual photoelectron lines [7]. The intensity of the line is then given by

$$I = n/S \tag{15.2}$$

and the atomic fraction c_x of a multi-element sample by

$$c_x = \frac{I_x/S_x}{\sum\limits_x I_x/S_x} \tag{15.3}$$

Although this works quite well in a number of examples, the absolute values can differ significantly from the given composition in other cases. As an example, we present the case of CdTe and CdS films, which are both expected to be very close to stoichiometry. Nevertheless, while the composition of clean CdTe surfaces determined by XPS using sensitivity factors reproduces this composition, an approximate 60 : 40 composition is reproducibly derived for CdS, which is not realistic. It is further explicitly pointed out that Equation 15.3 is only appropriate for homogeneous samples due to the dependence of the intensity on the inelastic mean path (Figure 15.2). When segregation of elements occurs at the surface, compositional profiles can be determined using angle-dependent measurements, which work well for flat surfaces and which can also be particularly useful for the determination of layer thicknesses [43]. An example for segregation of Sn at the surface of an *in situ* deposited Sn-doped In_2O_3 (ITO) sample is shown in

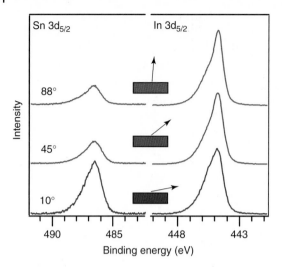

Figure 15.8 In 3d$_{5/2}$ and Sn 3d$_{5/2}$ core level spectra of an Sn-doped In$_2$O$_3$ sample prepared in situ by magnetron sputtering. The spectra are recorded at different take-off angles normalized to the intensity of the In 3d5/2 emission. An enrichment of Sn at the surface is evident.

Figure 15.8. The change in intensity ratio becomes apparent only at rather low emission angles, as the substrate also contains about 10 cation% Sn. Assuming a homogeneous composition, a Sn content of ~13 cation% is obtained using Equation 15.3 [18].

The surface sensitivity of XPS can affect the determination of composition in different ways. If a sample is covered by adsorbates, for example, the attenuation of the substrate signal depends on the mean free path λ of the photoelectrons in the adsorbate layer and thus on the kinetic energy. A composition analysis involving levels with strongly different energies is, therefore, only appropriate for clean surfaces. The composition analysis of very thin films is similarly affected. As the increase in the peak intensity with film thickness depends on λ, a film thickness of $>5 \lambda$ is required to obtain intensity saturation. Peaks with a lower kinetic energy (higher binding energy) have a smaller λ and reach the saturation intensity at a lower film thickness. Consequently, the intensity ratio overestimates the content of the element with the lower kinetic energy. This is exemplified using an experiment in which Cu$_2$S has been stepwise deposited *in situ* onto a clean TiO$_2$ surface (see Figure 15.9). For large deposition times, the stoichiometry of the deposited film is close to the expected value of 2 : 1. At low coverage, however, the Cu 2p intensity (BE $\sim$ 933 eV) increases faster than the S 2p intensity (BE $\sim$ 162 eV), leading to an apparent stoichiometry close to 4 : 1. A proper stoichiometry can thus only be determined if the film thickness is larger than 5 λ.

Compositional analysis of single crystal samples can be strongly affected by photoelectron diffraction. As (photo-)electrons are propagating as waves through

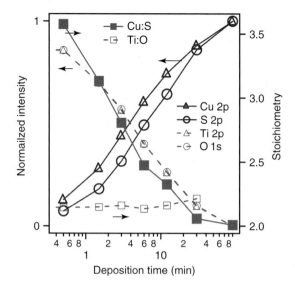

Figure 15.9 The evolution of Cu 2p and S 2p intensity of a Cu_2S film deposited onto a clean TiO_2 surface illustrates the difficulty in determining the composition of very thin films. The faster increase of the Cu 2p intensity compared to the S 2p intensity is related to the shorter inelastic mean free path of the Cu 2p photoelectrons. Due to the comparable binding energies of the Ti 2p (458 eV) and O 1s levels (530 eV), the substrate core levels are attenuated almost in parallel.

a material, they can be scattered and diffracted at the electron shells of the crystal lattice. X-ray photoelectron diffraction (XPD) can lead to considerable variation in peak intensities of single crystal surfaces with excitation energy and emission angle. XPD can be very useful in the determination of the structure of surfaces [44, 45], but compositional analysis can become impossible due to this effect.

15.3.1.3 Partial Density of States
The partial density of states describes the contribution of atomic orbitals of a compound to the valence band density of states. Its experimental determination using photoelectron spectroscopy is based on the dependence of the photoionization cross section on photon energy, which is presented, for example, in Ref. [46]. At low photon energies, as used in UPS, s- and p-orbitals have large cross sections, which decay rapidly with increasing photon energy. In contrast, the d-orbitals exhibit a slower decrease. Consequently, valence bands measured with UPS typically emphasize the s- and p-orbital contribution while the d-orbital contribution is emphasized at higher photon energies (>60 eV).

Figure 15.10 shows the deconvolution of the density of states of the chalcopyrite semiconductor $CuInSe_2$, which is an important thin film solar cell material. The valence band is mainly composed of Se 4p and Cu 3d orbitals. The contribution of the Cu 3d orbitals leads to the so-called p–d repulsion, which results in an

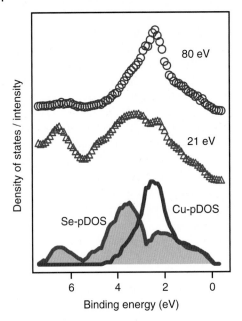

Figure 15.10 Valence band of vacuum-cleaved CuInSe$_2$ single crystal recorded with 21 and 80 eV synchrotron radiation and partial density of states extracted from energy-dependent valence band spectra [49].

upward shift of the valence band maximum and, consequently, to a smaller band gap compared to ZnSe with no metal-d contribution to the upper valence band [47]. The effect is of particular importance for solar cells, as surfaces and grain boundaries of CuInSe$_2$ thin films are Cu depleted, which leads to an increased band gap at these locations [48]. This is believed to contribute to the passive nature of the grain boundaries and, therefore, to the high efficiencies achieved with CuInSe$_2$ solar cells. The deconvolution of the density of states reveals an approximately equal contribution of Se-p and Cu-d states to the valence band maximum. The valence band recorded using 80 eV photon energy already resembles closely the Cu-d partial density of states, while the Se-p contribution is more strongly reflected in the spectrum recorded with 21 eV photon energy.

15.3.2
Analysis of Energy Shifts

The detailed analysis of core level binding energies in photoemission can provide detailed information on the chemical and electronic structure of the surface. However, as a large number of different effects can contribute to changes in binding energies, the identification of the origin of the observed shifts can be quite difficult. In the following we will give an overview and a few examples of the most important contributions to measured values of binding energies in photoelectron spectroscopy, which are relevant in the context of materials science and physical chemistry.

15.3.2.1 Chemical Shifts

Photoelectron spectroscopy is famous mostly because it can detect differences in chemical bonding of atoms and the literature is full of examples. The binding energy of a core electron in a solid roughly depends on the oxidation state of the atom. In most cases, a more positive oxidation state leads to higher binding energy. However, as there are also exceptions from this behavior, one has to restrict to reference databases for identification of chemical species [7, 13]. It is sometimes useful to distinguish between initial and final state contributions to the binding energy of core electrons [3, 4, 6]. The initial state is described by the occupied electronic level, whose energy depends on the local valence and core charge distribution at the atomic site. The energy of this level, however, cannot be measured as it changes with the removal of the electron. Final state contributions to the binding energy result from the screening of the positive charge of the core hole produced by photoemission by a polarization of the atoms' surrounding charges, leading to a reduction in the binding energy of the core level. As the screening occurs simultaneously with the excitation of the electron, the photoelectron contains in its kinetic energy the energy gain of the screening.

As an example for chemical shifts, Figure 15.11 shows the spectral changes recorded from a V_2O_5 thin film in the course of stepwise deposition of Na [50]. As a first step, a V_2O_5 thin film was deposited onto a highly oriented pyrolytic graphite substrate. No strong changes in chemical bonding are expected at the surface of the material due to its layered crystal structure. Surface core level binding energy shifts, which are well known from surface sensitive photoelectron core level spectra of three-dimensional materials [51–54], can therefore be neglected. Surface core level shifts of metals may be found at higher or lower binding energies depending on the center of the density of state of the surface layer compared to the bulk [51]. Semiconductors can show a variety of surface core level shifts depending on the details of the surface reconstruction [52–54]. Due to the absence of surface core level shifts for V_2O_5, the spectra show only a single oxygen and vanadium component, attributed to the oxidation states O^{2-} (a_1 in Figure 15.11) and V^{5+} (b_1) expected for bulk V_2O_5.

Upon deposition of Na, which is performed using commercially available alkali dispenser sources, significant changes of the spectra occur. For low Na deposition times, Na intercalates into the V_2O_5 donating its electron to the host lattice. This leads to a filling of unoccupied electronic states, which are mainly composed of V d-orbitals. The filling of unoccupied states leads to a shift of the Fermi level and, hence, to an increase in the V^{5+} and O^{2-} binding energies (for more details see Section 15.3.2.2). In addition a Na KLL line representing the intercalated Na atoms shows up (c_1). Already at low deposition times the V 2p level becomes asymmetric with an increased emission at the low binding energy side, which can be attributed to the formation of V^{4+}. The reduction of V continues with further Na deposition and additional peaks related to the formation of V^{3+}, V^{2+}, and V^{1+} occur (for a more detailed discussion please refer to Ref. [50]). For higher deposition times, Na remains mostly on the surface forming a Na_2O_2 surface phase, as indicated by additional Na and oxygen features (c_2 and a_2). The new oxygen component is

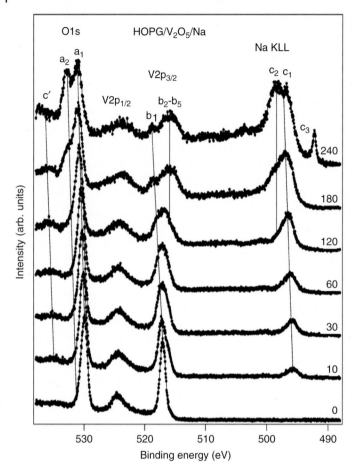

Figure 15.11 O 1s, V 2p core level, and Na KLL Auger transition recorded during stepwise deposition of Na onto a V_2O_5 thin film, which was deposited before onto highly oriented pyrolytic graphite (HOPG).

attributed to peroxide with oxygen in the formal oxidation state O^{1-}. For further Na deposition, metallic Na is also observed (c_3), as further reactions are kinetically hindered by the previously formed reaction layer.

As another example of chemical shifts, the photoemission spectrum of the C1s core level of the organic dye molecule PTCDA (3,4:9,10-perylenetetracarboxylic dianhydride) is displayed in Figure 15.12. Due to different electronegativities of the atoms, three distinct molecular carbon species can be discriminated in the spectrum. While the emissions of C–H and C–C species are found at similar energies, the carboxylic carbon atom occurs at a considerably higher binding energy due to the polarity of the C–O bonds. The additional intensities S_{C-C} and S_{C-O} are caused by inelastic scattered photoelectrons [55]. The shift between the

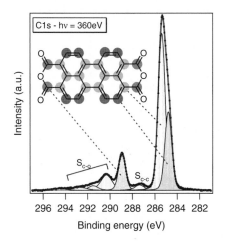

Figure 15.12 C 1s photoelectron spectrum of a freshly evaporated thin film of PTCDA. The different chemical environments of the carbon atoms give rise to distinct binding energies. In addition, satellite emissions are observed, which correspond to the excitation of the HOMO–LUMO gap of the molecule.

primary electron emissions and their satellites correlates quantitatively with the optical HOMO–LUMO gap of the PTCDA molecules.

Other examples of satellite lines, which appear close to photoelectron lines and thus can easily be mistaken for different chemical species, are plasmon excitations in degenerately doped semiconductors or vibrational losses. The plasmon energy in a solid depends directly on its free electron concentration [56]. For metals, surface and bulk plasmon excitations typically lead to satellite emission separated 5–10 eV from the main photoelectron line [7]. Multiple plasmon excitations also lead to multiple satellites, as exemplified, for example, by the free-electron like metals Na and Al [7]. In degenerately doped semiconductors like Sn-doped In_2O_3 (ITO) the carrier concentrations are 2–3 orders of magnitude smaller than in metals. As a consequence, the satellite emissions are observed only about 0.5 eV separated from the main photoelectron line [18]. Excitation of molecular vibrations can also cause the appearance of additional emissions. An example is provided by the methyl-terminated Si(111) surface, where the excitation of C-H stretching vibrations explains multiple satellite emissions separated by ∼0.4 eV [57].

15.3.2.2 Surface Potential Changes

The binding energies of solids in XPS and UPS are measured with respect to the Fermi energy of the spectrometer. If the sample is in electronic equilibrium and in electrical contact with the sample holder the Fermi energy at the surface of the sample is aligned with the Fermi level of the sample holder and, hence, the reference energy (Chapter 16). The reference energy can be easily measured by recording the topmost (lowest binding energy) emission of a clean metallic sample

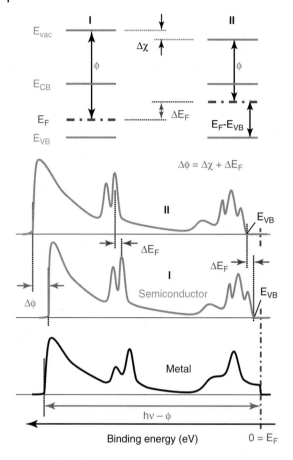

Figure 15.13 Changes in surface potential for semiconducting samples. The binding energy reference is given by the Fermi energy of the spectrometer, which can be directly measured from a metallic sample.

or by measuring the well-known core level binding energies of metals such Cu, Ag, or Au [7]. For non-metallic samples such as, for example, semiconductors, the Fermi energy can vary inside the energy gap [58]. Any variation of the Fermi level position inside the gap, caused, for example, by variation of the electrostatic potential at the surface, leads to a change in the Fermi level (ΔE_F) with respect to the occupied states, as illustrated in Figure 15.13. As the Fermi level of the sample is aligned with the spectrometer reference energy, the position of the Fermi energy on the binding energy scale of the experiment is not affected by changes in its position inside the gap. A change in the Fermi energy with respect to the band edges of the sample is instead observed by a change in the binding energies of all levels, that is, all valence, core, and Auger levels of the sample. Hence, if all levels exhibit parallel shifts upon a sample treatment, the shifts can be explained by changes in E_F.

A change in the Fermi level ΔE_F also changes the work function of the sample, which is given by the difference between the Fermi level and the vacuum level at the surface. The work function of a sample is, furthermore, affected by changes in the surface structure [59]. This can modify the surface dipole and hence the electron affinity χ, which is given by the difference between the vacuum level and the conduction band minimum. The change in the work function is, hence, given by the sum of the change in the Fermi level and the change in the electron affinity:

$$\Delta\phi = \Delta E_F + \Delta\chi \tag{15.4}$$

The energetic position of the valence band maximum (E_{VB}) is usually determined by a linear extrapolation of the leading edge of the valence band emission. More refined models, utilizing calculated density of states in the valence band, have also been described [60]. Values with an accuracy of 100 meV can only be obtained if high resolution photon sources such as UV lamps, X-ray monochromators, or synchrotron beam lines are used. A good indication for a reliable determination of E_{VB} is provided by the differences between E_{VB} and the different core level binding energies, which should be constant for a given material. This is very useful, as it allows one to follow changes in the Fermi energy by monitoring the binding energy shifts of the core levels. To separate from shifts caused by chemical changes, it is important to compare the changes for all core levels of a compound material.

The variation of the Fermi energy at a semiconductor surface can be related to band bending induced by adsorbates or interface formation, or to different doping of a material. Photoelectron spectroscopy offers two possibilities to distinguish between the different situations. First, depth profiling is partially possible by changing the photon energy. Even though this has become more useful with high energy XPS, the different information depth leads only to an asymmetric broadening of the peaks and not directly to a shift of the maximum. A direct identification of surface band bending is possible by exploiting surface photovoltage induced shifts (see also below) [61]. In the presence of a surface band bending, creation of electron–hole pairs leads to reduction of the band bending due to charge separation in the space charge field [58] and, hence, to a rigid shift of binding energies of all levels. Photovoltages are particularly expected at low temperatures due to the thermally activated nature of the current [58].

As an example of the variation of the Fermi level position at a semiconductor surface, the valence band maximum of *in situ* deposited ZnO films is displayed in Figure 15.14 [17]. The films were prepared by magnetron sputtering from ceramic ZnO and from Al-doped ZnO targets. Insertion of Al, which substitute for Zn lattice atoms, leads to degenerately doped films with electron concentrations of up to 10^{21} cm^{-3} [62]. For such highly doped films the Fermi level is expected to lie well above the conduction band minimum, in agreement with the values observed by XPS. Increasing the oxygen partial pressure during deposition leads to a considerable lowering of the Fermi level position [63]. This interpretation of the photoemission data as bulk Fermi level positions is only possible when no surface band bending is present at the ZnO surfaces. This can be deduced from the data themselves as the surface Fermi level positions are well in line with expected bulk positions, and

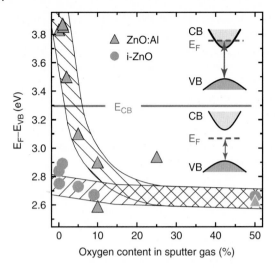

Figure 15.14 Surface Fermi level position of magnetron sputtered ZnO ($E_{gap} = 3.3$ eV) and Al-doped ZnO films [17].

also by theoretical calculations, which indicate that ZnO surfaces are free of surface states in the bulk band gap [64].

Similar changes in Fermi level position are observed for In_2O_3 and Sn-doped In_2O_3. In the latter case, it has been shown that the above-mentioned surface segregation of Sn (see Figure 15.8), is correlated with the surface Fermi level position [18]. This behavior can be explained by the defect chemistry of the material: in ITO, excess Sn donors are compensated by oxygen interstitials [65]. Reducing the material by heating in vacuum reduces the concentration of oxygen interstitials in the material. This increases the concentration of uncompensated Sn donors, which become unstable and segregate to the surface. The changes in Fermi level position and surface Sn concentration have also been recorded *in situ* using high-pressure photoemission [18]. It is remarkable that the reversible changes of the surface Sn concentration are observed on a timescale of 1 h down to temperatures of ~300 °C. Bulk diffusion of cations is not expected at these temperatures.

Changes in Fermi level position are also important in intercalation reactions of battery materials. Using *in situ* photoelectron spectroscopy, they can be monitored during intercalation and de-intercalation. The obtained results also enable one to distinguish between ionic and electronic contributions to the battery voltage [50, 66].

The determination of Schottky barrier heights (see Section 15.3.3) at metal/ semiconductor interfaces is directly possible by following the Fermi level at the substrate surface in the course of metal deposition [24, 67, 68]. A proper determination of the barrier height may require taking surface photovoltages into account, in particular if the experiments are carried out at low substrate temperatures [67, 68] or if the barrier heights are large [24]. The photovoltage can be induced by the photoemission photon source itself as inelastic scattering of primary photoelectrons creates electron–hole pairs [69, 70]. It is indicated by a deviation of core level or Fermi edge emission from reference binding energies (zero in the case of the Fermi energy) and its direction depends on band bending in the substrate. In principle, photovoltages can occur at all semiconductor surfaces and interfaces. If no metallic surface layers are present they are, however, difficult to identify. As already mentioned above, a temperature or intensity-dependent measurement can resolve this issue. Furthermore, cluster size effects can lead to comparable shifts of metallic binding energies [71] and are often difficult to distinguish from source-induced photovoltages for n-type semiconductors.

The photovoltage can be calculated using the general expression

$$U_{ph} = \frac{k_B T}{q} \times \left(\ln \frac{j_{ph}}{j_0} + 1 \right) \tag{15.5a}$$

$$j_0 = A^* T^2 \exp\left(-\Phi_B / k_B T\right) \tag{15.5b}$$

where j_{ph} and j_0 are the photocurrent density and the reverse saturation current density of the diode, respectively. Equation 15.5b is valid for semiconductor/metal interfaces when the current transport is dominated by thermionic emission over the barrier [72]. Here A^* is the effective Richardson constant and Φ_B the barrier height of the interface.

An example of a source-induced photovoltage at the p-WSe$_2$/In interface is presented in Figure 15.15 [67, 68]. Photovoltages are indicated by the deviation of the binding energy of the In 4d core level from its reference value of 16.7 eV. The experiment starts with a vacuum-cleaved single crystal WSe$_2$ surface, which exhibits flat band conditions as the surface is free of surface states. In a first step, indium is deposited onto the sample at liquid nitrogen temperature, which suppresses islands formation of the deposited In. The photovoltage for low In coverage is very large and amounts to ~0.8 eV. With increasing metal film thickness, indicated by the increase of the In peak intensity, the photovoltage gradually disappears. This is attributed to a short circuit between the sample surface and the sample holder formed by the growing In film [68, 73].

The sample is subsequently heated to room temperature, which leads to the clustering of the In film, as indicated by the decrease in the In intensity. Consequently, the metallic film is no longer continuous. This eliminates the short circuit leading to a re-appearance of the photovoltage. As predicted by Equation 15.5, the photovoltage is lower at higher temperature and the following cooling of the sample increases the photovoltage again. Using Equation 15.5, the temperature dependence of the photovoltage allows an independent determination of the barrier height [67, 68]. On the other hand, a quantitative evaluation of photovoltages allows one to verify that the transport across the interface is indeed accomplished by thermionic emission. In other cases it was possible to follow changes in the transport properties, caused, for example, by the intercalation of metal atoms [74]. It should be noted that, although there are also exceptions when the barrier height is very large [24], photovoltages induced by laboratory sources at room temperature can be neglected in most cases.

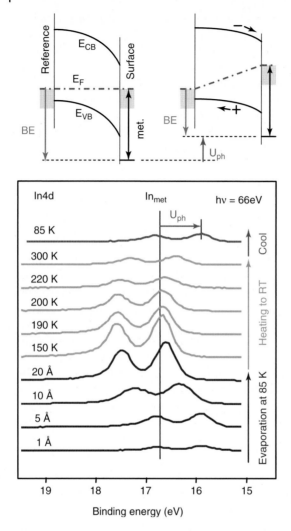

Figure 15.15 In 4d core level recorded during deposition and temperature cycling of a p-WSe$_2$/In interface experiment [67]. The deviation of the binding energy of the In 4d level from the indicated reference for metallic indium is caused by a surface photovoltage, which is induced by the photoemission light source (synchrotron radiation).

In the presence of photovoltages, the Fermi level at the surface is no longer aligned with the sample holder. This might also occur if non-conducting samples are investigated. Due to the emitted electrons, the sample surface will charge up to a certain amount, which can lead to energy shifts of several hundred volts. In principle, charge neutralizers (low-energy electron guns) are usually employed to overcome charging effects. However, an accurate determination of binding energies becomes very difficult and requires the use of spectator lines such as the typically adsorbed hydrocarbons or intentionally added silicone [7]. Analysis of Fermi level positions is no longer possible.

It is finally noted that changes in core level line shapes can also be caused by inhomogeneous surface potentials, whenever the surface potential varies along the surface. Band bending might vary across a surface due to inhomogeneous composition, structure, or adsorption. Charging and photovoltages [67] can also be inhomogeneous. The presence of inhomogeneous surface potentials can often be identified by comparing the line shapes of the different core levels of a compound.

15.3.3
Interface Analysis

Electronic properties of interfaces are particularly interesting for materials with band gaps between the highest occupied and the lowest unoccupied electronic states, like semiconductors and insulators. Interfaces are crucial for all electronic devices, for example, for the electrical contact. In many cases, interfaces between two different or differently doped materials provide the essential functional element. In solar cells, for example, a rectifying junction between an n- and a p-doped semiconductor is responsible for the photovoltaic effect. The n- and p-type semiconductors can be between the same material, as in silicon or GaAs, or between two different semiconductors, as in $Cu(In,Ga)Se_2$ and CdTe heterojunction solar cells [75]. In the dye sensitized (Grätzel) solar cell, the heterojunction is formed between a nanocrystalline TiO_2 film on a fluorine-doped SnO_2, an adsorbed monolayer of a dye molecule and an electrolyte solution [76]. This example illustrates how many different types of interfaces can be present in a single device: (i) an interface between two oxide semiconductors, (ii) an interface between an oxide semiconductor and an adsorbed organic molecule (the adsorption is accomplished by dipping in a solution where the solvent is also crucial), (iii) the interface between the adsorbed molecules and electrolyte, and (iv) the interface between the electrolyte and a metallic conductor, necessary for charge collection.

Energy band diagrams for two rectifying semiconductor contacts, a semiconductor/metal- and a semiconductor p/n-heterocontact, are shown in Figure 15.16. The important parameter of a semiconductor/metal interface is the Schottky barrier height Φ_B, which is given as the distance between the Fermi energy and the valence band maximum E_{VB} or the conduction band minimum E_{CB} for p-type or n-type semiconductors, respectively. Corresponding barrier heights for heterointerfaces are the valence and conduction band discontinuities ΔE_{VB} and ΔE_{CB}. Together with the doping profile, the barrier heights determine the potential distribution and thus the bending of energy bands due to the formation of space charge layers across an interface. In general, a localized charge transfer exists across the phase boundary due to the polarity of the interface bonds. This leads to an interface dipole, which shifts the energy bands of both materials relative to each other by a double layer potential drop δ. This is drawn as a step

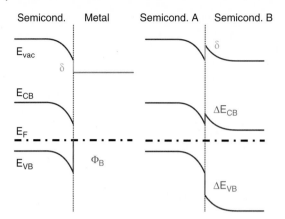

Figure 15.16 Energy band diagrams for semiconductor/metal- and semiconductor/ hetero-interfaces.

in the vacuum level at the phase boundary, which is used to indicate the spatial variation of the electric potential perpendicular to the interface.

Extensive research has been devoted in the past toward understanding the mechanisms governing barrier formation at interfaces in order to enable prediction and possible modification of barrier heights [77, 78]. Photoelectron spectroscopy has been extensively used in this context as it allows direct determination of barrier heights and interface dipole potentials from the analysis of work functions. More recently, photoemission has also been used to monitor changes in barrier heights at semiconductor/metal interfaces upon annealing in different oxygen partial pressures, which induces changes in interfacial defect concentration [24, 79].

The determination of Schottky barrier heights is straightforward from a stepwise deposition experiment. Once the valence band maximum binding energy of the clean substrate surface is known, one has to follow the changes in binding energy of the substrate core levels in dependence on deposition time. As an example, a detailed analysis of the interface formation between $(Ba,Sr)TiO_3$ and Pt is presented in Ref. [24]. Of course, chemical reactions might also occur during interface formation [80], which makes the determination of surface potential changes more difficult.

The determination of barrier heights at semiconductor heterointerfaces is also well described in the literature (see e.g., Refs. [17, 81, 82]). In principle, the valence band offset $\Delta E_{VB}(A, B)$ between two semiconductors A and B is determined according to

$$\Delta E_{VB}(A, B) = BE_{CL}^{VB}(A) - BE_{CL}^{VB}(A) - \Delta BE_{CL} \qquad (15.6)$$

where $BE_{CL}^{VB}(A, B)$ are the core level to valence band maximum binding energy differences of semiconductors A and B, and ΔBE_{CL} the binding energy difference of the core levels of the substrate and the overlayer determined during a stepwise deposition experiment. To extract the requested information the evolution of the

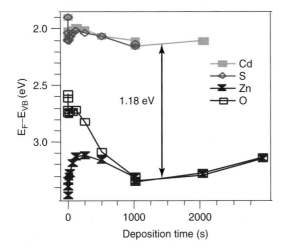

Figure 15.17 Evolution of valence band maxima recorded during deposition of ZnO onto CdS by magnetron sputtering [17]. A deposition time of 1000 s corresponds to a ZnO film thickness of ~2 nm.

core level binding energies of the substrates and the overlayer as a function of deposition time have to be displayed. It is convenient to subtract the $BE_{CL}^{VB}(A, B)$ values as done in Figure 15.17. From such a graph, the evolution of the valence band maxima of the substrate and overlayer can be obtained directly. The valence band discontinuity is then derived from the energy difference of the two valence band maxima.

In principle, the VBM of the substrate and the overlayer should change in parallel in the course of deposition. In addition, if compound materials are studied, the VBM positions can be derived from different core levels, which all should exhibit parallel shifts. This behavior is exemplified by the Cd 3d and S 2p lines of the CdS substrate in Figure 15.17 [17]. In contrast, the Zn 2p and O 1s lines show a strongly different evolution at low coverage, indicating a different chemical bonding between Zn and O in the first 2 nm of the film. This illustrates that it is often essential to follow the evolution of core levels in dependence on film thickness. It is also mandatory to consider all core levels of a material.

15.4
Case Studies: UPS

UPS can also be used to determine surface potentials and, partially, also chemical interactions at interfaces. Although also possible with XPS [83], the determination of work functions is usually performed by UPS, which is straightforward following the scheme in Figure 15.13. UPS is also of particular importance when the valence band density of states is the center of interest (see Section 15.3.1.3).

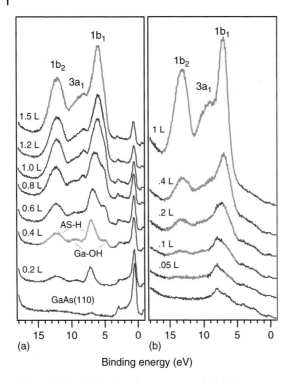

Figure 15.18 Valence band spectra recorded during low temperature adsorption of H_2O on vacuum-cleaved GaAs(110) (a) [33] and de-capped GaAs(100) (b). Water dosage is given in Langmuir (1L = 10^{-6} Torr · s).

In addition, energy-dependent and angle-resolved ultraviolet photoelectron spectroscopy (ARUPS) can be used to determine the energy band structure $E(k)$ of solids as energy and momentum are conserved in the photoemission process [2, 4, 84]. However, a detailed description of the procedure cannot be given here, however.

Valence band spectra recorded during the adsorption of water from the gas phase onto GaAs surfaces are displayed in Figure 15.18 as an example of UPS applied to semiconductor studies [33]. Water adsorption was performed by dosing purified water, using a leak valve, onto samples cooled to ~100 K. After a water dosage of >1 L the typical emission of adsorbed water molecules ($1b_1$, $3a_1$, and $1b_2$) in an ice layer was observed. On the GaAs(110) surface, which was prepared by cleaving an oriented single crystal in vacuum prior to the adsorption experiment, different emission features were observed for lower water dosage. These can be attributed to dissociated water species, leading to the formation of surface Ga-OH and As-H species [33]. No dissociative adsorption was observed on GaAs(100), which was prepared by de-capping an As-capped GaAs(100) wafer. The different adsorption behavior can be explained by the different surface atomic structures

[85]. On GaAs(110), Ga and As dangling bonds exist on neighboring surface atoms. This makes water dissociation feasible as the –OH and –H fractions of the H_2O molecule can be bound on the surface in the close vicinity. This is not possible on GaAs(100), as a single substrate surface species dominates for this surface orientation.

15.5
Conclusions and Perspectives

Photoelectron spectroscopy is a widely used technique for the chemical composition and bonding analysis of solids, which provides valuable information on the surface properties after sample preparation and/or treatment. Typical detection limits are a small percentage of a monolayer. The evolution of intensity and line shape in the course of surface coverage, for example, by adsorption or film deposition, gives detailed insights into the morphological properties of the deposit and into chemical interface reactions. Additional information on the electronic surface properties is obtained when the binding energies are referenced to the Fermi energy of the material, which is possible for electrically conducting samples such as metals or semiconductors. With the Fermi energy as reference, work functions and electrostatic surface potentials can be accessed. These are sensitive to changes in composition and chemical bonding too small to be observed directly as spectroscopic features. Electronic properties relevant for charge transport across interfaces, like energy band alignment, can also be analyzed.

Understanding the properties of materials, surfaces, and interfaces, and even more their function in a particular device, requires not only careful data acquisition and analysis, but also substantial control of suitable sample preparation. Most important for this task is the development of combinations of photoelectron spectrometers and up-to-date sample preparation to avoid the contamination of surfaces and interfaces which can strongly affect the electronic surface properties. To figure out the intrinsic properties of surfaces and interfaces can be particularly challenging if non-vacuum-based sample preparation is involved. Besides sample preparation, post-deposition or device operation induced changes of materials, surfaces, and interfaces are important. So far only very few studies of such effects have been performed, which is mostly related to the limitation caused by the high surface sensitivity and the low pressure environment of the technique.

Future developments of photoelectron spectroscopy include increase in lateral resolution which may eventually allow the assessment of chemical composition and electric potential gradients at grain boundaries. High-energy photoelectron spectroscopy, where the probed electronic structure is less affected by surfaces, may become particularly useful for the study of buried interfaces and the effect of post-deposition treatments. In addition, high-pressure photoelectron spectroscopy is an important step towards understanding surface properties under real processing and operation conditions.

15.6
Supplementary Material

15.6.1
Photon Sources

Photon sources are essential parts of photoelectron spectroscopy. A suitable light source has to fulfill three basic premises: the photon energy has to be higher than the materials' work function to emit photoelectrons, the radiation has to be monochromatic, at least to an extent that different core levels and/or chemical oxidation states can be resolved, and the intensity or flux density or brilliance has to be high enough to produce a satisfactory count rate and signal-to-noise ratio.

X-ray sources are typically implemented either in a dual anode or in a single anode monochromator configuration (see Figure 15.19). The water-cooled anodes

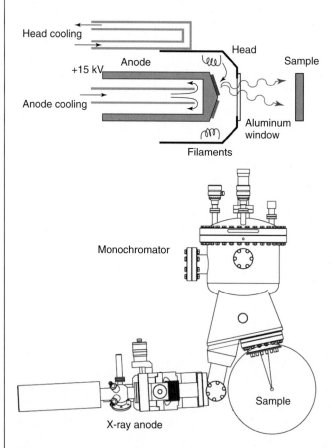

Figure 15.19 General configurations of a dual anode (a) and a monochromator X-ray source (b).

Table 15.1 Selected characteristic X-ray lines.

Anode material	Line (eV)	FWHM (eV)
Mg K_α	1253.6	0.7
Al K_α	1486.6	0.85
Al $K_{\alpha'}$ (monochromated)	1486.44	0.3
Zr M_ξ	151.4	0.68
Zr L_α	2042.4	1.7
Ag L_α	2984.3	2.6
Ag L_α' (monochromated)	2984.3	0.6
Ti K_α	4510.0	2.0
Cu K_α	8048.0	2.6

are coated with different materials giving rise to characteristic X-ray emission lines (see Table 15.1). The electrons emitted from the filament are accelerated by high voltage to the anode. Besides the characteristic radiation, unspecific and disturbing bremsstrahlung background is also emitted, that is mostly suppressed by an aluminum window. In the monochromator set-up, the X-rays are emitted and reflected by a monochromator crystal, which also focuses the X-rays onto the sample. The commonly used monochromator design is of Rowland circle geometry. The energy width produced can be below 20 meV. The intensity of the X-rays is by about 1–2 orders of magnitude lower than for a non-monochromated dual-anode source.

The lowest photon energies available from X-ray sources are around 150 eV. For this reason, and because of the comparably large line width, X-ray sources are not suitable for high resolution valence band spectroscopy. UV sources using gas discharge are more suitable for this task. The energy widths of gas discharge lines are a couple of millielectronvolts. Standard UV sources still work according to the gas discharge principle. High intensity sources use the electron cyclotron resonance principle (without filament) or the duoplasmatron principle (with filament and magnetically confined plasma).

15.6.2
Energy Analyzers

A modern hemispherical electron energy analyzer consists of three elements: the lens, the analyzer itself, consisting of an outer and an inner hemisphere, the entrance/exit slit assembly, and, finally, the electron detector (see Figure 15.20). The hemispheres are the energy dispersive element. For electron detection the inner hemisphere is on positive and the outer on negative potential. This voltage determines the energy, with which the electrons can be transmitted through the analyzer on the medium radius, the so-called pass energy. The pass energy, the mean radius, the mechanical precision of the hemispheres and the entrance slit,

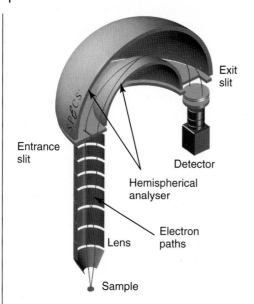

Figure 15.20 Modern hemispherical electron energy analyzer with lens, outer and an inner hemisphere, entrance and
exit slit, and the electron detector.

the electrical stability of the voltages and the size of the detector acceptance area
determine the energy resolution. Nowadays the energy resolution of state-of-the-art
electron analyzers can reach values <1 meV. To keep the energy resolution and
the transmission for all kinetic electron energies constant, the pass energy has
to be kept constant and the electrons have to be retarded or accelerated to this
pass energy. This mode is called fixed analyzer transmission (FAT) or constant
analyzer energy (CAE) mode. Most electron analyzers can also detect ions instead
of electrons. For this purpose the power supplies are bipolar and can reverse the
potentials at the hemispheres.

At the analyzer exit, in front of the detector, electrons of slightly different pass
energies are dispersed along the gap between the outer and inner hemisphere, so
that different energies or an energy window of the spectrum (about 10% of the pass
energy) can be detected at the same time. Perpendicular to this direction, electrons
entering the analyzer at different positions of the entrance slit can be inversely
detected at different positions at the detector in the non-dispersive direction.
This means that, at the detector, an inverse image stripe of the entrance slit is
produced. The distribution of the electrons at the entrance slit is determined by
the lens. In modern analyzers the lens is switchable between various modes with
different spatial and angular resolution, giving additional flexibility to the analyzer.
Hemispherical analyzers, as described above, can meanwhile also handle voltages
(and, therefore, electron kinetic energies) up to 15 keV, so that with a suitable
X-ray source (synchrotron) hard X-ray XPS studies can be made to be less surface
sensitive.

15.6.3
Electron Detectors

One major advantage of modern hemispherical analyzers is the exchangeability of different types of electron and ion sensitive detectors. The classical detector choice for an electron analyzer is the channeltron detector. Today, arrays of channeltrons, filling the homogeneous part of the analyzer gap between inner and outer hemisphere in a dispersive direction, are used for faster spectra recording.

Using channeltron detector systems the sample has to be tilted with respect to the analyzer to record angular-dependent measurements such as, for example, in band structure or photoelectron diffraction experiments. Angular-dependent measurements are performed much faster using two-dimensional detectors. Electrons transmitted through a hemispherical analyzer hit a channelplate (MCP) exciting secondary electrons. The position of the electron count on the detector contains in the dispersive direction the energy information, and in the non-dispersive direction either the emission angle information or the spatial information, depending on the lens mode used. The secondary electrons are accelerated to a phosphorus screen leading to light emission that finally can be recorded through a vacuum viewport using a CCD camera. Due to the smaller pixel size in comparison to the channeltron size, the energy resolution comes down to $<1\,meV$ and the angular resolution is $<0.1°$. On the other hand, the dynamic range of a 2D-CCD detector is a factor of about 1000 smaller than that of channeltron detectors.

In delay line detectors the electrons also hit an MCP stack, but the secondary electron cloud emitted to the back side impinges on a crossed coil (one for the x- and one for the y-direction) assembly. The run time difference of the signal counted to the two x and the two y contacts gives the position of the electron count directly in electronic form. In addition to CCD detectors, time-dependent measurements in the nanosecond to picosecond range are also possible. This makes the detector also suitable for time-of-flight spectrometers. Nearly no background noise is produced and the dynamic range is also improved by 1–2 orders of magnitude compared to CCD detection, making it suitable for small and high count rates.

References

1. Ertl, G. and Küppers, J. (1985) *Low Energy Electrons and Surface Chemistry*, Verlag Chemie, Weinheim.
2. Cardona, M. and Ley, L. (eds) (1978 1979) *Photoemission in Solids*, vol. I & II, Springer-Verlag, Berlin.
3. Hüfner, S. (1995) *Photoelectron Spectroscopy*, Springer-Verlag, Berlin.
4. Reinert, F. and Hüfner, S. (2005) *New J. Phys.*, **7**, 97.
5. Briggs, D. and Seah, M.P. (1983) *Practical Surface Analysis by Auger and X-Ray Photoelectron Spectroscopy*, John Wiley & Sons, Inc., New York.
6. Egelhoff, W.F. Jr. (1987) *Surf. Sci. Rep.*, **6**, 253.

7. Moulder, J.F., Stickle, W.F., Sobol, P.E., and Bomben, K.D. (1995) *Handbook of X-ray Photoelectron Spectroscopy*, Physical Electronics, Inc., Eden Prairie.

8. Einstein, A. (1905) *Ann. Phys.*, **17**, 132.

9. Eberhardt, W. (ed.) (1995) *Applications of Synchrotron Radiation*, Springer-Verlag, Berlin.

10. Bluhm, H., Hävecker, M., Knop-Gericke, A., Kiskinova, M., Schlögl, R., and Salmeron, M. (2007) *MRS Bull.*, **34**, 1022.

11. Rhodin, T.N. and Gadzuk, J.W. (1979) in *The Nature of the Surface Chemical Bond* (eds T.N. Rhodin and G. Ertl), North-Holland, Amsterdam, p. 113.

12. Tanuma, S. Powell, C.J., and Penn, D.R. (1991) *Surf. Interface Anal.* **17**, 911 & 927.

13. Wagner, C.D., Naumkin, A.V., Kraut-Vass, A., Allison, J.W., Powell, C.J., and Rumble, J.R. Jr. NIST X-ray Photoelectron Spectroscopy Database, *http://srdata.nist.gov/xps/* (accessed 6 June 2000).

14. Klein, A., Dieker, H., Späth, B., Fons, P., Kolobov, A., Steimer, C., and Wuttig, M. (2008) *Phys. Rev. Lett.*, **100**, 016402.

15. Weiss, W. and Ranke, W. (2002) *Prog. Surf. Sci.*, **70**, 1.

16. Kunat, M., Girol, S.G., Becker, T., Burghaus, U., and Wöll, C. (2002) *Phys. Rev. B*, **66**, 081402.

17. Klein, A. and Säuberlich, F. (2008) in *Transparent Conductive Zinc Oxide: Basics and Applications in Thin Film Solar Cells* (eds K. Ellmer, A. Klein, and B. Rech), Springer-Verlag, Berlin, p. 125.

18. Gassenbauer, Y., Schafranek, R., Klein, A., Zafeiratos, S., Hävecker, M., Knop-Gericke, A., and Schlögl, R. (2006) *Phys. Rev. B*, **73**, 245312.

19. Göpel, W., Rocker, G., and Feierabend, R. (1983) *Phys. Rev. B*, **28**, 3427.

20. Fritsche, J., Schulmeyer, T., Kraft, D., Thißen, A., Klein, A., and Jaegermann, W. (2002) *Appl. Phys. Lett.*, **81**, 2297.

21. Fritsche, J., Klein, A., and Jaegermann, W. (2005) *Adv. Eng. Mater.*, **7**, 914.

22. Vitomirov, I.M., Raisanen, A., Finnefrock, A.C., Viturro, R.E., Brillson, L.J., Kirchner, P.D., Pettit, G.D., and Woodall, J.M. (1992) *Phys. Rev. B*, **46**, 13293.

23. Schulmeyer, T., Hunger, R., Klein, A., Jaegermann, W., and Niki, S. (2004) *Appl. Phys. Lett.*, **84**, 3067.

24. Schafranek, R., Payan, S., Maglione, M., and Klein, A. (2008) *Phys. Rev. B*, **77**, 195310.

25. Jaegermann, W. and Mayer, T. (2004) *Sol. Energy Mat. Sol. Cells*, **83**, 371.

26. Mayer, T., Lebedev, M., Hunger, R., and Jaegermann, W. (2005) *Appl. Surf. Sci.*, **252**, 31.

27. Beerbom, M., Mayer, T., and Jaegermann, W. (2000) *J. Phys. Chem. B*, **104**, 8503.

28. Mayer, T., Lebedev, M.V., Hunger, R., and Jaegermann, W. (2006) *J. Phys. Chem. B*, **110**, 2293.

29. Jaegermann, W. and Mayer, T. (1995) *Surf. Sci.*, **335**, 343.

30. Lebedev, M.V., Ensling, D., Hunger, R., Mayer, T., and Jaegermann, W. (2004) *Appl. Surf. Sci.*, **229**, 226.

31. Hunger, R., Schulmeyer, T., Klein, A., Jaegermann, W., Lebedev, M., Sakurai, K., and Niki, S. (2005) *Thin Solid Films*, **480–481**, 218.

32. Ensling, D., Hunger, R., Kraft, D., Mayer, T., Jaegermann, W., Rodriguez-Girones, M., Ichizli, V., and Hartnagel, H.L. (2003) *Nucl. Instrum. Meth. B*, **200**, 432.

33. Henrion, O., Klein, A., Pettenkofer, C., and Jaegermann, W. (1996) *Surf. Sci. Lett.*, **366**, L685.

34. Schwanitz, K., Mankel, E., Hunger, R., Mayer, T., and Jaegermann, W. (2007) *Chimia*, **61**, 796.

35. Jaegermann, W. (1996) in *Modern Aspects of Electrochemistry* (ed. R.E. White), Plenum Press, New York, p. 1.

36. Mayer, T. and Jaegermann, W. (2000) *J. Phys. Chem. B*, **104**, 5945.

37. Siegbahn, H., Svensson, S., and Lundholm, M. (1981) *J. Electron. Spectrosc. Relat. Phenom.*, **24**, 205.

38. Morgner, H. (1994) *J. Electron. Spectrosc. Relat. Phenom.*, **68**, 771.

39. Weber, R., Winter, B., Schmidt, P.M., Widdra, W., Hertel, I.V., Dittmar, M., and Faubel, M. (2004) *J. Phys. Chem. B*, **108**, 4729.

40. Kowalczyk, S.P., McFeely, F.R., Ley, L., Gritsyna, V.T., and Shirley, D.A. (1977) *Solid State Commun.*, **23**, 161.

41. Tougaard, S. (1986) *Phys. Rev. B*, **34**, 6779.

42. Doniach, S. and Sunjic, M. (1970) *J. Phys.*, **C3**, 285.

43. Fadley, C.S. (1984) *Prog. Surf. Sci.*, **16**, 275.

44. Chambers, S.A. (1992) *Surf. Sci. Rep.*, **16**, 261.

45. Fadley, C.S. (1993) *Surf. Sci. Rep.*, **19**, 231.

46. Yeh, J.J. and Lindau, I. (1985) *At. Data Nucl. Data Tables*, **32**, 2.

47. Jaffe, J.E. and Zunger, A. (1984) *Phys. Rev. B*, **29**, 1882.

48. Persson, C. and Zunger, A. (2003) *Phys. Rev. Lett.*, **91**, 266401.

49. Löher, T., Klein, A., Pettenkofer, C., and Jaegermann, W. (1997) *J. Appl. Phys.*, **81**, 7806.

50. Wu, Q.H., Thissen, A., and Jaegermann, W. (2004) *Solid State Ionics*, **167**, 155.

51. Riffe, D.M., Wertheim, G.K., and Citrin, P.H. (1989) *Phys. Rev. Lett.*, **63**, 1976.

52. Paggel, J.J., Theis, W., Horn, K., Jung, C., Hellwig, C., and Petersen, H. (1994) *Phys. Rev. B*, **50**, 18686.

53. Nakamura, K., Mano, T., Oshima, M., Yeom, H.W., and Ono, K. (2007) *J. Appl. Phys.*, **101**, 043516-1–043516-5.

54. McLean, A.B. (1989) *Surf. Sci.*, **220**, L671.

55. Jung, M., Baston, U., Porwol, T., Freund, H.-J., and Umbach, E. (2004) The XPS Peak Structure of Condensed Aromatic Anhydrides and Imides, *http://arxiv.org/abs/cond-mat/0408663* (accessed on 30 August 2004).

56. Kittel, C. (1996) *Introduction to Solid State Physics*, John Wiley & Sons, Ltd, Chichester.

57. Hunger, R., Fritsche, R., Jaeckel, B., Jaegermann, W., Webb, L.J., and Lewis, N.S. (2005) *Phys. Rev. B*, **72**, 045317.

58. Sze, S.M. (1981) *Physics of Semiconductor Devices*, John Wiley & Sons, Inc., New York.

59. Zangwill, A. (1988) *Physics at Surfaces*, Cambridge University Press, Cambridge.

60. Kraut, E.A., Grant, R.W., Waldrop, J.R., and Kowalczyk, S.P. (1983) *Phys. Rev. B*, **28**, 1965.

61. Jacobi, K., Myler, U., and Althainz, P. (1990) *Phys. Rev. B*, **41**, 10721.

62. Ellmer, K. (2000) *J. Phys. D: Appl. Phys.*, **33**, R17.

63. Erhart, P., Klein, A., and Albe, K. (2005) *Phys. Rev. B*, **72**, 085213.

64. Ivanov, I. and Pollmann, J. (1981) *Phys. Rev. B*, **24**, 7275.

65. Hwang, J.-H., Edwards, D.D., Kammler, D.R., and Mason, T.O. (2000) *Solid State Ionics*, **129**, 135.

66. Tonti, D., Pettenkofer, C., and Jaegermann, W. (2004) *J. Phys. Chem. B*, **108**, 16093.

67. Schlaf, R., Klein, A., Pettenkofer, C., and Jaegermann, W. (1993) *Phys. Rev. B*, **48**, 14242.

68. Klein, A., Tomm, Y., Schlaf, R., Pettenkofer, C., Jaegermann, W., Lux-Steiner, M.C., and Bucher, E. (1998) *Sol. Energy Mat. Sol. Cells*, **51**, 181.

69. Alonso, M., Cimino, R., and Horn, K. (1990) *Phys. Rev. Lett.*, **64**, 1947.

70. Hecht, M.H. (1990) *Phys. Rev. B*, **41**, 7918.

71. Wertheim, G.K. (1989) *Z. Phys. D*, **12**, 319.

72. Sze, S.M. (1985) *Semiconductor Devices*, John Wiley & Sons, Inc., New York.

73. Horn, K., Alonso, M., and Cimino, R. (1992) *Appl. Surf. Sci.*, **56–58**, 271.

74. Schellenberger, A., Schlaf, R., Pettenkofer, C., and Jaegermann, W. (1992) *Phys. Rev. B*, **45**, 3538.

75. Luque, A. and Hegedus, S. (eds) (2003) *Handbook of Photovoltaic Science and Engineering*, John Wiley & Sons, Ltd., Chichester.

76. O'Reagan, B. and Grätzel, M. (1991) *Nature*, **353**, 737.

77. Mönch, W. (2003) *Electronic Properties of Semiconductor Interfaces*, Springer-Verlag, Heidelberg.

78. Franciosi, A., and Van de Walle, C.G. (1996) *Surf. Sci. Rep.*, **25**, 1.

79. Mosbacker, H.L., Strzhemechny, Y.M., White, B.D., Smith, P.E., Look, D.C., Reynolds, D.C., Litton, C.W., and Brillson, L.J. (2005) *Appl. Phys. Lett.*, **87**, 012102.

80. McGilp, J.F. (1984) *J. Phys. C*, **17**, 2249.

81. Yu, E.T., McCaldin, J.O., and McGill, T.C. (1992) *Solid State Phys.*, **46**, 1.

82. Kraut, E.A., Grant, R.W., Waldrop, J.R., and Kowalczyk, S.P. (1980) *Phys. Rev. Lett.*, **44**, 1620.

83. Schlaf, R., Murata, H., and Kafafi, Z.H. (2001) *J. Electron. Spectrosc. Relat. Phenom.*, **120**, 149.

84. Himpsel, F.J. (1983) *Adv. Phys.*, **32**, 1.

85. Moll, M., Kley, A., Pehlke, E., and Scheffler, M. (1996) *Phys. Rev. B*, **54**, 8844.

16
Photoelectron Microscopy: Imaging Tools for the Study of Surface Reactions with Temporal and Spatial Resolution

Eva Mutoro, Bjoern Luerßen, Sebastian Günther, and Jürgen Janek

■ **Method Summary**

Acronyms

Photoelectron microscopy (PEM) (synchrotron-based and non-synchrotron-based) especially:

- Scanning photoelectron microscopy (SPEM)
- Photoelectron emission microscopy (PEEM), UV-PEEM (Ultraviolet-PEEM), XPEEM (X-ray-PEEM), TOF-PEEM (Time of flight-PEEM).
- X-ray photoelectron spectroscopy (XPS), Ultraviolet photoelectron spectroscopy (UPS)
- Electron spectroscopy for chemical analysis (ESCA), µ-ESCA (ESCA from micro-spot)
- Low energy electron microscopy (LEEM)
- Low energy electron diffraction (LEED).

Benefits (Information Available)

- Technique is surface sensitive with information depth 1 nm–10 nm depending on electron kinetic energy and geometry (XPS, UPS)
- Information from subsurface species or buried layers (imaging in synchrotron-based PEEM)

Methods in Physical Chemistry, First Edition. Edited by Rolf Schäfer and Peter C. Schmidt.
© 2012 Wiley-VCH Verlag GmbH & Co. KGaA. Published 2012 by Wiley-VCH Verlag GmbH & Co. KGaA.

- *In situ* analysis of catalytic and/or electrochemical reactions at surfaces (UV-PEEM, XPEEM, SPEM)
- Combination of PEEM and LEEM: correlation of chemical information with structural surface changes (e.g., reconstructions) with lateral resolution
- Identification of surface phases with lateral resolution of 100 nm (SPEM), 30 nm (XPEEM), ≤10 nm (LEEM)
- μ-ESCA: spectral information from μ-spot (ø ~ 100 nm (SPEM), ≥2 μm (XPEEM))
- μ-LEED: structural information (LEED pattern) from small surface phases (ø ~ ≥2 μm (LEEM)), dark-field imaging (lateral resolution >10 nm)
- Spectromicroscopy and microspectroscopy, subsequent acquisition of image-stacks (XPEEM), parallel acquisition in multichannel-detection mode (SPEM).

Limitations (Information Not Available)

- Generally acquisition time limits are severe; time required per image acquisition: milliseconds to seconds (UV-PEEM, LEEM), seconds to minutes (XPEEM), several minutes (SPEM)
- Exception: very fast time resolution (<125 ps) possible with pulsed sources and TOF-PEEM, even femtosecond-range obtained with conventional PEEM
- Time required for μ-spectroscopy: approximately seconds (multichannel detection mode in SPEM or μ-spot mode in PEEM), seconds to minutes (μ-ESCA in SPEM (depending on spectrum)), several minutes (XPEEM spectromicroscopy imaging XPS)
- Techniques restricted to $p \leq 1 \cdot 10^{-5}$ mbar (if not differentially pumped)
- Due to large topography contrast microscopies are generally restricted to samples with flat surfaces or other well defined surface geometries (SPEM, PEEM, LEEM)
- High level performance with conducting samples only (PEEM, LEEM)

16.1
Introduction

The study of chemical reactions *in situ* requires analytical methods with sufficient temporal resolution; and as many chemical reactions – especially those in the solid state – occur under spatially inhomogeneous conditions, *in situ* studies require both temporal and spatial resolution. Even despite significant instrumental developments, this requirement is still a challenge today – at least for reactions in the interior of solids and for surface reactions under atmospheric conditions. Here, we focus on some very specific photoelectron techniques which offer valuable information on the kinetics of surface reactions.

Photoelectron-based analytical methods offer information on electronic states and, therefore, have become indispensable in (physical) chemistry. Numerous books and reviews have appeared, addressing both instrumental aspects and possible fields of applications (see [1] for comprehensive references). A survey of all currently available varieties of photoelectron spectroscopy and microscopy is far beyond the scope of this chapter. Therefore, we focus on one specific development within physical chemistry, that is, the use of photoelectrons for the generation of spatially and time-resolved information from inhomogeneous or heterogeneous surfaces.

This chapter is primarily written as an introductory text for interested advanced students in physical chemistry. As typical users of photoelectron microscopy (PEM) and scanning spectroscopy we rather concentrate on successful applications of these methods than on details of instrumentation or data analysis. Readers with more interest in these aspects are referred to the literature. Recent reviews or comprehensive descriptions of PEM and scanning microscopy can be found in [2, 3].

We have structured the subject as follows: in Section 16.2 we briefly introduce some basics of photoelectron spectroscopy before we present PEM as the major subject of this chapter in Section 16.3. Here, we focus on scanning photoelectron microscopy (SPEM) and photoelectron emission microscopy (PEEM) in Sections 16.3.2 and 16.3.3, respectively. However, the main part of the chapter (Section 16.4) presents surface reactions (Section 16.4.1) and high temperature electrode processes (Section 16.4.2) as examples for the application of PEM.

16.2
Basic Principles: Photoelectron-Based Analytical Techniques

All techniques designed for the analysis of a surface of unknown composition are based on the general principle "incoming species – interaction with the surface – outcoming species" (Figure 16.1).

Taking the typical binding energies of electrons in atoms or molecules as a rough measure, there are many ways to generate electrons from a sample surface by irradiation or particle bombardment. In essence, the formation of secondary electrons is a ubiquitous phenomenon in virtually all types of excitation of a solid.

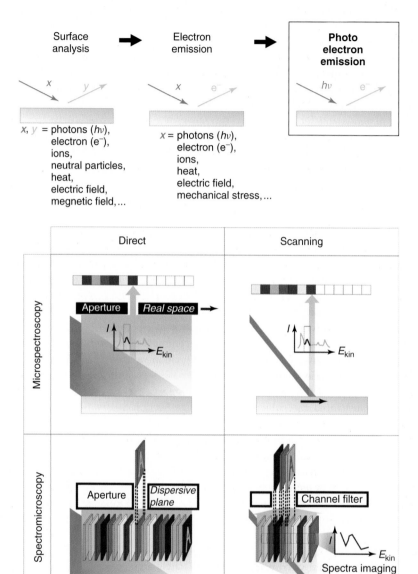

Figure 16.1 Relation of photoelectron microscopy to general surface analytical techniques and the differences between microspectroscopy and spectromicroscopy.

In scanning electron microscopy this effect is used to extract information on the surface morphology ("electrons in, electrons out"). In photoelectron spectroscopy we use photons in different energy ranges to cause the emission of electrons (photoelectrons, photoelectric effect) from a solid sample.

This process is today known as the *(outer) photo effect* for the case of electron emission from a metal surface, was first observed by *Alexandre Edmond Becquerel* (the father of the better known *Antoine Henri Becquerel*) in 1839 and was fully described and explained by *Albert Einstein* in 1905.

The photoelectric effect was one of the key experiments in the development of early quantum physics, and is a true classic in textbook physics: the sum of the kinetic energy of the emitted photoelectrons E_{kin} and the work function Φ (i.e., the minimum energy required to remove an electron out of the solid, see Section 16.6.3) equals the photon energy:

$$h\nu = E_{kin} + \Phi \tag{16.1}$$

It is at this point instructive to compare the photoelectric effect with the generation of secondary ions in secondary ion mass spectrometry (SIMS), see Chapter 17. Here, (primary) ions (e.g., Bi^+ or Ga^+) are used to generate secondary ions from the target, that is, the sample surface. These secondary ions, their mass and their count rate, provide information on the surface composition. However, we do not necessarily need primary ions to generate secondary ions. These will also be emitted, of course with different rate, for example, upon irradiation with energetic photons or neutral species.

In general, we can use two different routes to obtain local information via photoelectron emission (see Figure 16.1): we either collect spectroscopic information from a certain area of the sample by scanning the surface with a focused photon beam or we use suitable electron optics in order to create a microscopic (and energy-filtered) image of the sample surface.

Today, both routes are often combined (see Figure 16.1): In *microspectroscopy* the image (the display of local information) is composed of spectroscopic information gathered from small areas of the sample. This can be done by microspot illumination (as in SPEM) or by using small spot apertures in emission microscopes, that is, with full-field illumination. In *spectromicroscopy* an energy-filtered image is created, thus spectroscopic information is gained out of a series of gray-level images taken at different energies. This can either be done in a scanning microscope or an imaging microscope by collecting exclusively electrons with a pre-defined energy.

Various types of photoelectron microscopes have been constructed, offering different information and different specific advantages. A reasonable classification can be made by distinguishing the photoelectron energy and the collection mode (X-ray or UV photoelectrons, use of filter), the operation mode of the microscope (PEM working in UPS, X-ray photoelectron spectroscopy (XPS), or X-ray absorption spectroscopy (XAS) mode), the source–sample–detector geometry, and the concepts used for achieving spatial resolution [2, 4].

In the following we will concentrate on PEEM without energy filtering as a typical local laboratory experiment and SPEM as a synchrotron-based experiment. For a better understanding of these spatially resolving techniques we briefly explain the most common photoelectron spectroscopies (UPS and XPS) in the following two subsections.

16.3
Experimental Methods[1)]

16.3.1
Ultraviolet Photoelectron Spectroscopy and X-Ray Photoelectron Spectroscopy

In general, plain photoelectron spectroscopy follows a straightforward scheme: The photons, as exciting species, are (more or less) focused on the sample surface, and the energy of the emitted photoelectrons is analyzed by using a suitable detector which counts the number of emitted photoelectrons as a function of their kinetic energy. Usually a concentric hemispherical analyzer (CHA) is applied, which uses an electric field to differentiate between electrons depending on their kinetic energies. Often, these kinetic energies are used in first raw representations of spectra. In most cases the kinetic energies are converted into photoelectron binding energies (E_B) according to the following equation (see Figure 16.3 for a graphical presentation; see also Section 15.2).

$$|E_B| = h\nu - E_{kin} - \Phi \tag{16.2}$$

It is important to note that Figure 16.3 displays the electron spectrum referenced to the vacuum level in front of the sample surface. In a real experiment the electron kinetic energies reference to the vacuum level in front of the electron analyzer lens system. As a result, Φ in formula 16.2 denotes the work function of the analyzer rather than one of the sample surface. Both values for Φ can be separately extracted from photoemission data [5].

The full range of energies used in photoelectron spectroscopy is often called the vacuum ultraviolet range (VUV, 6 eV < $h\nu$ < 6 keV), being a technical definition to account for the fact that all applications in this range require vacuum conditions. Alternatively, virtually the same range (10 eV < $h\nu$ < 6 keV) is also often referred to as the extreme ultraviolet (XUV or EUV) range. The interval between 100 eV and 1000 eV is usually called the "soft" X-ray range. Photons in the energy range 0 eV–50 eV are used for UPS, for example, applying emission lines of helium discharge lamps (He I: 21.1 eV, He II: 42.8 eV). XPS (>1000 eV) employs X-ray tubes. Since the introduction of tunable photon sources (see Section 16.6.2) this technical distinction starts to lose relevance [6].

Both UPS and XPS are highly sensitive to the surface of the irradiated samples – or better, deliver primarily information on surface states, which can be explained by the following arguments: The incoming photons (let us assume soft X-rays in the case of XPS) penetrate the solid during the excitation process to a depth of a few micrometers, and lead to the generation of excited electrons. It is

---

[1)] See also Section 16.3, page 479

trivial that only those electrons which leave the solid after the excitation can be detected by our analyzer, presuming that they are emitted in the direction of the detector optics.

Photoelectrons that are not created in the topmost layers of the sample lose energy by inelastic scattering (described by the *inelastic mean free path* $\lambda(E_{kin})$) or may change the direction of their momentum by elastic scattering (followed later by inelastic scattering). The nearer to the surface the excited atoms are located, the less photoelectrons we lose by these scattering processes. The standard description of these processes is part of the "three-step model", which defines the *effective mean free path* as a result of combined inelastic and elastic scattering [7, 8]. The effective mean free path depends strongly on the kinetic energy of the generated photoelectrons and on the properties of the sample material. In general, the effective mean free path has a minimum in the order of 1 nm or less for energies slightly beyond the UV region (roughly around 70 eV). With decreasing (UV region, roughly up to 10 nm) or increasing energies (soft X-rays, roughly up to 2 nm) the mean free path increases (see e.g., [9]). In summary, the probability for the escape of photoelectrons out of the solid sample decreases drastically with increasing depth d and equals approximately zero for $d \approx 5 \cdot \lambda$, thus only photoelectrons from a depth less than $d \approx \lambda$ contribute to the photoelectron spectrum. The surface sensitivity can be further increased if the detector is positioned at grazing emission (i.e. under an azimuthal angle θ approaching 90°).

What may appear to be simply a complication is in fact the basis for more advanced modifications of photoelectron spectroscopy. As indicated schematically in Figure 16.2, the angle-resolved detection of photoelectrons (in the case of UPS abbreviated as ARUPS) offers a wealth of additional information on the electronic structure of a material.

16.3.1.1 UPS

UPS uses photons in the energy range of approximately 3 eV–100 eV, usually generated by a noble gas discharge lamp (for example, a He lamp emits photons

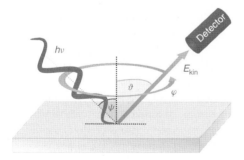

Figure 16.2 Scheme of the photoelectron experiment indicating the angles of incoming photons and outgoing electrons relative to the surface normal.

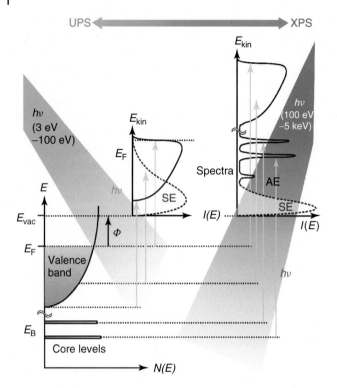

Figure 16.3 Schematical illustration of UV photoelectron spectroscopy (UPS) and X-ray photoelectron spectroscopy (XPS), SE = secondary electrons, AE = Auger electrons. Explanation see text.

with an energy of 21.2 eV). A schematical UPS spectrum is shown in Figure 16.3. A measured spectrum always shows the number of photoelectrons ("counts", intensity) as a function of their kinetic energy (E_{kin}). The energy of the incoming photons is only sufficient to generate photoelectrons from the outermost electron states, the valence shells, near the Fermi level (E_F) of the material. Therefore, UPS is highly useful for acquiring information about the electronic structure (occupied states) of a sample.

In inverse photoelectron spectroscopy (IPES) the surface is illuminated with electrons of a certain kinetic energy (E_{kin}). The de-excitation of the electrons into unoccupied states of a solid causes the emission of photons which are energy-filtered and collected. The energy of the emitted photon reflects the difference between the kinetic energy of the incoming electron and the empty electron state which is occupied during the de-excitation process. To put it simply, IPES probes unoccupied electron states. Together UPS and IPES can help to fully explore the band structure of a solid.

16.3.1.2 **XPS**

In XPS photon energies between 100 eV and 5000 eV are used in order to excite electrons especially from lower occupied states and the core levels (electrons on inner shells). Figure 16.3 shows schematically the main differences between the XPS and UPS and the resulting spectra.

The higher the photon energies, the more we excite electrons from energetically "deeper" inner shells ("core electrons"). Since each element is characterized by distinct energies of its core levels, XP spectra can be used for chemical analysis of the irradiated matter. In addition, the exact energies of the core levels depend on the chemical binding states of the excited atoms, and thus, the resulting spectrum provides information on the chemical identity of the sample. In many standard cases we are then able to interpret an XP spectrum in terms of relative concentrations of atoms in different valence states. However, in many other cases of less studied materials and elements the identification of a chemical state is less straightforward. The photoelectron spectroscopy of core levels can thus be used as an analytical tool, and it is often referred to as electron spectroscopy for chemical analysis (ESCA) [10].

Summarizing, a full XP spectrum can be divided into three parts: at low kinetic energies (= high binding energies) we observe the emission of "true" secondary electrons at a kinetic energy of a few eV with a peak as broad as some 10 eV. In principle, all emitted electrons, except the ones originating from the primary core hole formation process, can be regarded as "secondaries". However, historically only these low-energetic electrons produced in a cascade process are considered in this way. At higher kinetic energies peaks appear which correspond to photoelectrons being excited from atomic core levels. Finally, electrons ejected from the valence band close to the Fermi edge are detected.

It has to be mentioned that the photon irradiation also leads to the ejection of so-called Auger electrons. In this double ionization process the photon creates a primary electron hole which is subsequently filled by an energetically higher-lying electron. The energy gained by the difference between the higher-lying electron level and the level of the primary electron hole can be transferred by a radiationless process to a third electron which is then ejected from the solid leaving two holes, that is, a doubly ionized atom. The signatures of the peaks of Auger electrons are also found in an XP spectrum. Their energies are also element-specific and can be used for an elemental analysis. Because the difference between two levels is transferred to the emitted Auger electron, the kinetic energy of this ejected electron is independent of the energy of the excitation source. The only requirement for the ejection of Auger electrons is the presence of a primary electron hole. Since this hole can be also generated by electron irradiation (of typically 3 keV–5 keV), Auger electron spectroscopy (AES) can be performed using an electron source instead of a photon source, which is considerably cheaper. Today, AES is one the most widespread surface-analytical methods.

As other chapters in this book this will focus in particular on photoelectron spectroscopy, we restrict ourselves to these short introductory remarks. The interested reader will find helpful presentation of UPS, XPS, and related spectroscopic techniques in, for example [1, 5, 6].

16.3.2
Scanning Photoelectron Microscopy

The construction of a scanning photoelectron microscope is depicted in Figure 16.4, essentially representing the design of the μ-ESCA beamline at the synchrotron ELETTRA in Trieste, Italy. The photon beam is extracted from the synchrotron source via a so-called undulator (see Section 16.6.2). In a first step, the photon beam passes a monochromator which filters the photon beam. In a second step, the monochromatic photon beam is then focused by a Fresnel zone plate and an aperture to the sample surface with a final width of 0.15 μm [11, 12]. The emitted photoelectrons are energy-filtered and collected in a hemispherical energy analyzer which is positioned at an angle of 60° relative to the photon beam.

In the depicted case we obtain a photon flux of approximately 10^{10} per s and per $(100 \times 100)\,nm^2$ sample surface area. If we compare this with the typical photon flux of 10^8 per s and per cm^2 sample surface area, we find that the synchrotron light source offers a photon irradiation which is stronger by a factor of at least 10^{12} than in the case of a local laboratory photon source. It is primarily this high photon intensity which allows the collection of spatially resolved spectral information in reasonable time intervals.

Two experimental modes can be used: (i) the photon beam is focused to a defined position on the surface and an XPS spectrum is acquired with high energy resolution (typically 0.4 eV). Thanks to the intensity of the photon source, the acquisition of a single spectrum takes in the order of minutes for elements with a reasonably high concentration in the sample. (ii) In the second mode of operation the sample is scanned with respect to the photon beam, and the analyzer is set to accept photoelectrons of a specific kinetic energy corresponding to the electron emission from an element of interest. After each scanning step the photoelectrons within the chosen energy range are counted, and we obtain an XPS map of the

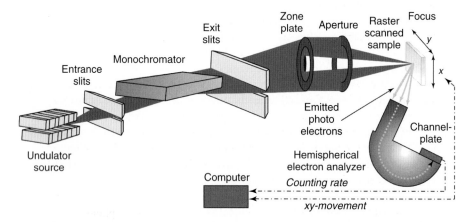

Figure 16.4 Construction of the scanning photoelectron microscope (SPEM, μ-ESCA beamline) at ELETTRA (sincrotrone, Trieste, Italy).

surface reflecting the elemental concentration on the probed surface. The total time required to generate such an image (XPS map) depends on the chosen acquisition time (needed to provide sufficient signal to noise) at each single spot and on the number of scan steps (pixels) of the constructed image.

We will see in the next section that SPEM and PEEM are kind of complementary: SPEM offers chemical information with high spatial and medium time resolution; only slow surface processes can be imaged as a function of time. In contrast, PEEM (without an energy filter or analyzer) does not offer detailed chemical information but provides a high spatial and time resolution in the imaging mode.

16.3.3
Photoelectron Emission Microscopy

PEEM is a special type of PEM making use of a so-called immersion or cathode lens. In contrast to SPEM it is a direct parallel and not a scanning microscopy method. Various types of PEEM systems exist [2], here, our main attention is focused on UV-PEEM set to purely imaging mode without energy filtering of the acquired photoelectrons. For a more detailed insight into UV-PEEM [2, 13] and PEEM using X-ray excitation (XPEEM) [14] we refer to the literature.

PEEM is a non-destructive method for the majority of materials, but the most important advantage of PEEM is the possibility to observe dynamic processes in real time. Furthermore, PEEM offers (depending on the instrument) zoom ranges with a field of view between 1 mm and some μm and spatial resolution in the nm range [15, 16]. The depth of information [17] depends on the light which irradiates the surface. Using light in the UV-range the generated image is dominated by secondary electrons. Note that the secondary electron yield indirectly reflects the bulk properties of the material. High temperature studies using PEEM present no major problems, and imaging up to temperatures of 2000 K has been reported. As all microscopic techniques which are based on the analysis of emitted or scattered electrons, PEEM requires vacuum conditions (UHV). In practice, the pressure within a PEEM chamber should not exceed 10^{-3} Pa in order to obtain reasonable experimental information, but the upper pressure limit can also be extended by differential pumping.

The contrast in PEEM images can be formed by quite a number of different mechanisms, which creates not only opportunities but also issues in interpretation. The correct interpretation of an UV-PEEM image is not necessarily simple, as no direct "chemical" information is obtained, and may even be impossible without additional information on the sample. However, the high information content of the contrast is of course often highly valuable, in particular once PEEM is combined with other surface-analytical techniques.

16.3.3.1 Construction and Function
In a typical PEEM experiment the sample surface is irradiated with photons at a constant angle ψ with respect to the surface normal (Figure 16.5). The most common laboratory UV light sources are mercury discharge lamps (typically 100 W,

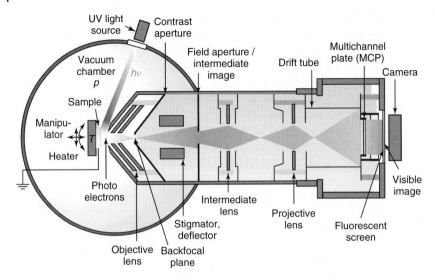

Figure 16.5 Construction of a UV photoelectron emission microscope.

$I_{max} \approx 4.9$ eV) and deuterium lamps (typically 200 W, $I_{max} \approx 6.4$ eV). The emitted photoelectrons are then accelerated by a high electric field between the sample and the imaging column. (In the case of the depicted instrument the sample is grounded and the imaging column biased.) The objective–also called cathode lens or electrostatic immersion lens–consists of the sample itself, a biased aperture and a subsequent lens [18]. The first real intermediate image is formed in the back-focal plane behind the objective. The contrast aperture for manipulation of the angular distribution of the photoelectrons is also inserted in this plane. In some PEEMs a stigmator is placed between the objective and the intermediate lens to compensate lens aberration. A second aperture, the field aperture, cuts the outer part of the real space image and is positioned at the first real intermediate image. After the photoelectrons pass the intermediate and the projective lens, they are focused onto the multi-channel plate (MCP). Depending on the settings of the intermediate lens either one or two real images are formed in the imaging column. After amplification of the electrons (MCP) they are converted into a visible image on a phosphorus screen and can be monitored with a CCD camera.

It should be mentioned, that spectromicroscopy systems need an additional energy filter. It is possible to place a retarding field analyzer in front of the MCP acting as a high pass filter (i.e., only electrons with kinetic energies above a certain value can pass). By subtracting two images–one with photoelectrons collected with a kinetic energy below the threshold of the core level and the other with photoelectrons above this value–one can obtain chemical information. However, a much better signal-to-noise ratio can be achieved with a band pass filter by inserting an appropriate imaging energy filter in front of the projector lens. In such elaborate systems with band pass energy filtering the intermediate and projective lenses are lens systems rather than single lenses [19].

16.3.3.2 Contrast Mechanisms

Contrast in UV-PEEM images can be caused by various effects:

- **Work function contrast**: if a sample is irradiated with photons from a laboratory UV discharge lamp, so-called threshold excitation can be performed and the PEEM can produce contrast due to differences in the local work function of the sample. The sample emits photoelectrons and the image becomes light if the energy of the photons is higher than the work function (Figure 16.6a). In the simplest cases, differences in the local work function of a sample surface can be caused, for example, by the presence of different materials or different phases of the same compound. Spatially inhomogeneous electronic doping levels also cause differences in the work function. Adsorption on the surface changes the surface dipole and thus the work function. Differently orientated faces of a crystal also show a slightly different work function, so that PEEM offers information on the local crystal orientation of polycrystalline materials (see Section 16.6.1 [20]). However, by using only one energy for excitation it is not always possible to draw a conclusion on the work function from the luminosity of two materials.
- **Material contrast**: two materials, even having the same work function, can show a contrast in a PEEM image due to a different total secondary electron yield of both materials (Figure 16.6b) [21].
- **Surface topography contrast**: low energy photoelectrons leave the sample mainly in the direction of the surface normal before they are accelerated by the external field. This leads to a topographic contrast at edges and corners (Figure 16.6c). This contrast also depends on the position of the aperture stop—edges can appear darker or brighter than flat regions depending on whether the deflected electrons are suppressed or not. In general, any non-planarity leads to a deformation of the electric field, and thus to a contrast in the PEEM image. The contrast due to topography complicates the determination of the size of three-dimensional objects.
- **Shadowing effects**: The photons irradiate the surface at a constant angle (e.g. 75° with respect to the surface normal) and thus, hillocks, pits, and grooves cause shadows on the surface. In these shadows the photon irradiation is reduced

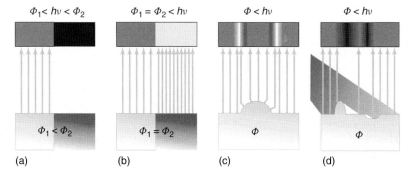

Figure 16.6 Contrast mechanisms in PEEM: (a) work function, (b) material, (c) topography, and (d) shadowing effects.

and the photoelectron yield is lowered and darker regions appear (Figure 16.6d). As the position of the light source is known, the identification of shadows is usually straightforward. Once shadows are clearly identified, their length offers approximate information on the height or depth of surface defects.

- **Magnetism** [23]: using UV light sources two different magnetic contrast mechanisms can occur–a Lorentz contrast and a contrast. Lorentz contrast microscopy uses the effect that the local magnetic field causes deviation of the trajectory of the emitted photoelectrons. By placing a contrast aperture at an appropriate position in the imaging column, electron beams emitted from differently oriented magnetic domains are cut differently due to their changed trajectory along the electron optical axis. Thus, magnetic contrast is obtained. In addition, using polarized light PEEM can generate a magnetic contrast. Using polarized X-ray sources we can use X-ray magnetic circular dichroism (XMCD) for the investigation of ferromagnetic materials and their domain structures. X-ray magnetic linear dichroism (XMLD) (using linearly polarized light) can be used for the imaging of antiferromagnetic materials. A review explaining the technique in more detail is given by Stöhr [23].

- **Photon energy**: the variation in the photon energy offers *a priori* a simple way to modify the contrast. At low energies the light can be tuned for threshold excitation (see e.g., Figure 16.1a,b). When using light in the soft X-ray regime the energy can be tuned through an absorption edge while all photoelectrons or only secondary electrons are collected. This electron intensity is strongly related to the absorption coefficient which is element specific [24]. Adjusting the energy of the illuminating light below and above a certain core level results in an intensity variation due to electron emission related to the photoionization of the specific core level. Thus, contrast is obtained which is related to specific core level excitation. For example, using the Ag 3d core level, an image can be obtained which is bright where Ag is present in the sample. That is, the technique is element specific although energy filtering in the microscope is not necessary, which makes this type of microscopy very popular for PEEM. Especially, using the information provided by the energy structure of the absorption edge ((NEXAFS (near-edge X-ray absorption fine structure) or XANES (X-ray absorption near-edge structure)) makes this spectroscopy very powerful [25, 26]. Combining it with polarised light illumination provides additional information, which can be used, for example, for imaging of magnetic and antiferromagnetic domains.

16.4
Case Studies (Selected Applications)

16.4.1
Chemical Waves in the NO + H$_2$ Reaction on Rh(110)

An example where PEM and related techniques are used very effectively is the characterization of chemically driven self-organization processes on surfaces. In this interdisciplinary field nonlinear dynamics, surface science, and catalysis meet [27]. Microscopy with sufficient temporal and spatial resolution is required, together with chemical or structural sensitivity in order to observe spatiotemporal patterns on surfaces. The following example of the catalytic $NO + H_2$ reaction on a $Rh(110)$ single crystal surface highlights the advantage of combining laboratory UV-PEEM with XPEM (in our case a SPEM) and surface structure sensitive low energy electron microscopy (LEEM).

The $NO + H_2$ reaction leads to a variety of chemical patterns on $Rh(110)$, as there are fronts [28], traveling wave fragments [29], target patterns [30], and spirals of various shapes [28]. Figure 16.7 shows a laboratory UV-PEEM image of a target pattern.

Thanks to the fast data acquisition of the technique, it is easy to follow how the target patterns evolve in time: The bright concentric rings travel over the surface, expanding radially. Once the rings have grown sufficiently, a new ring is formed in the center of the pattern, filling the area left by the expanding rings. Since PEEM primarily images the work function, we obtain only indirect information on the surface phases. An additional microscopic technique is required in order to clarify the chemical differences between bright and dark areas in Figure 16.7, and thus, to be able to establish a model of the surface processes responsible for the pattern formation.

Before we discuss the experimental details in more depth we have to introduce the effects that drive the observed chemical waves: the observed patterns are the effect of NO and H_2 which react catalytically on the $Rh(110)$ single crystal surface to give N_2 and H_2O. Technically, this type of reaction is important for the design of the automotive catalytic converter, where Rh particles are used to catalytically reduce NO to N_2 [31]. However, the chemical patterns we address here

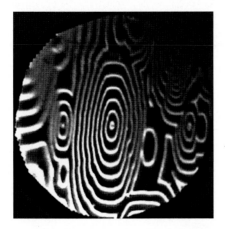

Figure 16.7 PEEM image of a target pattern generated on the $Rh(110)$ surface by the catalytic $NO + H_2$ reaction. The diameter of the imaged area is about $550\,\mu m$. Experimental conditions: $T = 530$ K, $p(NO) = 1.7 \cdot 10^{-7}$ mbar, $p(H_2) = 7 \cdot 10^{-7}$ mbar.

take place at $p \approx 10^{-6}$ mbar, a pressure far lower than the pressure (in the bar range) during technically important processes. On a heterogeneous catalyst (in our case the Rh(110) single crystal) gas molecules can adsorb, eventually dissociate, travel laterally, and react with other adsorbed gas molecules or atoms. The reaction products may leave the surface and again free the catalyst surface to host new incoming gas particles. Qualitatively, it can be expected that the catalytic reactivity of a surface changes after adsorbing a different amount of gas particles on it or by changing the type of adsorbates. Indeed, the difference in local reactivity, together with the fact that the adsorbed gas particles can laterally diffuse on the surface, drive the observed chemical patterns. Therefore, such systems are named reaction–diffusion systems [27]. They can generate target patterns or moving spirals as chemical waves if they are excitable [32, 33]. During the so-called excitation cycle the system can be excited from its starting, inactive state into an active state. Subsequently, the system gradually loses its activity until it reaches its starting, inactive state, from which it can be excited again. The proposed excitation cycle for the catalytic reaction of NO and H_2 on Rh(110) [34] which triggers the observed chemical waves is depicted in Figure 16.8.

On the starting (nearly) oxygen-saturated surface (state 1) H_2 is not able to adsorb, except in the vicinity of a surface defect which will trigger the excitation cycle. There, hydrogen reacts with the adsorbed oxygen forming H_2O, which immediately desorbs, producing a locally adsorbate "free" surface (state 2). On this "free" and catalytically active surface, H_2 and NO can adsorb and dissociate. Since the adsorbed oxygen is further reacted-off by H_{ad}, N_{ad} accumulates leading to a N-saturated surface (state 3). On the N-covered surface less hydrogen can dissociate while NO is still able to adsorb and dissociate leading to an increase in the oxygen

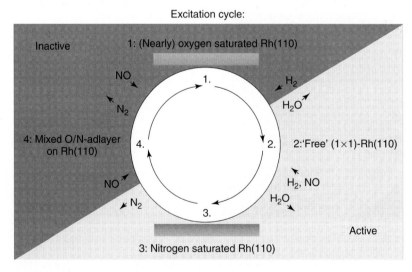

Figure 16.8 The excitation cycle that triggers a pulse in the catalytic NO + H_2 reaction on Rh(110).

coverage and thus, an increase in the total adsorbate coverage (state 4). Parallel to the build-up of a high total coverage the surface loses its catalytic activity. Due to lateral interactions between adsorbed nitrogen and oxygen the N_{ad} is destabilized and slowly desorbs as N_2. This leads to a gradual enrichment of adsorbed oxygen until the initial state 1 is reached. The sequence described is called the excitation cycle of the system. If the surface is locally excited it can perform one loop in the excitation cycle until the next excitation event occurs. Since the adsorbates are mobile (hydrogen, especially, is highly mobile) a triggering pulse can travel over the surface always igniting locally a loop in the excitation cycle along the propagating pulse.

By adsorbing a defined amount of gas molecules, laterally homogeneous adsorption layers can be prepared and thus, the adsorbate phases belonging to the different states of the excitation cycle are well characterized ([35] and references therein). Table 16.1 compiles the known coverage of the corresponding adsorbate layers. Here, the unit 1 ML indicates that each surface atom in the (110) plane of the Rh crystal is covered by one atom of the indicated type. Due to lateral interactions the adsorbed gas particles do not populate every site, but arrange in a highly ordered way which leads to a periodicity larger than the (1 × 1) unit cell of the (110) surface. Moreover, in several adsorbate layers even the Rh surface atoms themselves rearrange differently from the (1 × 1) registry of the bulk terminated (110) surface. The resulting symmetry of the unit cell of the corresponding adsorbate layer is also indicated in Table 16.1.

Having these results in mind, PEM can be used to prove that the proposed excitation cycle correctly describes the surface processes in the pulses of the observed chemical waves on Rh(110). This is possible if either the coverage of the corresponding adsorbate layers or their corresponding ordered surface structures can be imaged with sufficient lateral resolution and resolution in time (remember that the pulses are traveling!).

Figure 16.9 shows the result of an experiment addressing the first possibility: Here, a traveling pulse was imaged using a SPEM by tuning the kinetic energy of the acquired photoelectrons first to the N 1s and secondly to the O 1s core levels (Figure 16.9a). After performing a separate calibration experiment, the gray level intensity of the O 1s and N 1s image was related to the true adsorbate coverage,

Table 16.1 Coverage and structure of different adsorbate layers on the Rh(110) surface. Data are compiled in [35].

State	Coverage θ (ML)	Unit cell
1. (Nearly) oxygen-saturated Rh(110)	$\theta(O) = 0.66$	c(2 × 6)
2. "Adsorbate free" Rh(110)	$\theta = 0$	(1 × 1)
3. (a) Low N_{ad}-covered Rh(110)	$\theta(N) = 0.33$	(3 × 1)
(b) N_{ad}-saturated Rh(110)	$\theta(O) = 0.50$	(2 × 1)
4. Mixed oxygen and nitrogen-covered Rh(110)	$\theta(N) = 0.25$	c(2 × 4)
	$\theta(O) = 0.50$	

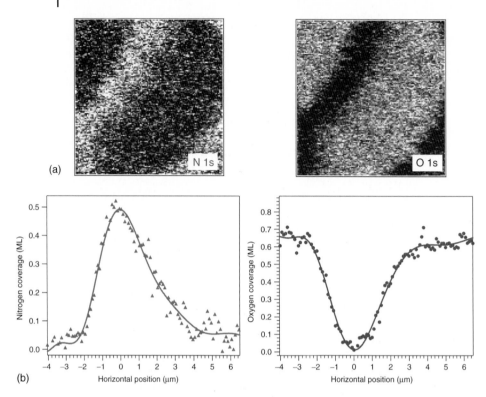

Figure 16.9 (a) N 1s and O 1s images of a pulse of a chemical wave (SPEM). (b) Calibrated gray scales obtained from the images show the corresponding adsorbate coverage and thus the correct gradient within the pulse. Image size: $(12.8 \, \mu m)^2$, experimental conditions: $T = 530$ K, $p(NO) = 2.0 \cdot 10^{-7}$ mbar, $p(H_2) = 6.0 \cdot 10^{-7}$ mbar.

which is plotted in Figure 16.9b versus the relative position in the pulse. Since the count rate in X-ray spectroscopy using a hemispherical analyzer is low, imaging X-ray-induced photoelectrons in SPEM is slow. Therefore, the traveling pulses had to be virtually slowed down by applying an offset ramp to the scanning of the images, so that the illuminating X-ray beam followed the pulses.

The results in Figure 16.9b clearly confirm that the proposed coverages of the excitation cycle (Figure 16.8) are reached within the pulse [36], and the adsorbate concentration gradient in the pulse can be extracted. One observes a steep increase in the N coverage, followed by a slow decrease while the O coverage evolves, and *vice versa* starting with a steep decrease followed by a slow increase. The measured adsorbate concentration gradients are compatible with the intuitive expectation and with computational simulations of the pulse [37]. By changing the kinetic energy during acquisition of an image from the N 1s to the O 1s core level it can be determined whether there is a spatial shift between the decrease in the oxygen and the increase in the nitrogen coverage. Such a shift would be detected if a significantly large area in the pulse was adsorbate free Rh(110) (state 2). Within our resolution limit of ≈1 μm, which is mostly influenced by the acquisition time (the pulse is moving) rather than by the lateral resolution of the instrument, such a shift is not detectable. Therefore, the (almost) adsorbate-free zone cannot exceed a lateral area of 1 μm.

Using a third microscopy, LEEM can be used to image the expected sequence of ordered adsorbate structures in the pulse i.e., the sequence (3 × 1) → (2 × 1) → c(2 × 4) → c(2 × 6), see Table 16.1. In LEEM the specimen is illuminated with plane-parallel electrons and the reflected electron beam can be used to image its surface. In contrast to X-PEM, LEEM does not usually suffer from a low count rate and thus this microscopy is much faster. In the so-called μ-LEED mode of the LEEM instrument, exclusively electrons leaving the surface from a ≈2 μm spot can be accepted for imaging. Upon imaging the focal plane of the electron image, the diffraction pattern from the masked part of the surface is obtained. As an example, Figure 16.10 shows the low energy electron diffraction (LEED) patterns obtained from a clean and a nitrogen-saturated Rh(110) surface. The pattern of the clean

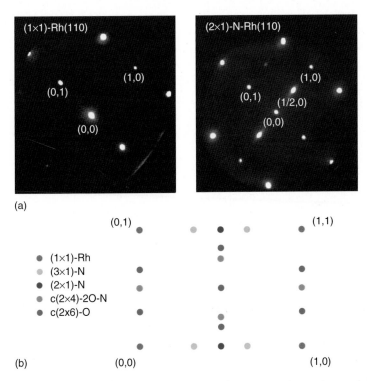

(a)

(b)

Figure 16.10 (a) LEED patterns obtained from the clean (1 × 1)- and the nitrogen covered (2 × 1)-N-Rh(110) surface. (b) Indication of the spot positions of the corresponding LEED patterns according to the (1 × 1) substrate lattice, the (3 × 1) and the (2 × 1) nitrogen layer, the mixed nitrogen-oxygen c(2 × 4)-2O-N-layer and the (almost) oxygen-saturated Rh(110) surface covered by a c(2 × 6)-O adsorbate layer.

surface consists of a (1 × 1) unit cell, reflecting the (primitive) grid of the (110) surface. Since the unit cell of the ordered nitrogen layer is doubled in size, that is, it follows a periodicity with a (2 × 1) unit cell, in reciprocal space 1/2 order spots occur. Therefore, the corresponding LEED pattern shows additional spots at the (1/2,0) and related positions. In Figure 16.10b the spot positions of the diffraction patterns are indicated for all surface phases listed in Table 16.1.

When recording the LEED pattern in the described μ-spot mode as a function of time, one sees how the ordered structures evolve each time a pulse enters the analysis area. Knowing the traveling velocity of the pulse, one can relate the time to a distance coordinate [38]. Figure 16.11 displays the intensity evolution of LEED spots characteristic for the indicated ordered phases versus the relative distance coordinate dx in the pulse.

Since every spot is extracted from the same image of the LEED pattern the evolution of the spot intensities is already synchronized. It is clearly visible that the pulse starts with a (3 × 1)-N phase, directly followed by a (2 × 1)-N-saturated surface. After this, the mixed c(2 × 4)-2O-N layer immediately appears once the (2 × 1) ordered surface decays. Finally, the c(2 × 4) intensity slowly decays while the surface becomes increasingly oxygen-enriched and the c(2 × 6)-O layer is built up. After completion of this surface phase the next pulse enters the analysis area and the next loop in the excitation cycle is observed. Again, a clear (1 × 1)

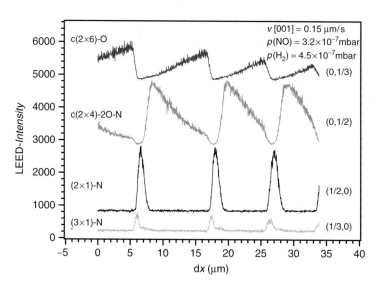

Figure 16.11 Intensity evolution of several spots of a μ-LEED measurement (diameter of the analysis area approximately 2 μm) of traveling pulses in a chemical wave characteristic for several surface phases. For the (3 × 1)-N adsorbate layer the (1/3,0) spot was chosen, the (1/2,0)-spot belongs to the (2 × 1)-N phase, the (0,1/2) spot indicates the presence of the mixed c(2 × 4)-2O-N layer and the (0,1/3) spot is characteristic for the c(2 × 6)-O surface. Experimental conditions: $T = 530$ K, $p(NO) = 3.2 \cdot 10^{-7}$ mbar, $p(H_2) = 4.5 \cdot 10^{-7}$ mbar.

LEED pattern of an adsorbate-free Rh(110) surface is not observed, so that we can conclude that the area belonging to this phase is significantly narrower than $2\,\mu m$. Indeed, mathematical simulations revealed that the width of the front area of the pulse in the chemical wave with an adsorbate coverage less than 0.1 ML stays well below $1\,\mu m$ [39].

Finally, so-called dark field imaging in LEEM can be used to image the lateral distribution of the different adsorbate phases. In this imaging mode only electrons originating from a special LEED spot are used for imaging. For this purpose an aperture can be inserted at the position of the focal plane of the electron image in the microscope in such a way that it masks out all electrons except those traveling through the desired diffraction spot. For example, by positioning the aperture on the (1/3,0) spot of the LEED pattern (see Figure 16.10), only electrons from the (1/3,0) spot are used to image the surface. Thus, the resulting image appears bright, where the (3×1)-N surface phase is present. The other phases can be imaged accordingly, using the (1/2,0) spot for the (2×1)-N, the (0,1/2) spot for the $c(2 \times 4)$-2O-N and the (0,1/3) spot for the $c(2 \times 6)$-O surface phase. In Figure 16.12 the corresponding four images area displayed which have been subsequently recorded after spiral waves were ignited on Rh(110) [38].

As a result, in each panel the spatial distribution of the ordered surface phases inside the chemical wave is imaged: (i) (3×1)-N, (ii) (2×1)-N, (iii) $c(2 \times 4)$-2O-N, and (iv) the $c(2 \times 6)$-O layer. Again the images clearly demonstrate that the ordered phases proposed in the excitation cycle of Figure 16.8 are found inside the pulse of the $NO + H_2$ chemical waves on Rh(110).

The presented example shows very well that XPEM has an excellent capability for local determination of adsorbate coverages and imaging non-homogeneous adsorbate distributions on surfaces. On the other hand, LEEM provides information on ordered surface phases which can even be imaged. Thus, the combination of both techniques enables one to study even complex surface reactions and extract quantitative results with high spatial and time resolution.

16.4.2
Electrode Reactions: Photoelectron Microscopy and Electrochemistry

For a long time after its development PEEM was mainly applied to purely chemical surface science problems, as discussed in the previous example. In recent years the application of PEEM to electrochemical surface reactions has also been successfully established. In the following, we show how the processes during polarization of an electrode system on a solid electrolyte can be observed *in situ* with PEEM and SPEM in a high temperature electrochemical cell.

In order to introduce this topic for readers with different backgrounds, we have to start with the original motivation for the application of PEEM: The importance of heterogeneous catalysts for a multitude of chemical processes (often with technical relevance) raised quite early the question whether it is possible to influence the activity of the catalyst by electrochemical polarization.

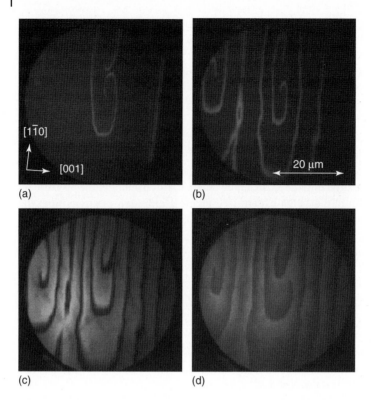

Figure 16.12 Dark field LEED image of the same area as in Figure 16.11 after spiral waves have been ignited on Rh(110). The subsequently acquired images were obtained using electrons from: (a) the (1/3,0) spot, (b) the (1/2,0) spot, (c) the (0,1/2) spot, and (d) the (0,1/3) spot. Experimental conditions: $T = 550$ K, $p(NO) = 3.5 \cdot 10^{-7}$ mbar, $p(H_2) = 2.0 \cdot 10^{-6}$ mbar.

First experimental studies of the ethylene oxidation on a platinum catalyst film, which was prepared as an electrode in an electrochemical cell Pt/oxygen ion solid electrolyte/Pt, were performed in the early 1980s by Vayenas and coworkers. They applied an anodic voltage to the catalytically active platinum electrode and observed a reversible increase in the catalytic reaction rate which was much higher (by a factor up to 10^5) than one would expect simply due to the transport of oxygen ions through the solid electrolyte to the catalyst surface. This unexpected phenomenon was called "NEMCA" (non-Faradaic electrochemical modification of catalysis), nowadays also often referred to as "EPOC" (electrochemical promotion of catalysis). The term "non-Faradaic" expresses the fact that each ion pumped to the catalyst electrode triggers (many) more than one surface reaction, that is, it is in fact a "promotion" effect. Despite extensive work by Vayenas and others [40], the reason for the rate increase and the exact mechanism of the change in the catalytic activity have not been fully clarified. Up to now, there are different attempts to explain the NEMCA effect including a highly catalytically active oxygen species, the so-called "backspillover" oxygen by Vayenas *et al.* [40], a chain reaction mechanism by Sobyanin *et al.*, and an ignition effect by Imbihl *et al.* [41]. However, all hypoteses to explain the NEMCA effect so far involve an electrochemically generated (charged) oxygen surface species, which we will call in the following "spillover" oxygen. The spill-over generation and its spreading over the active surface forms an attractive application of PEEM, as outlined below.

Figure 16.13 shows schematically the processes which occur duing the electrochemical polarization of an electrode prepared as the catalyst film. During anodic (positive) polarization of the catalytically active metal (working electrode) oxygen is transported from the back (counter electrode, porous platinum) to the so-called "three phase boundary" where solid electrolyte, metal electrode, and gas phase meet. It was the aim of our PEEM experiments to study the electrochemical oxygen generation and subsequent diffusion process *in situ* on a well-defined model system. Therefore, we used a dense platinum film (prepared by pulsed laser deposition) as working electrode and a yttria-stabilized zirconia (YSZ) single crystal as oxygen ion conductor. The counter electrode was prepared with porous platinum paste. With this set-up we planned to answer two major questions:

1) Is the oxygen (the spill-over species?) exclusively generated at the three phase boundary?
2) Is it possible to monitor the electrochemically driven oxygen diffusion on the platinum electrode *in situ*?

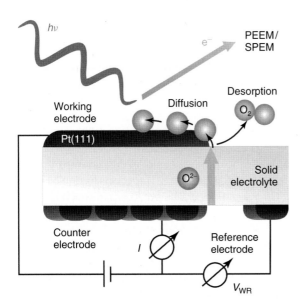

Figure 16.13 Schematical processes during electrochemical polarization of a Pt/oxygen solid electrolyte/Pt cell. The upper electrode is the anode.

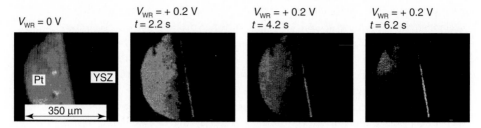

Figure 16.14 Sequence of PEEM images during the anodic polarization of a dense platinum electrode deposited on a YSZ single crystal (see Figure 16.13 for the cell arrangement).

Figure 16.14 shows a sequence of PEEM images taken during the anodic polarization of the working electrode [42]. Due to the difference in the secondary electron yield we can identify Pt as a bright semicircle on the left half of the first (left) image. Electron removal from the electronically insulating oxygen ion conductor YSZ (large bandgap of $\sim$5.2 eV) requires more energy than electron emission from the Pt electrode and thus appears dark.

As soon as the anodic voltage $V_{WR} = +0.2$ V is applied to the platinum electrode, a dark area begins to cover the electrode. After approximately 6 s more or less the whole visible part of the electrode has darkened, only a small stripe (exactly at the boundary between platinum and YSZ) remains bright. According to the functional principle of the PEEM we can interpret the result as follows: when the PEEM image darkens less photoelectrons are registered, because the work function of the sample has increased. The work function of metals is usually increased by adsorbed (electronegative) gas molecules on the surface (see Section 16.6.3). This process starts at the edge of the platinum electrode. In terms of the spill-over model this means that a mobile surface species (oxygen) is formed at the three phase boundary and diffuses on the platinum electrode which is finally covered completely with oxygen.

We emphasize that no information on the identity of the surface species is obtained, beyond the simple fact of the induced work function change. But quantitative information can be obtained on the kinetics of the spreading process by a numerical analysis of the gray tones [42]. Topographic effects have to be taken into account as a possible (static) disturbance. The bright stripe which remains visible after the diffusion experiment (final image of Figure 16.14) cannot be explained unambiguously. It might be a topographic effect at the edge of the electrode, but an oxygen-free surface part (for whatever reason) cannot be ruled out.

In contrast to PEEM, XPS provides information on the chemical nature (determination of the element and its oxidation state). Therefore, we transferred the electrochemical experiment to a scanning photoelectron microscope (μ-ESCA beamline at ELETTRA) where element-specific X-ray photoelectron images can be taken [43]. As the sampling rate for the spectral information is much lower than the rate for the collection of PEEM images, we used a "trick" to image the spreading of spill-over oxygen: First, we chose a hole in the otherwise dense platinum electrode (resulting from the PLD preparation process; dark spot in the middle of the images in Figure 16.15). Then we exposed the platinum to ethylene at elevated temperatures ($p(C_2H_4) \approx 1 \cdot 10^{-6}$mbar) leading to its decomposition and full coverage of the platinum with a thin film of carbon.

Upon anodic polarization of the working electrode ($V_{WR} = +0.5$ V) oxygen appears at the three phase boundary (tpb), that is, at the edges of the hole and reacts with the carbon film on the platinum, thus forming CO_2 which then desorbs into the gas phase. Slowly and continuously the carbon film was cleaned from the electrode, and a moving reaction front resulted. We could control this process via the applied potential. When we turned the potential off the oxygen supply stopped, and the front between the carbon-free and carbon-covered areas was stopped. Figure 16.15a illustrates schematically this electrochemically controlled surface reaction.

In Figure 16.15b two sets of SPEM images–all under an anodic polarization of $V_{WR} = +0.5$ V–are shown: three Pt 4f (upper row) and the three corresponding C 1s images (lower row). The brighter the pixels, the higher the surface concentration of platinum with respect to carbon. We notice that at time t_1 only a few areas are bright in the platinum image, whereas the carbon image is more or less homogeneously bright. Later, larger areas of the platinum surface appear bright, as more and more carbon is removed by reacting with the electrochemically generated oxygen. Therefore, the corresponding parts in the carbon images are dark.

This SPEM experiment clearly proves that the electrochemically generated oxygen is generated at the three phase boundary–which was already indicated by the PEEM experiment–and diffuses on the electrode surface from the three phase boundary to the reaction front.

Further details of these experiments and additional SPEM experiments (in particular spectra with spatial resolution) can be found in [42–47].

16.4.3
Recent Developments in Instrumentation and Application

Recent instrumental developments range from new aberration corrections [48], time-of-flight (TOF) detectors to photoelectron spin polarization analysis, making PEEM to a powerful tool for the solution of advanced analytical problems.

An interesting development is the use of a *free electron laser* (FEL) in the UV range as excitation source for a UV-FEL-PEEM, essentially employing a standard PEEM [49, 50]. A FEL offers two important advantages compared to common laboratory UV sources: it is extremely intense and the laser energy can be tuned, resulting in better spatial and temporal resolution. Thus, the investigation of fast surface processes on the nanometer scale become possible. At present, FEL are only available at synchrotrons (see Section 16.6.2), and thus, the UV-FEL PEEM cannot be considered as a routine method. However, once available it offers a multitude of experimental opportunities. Recent developments in FEL using linear accelerators [51, 52] will provide pulsed light which is orders of magnitude more brilliant than

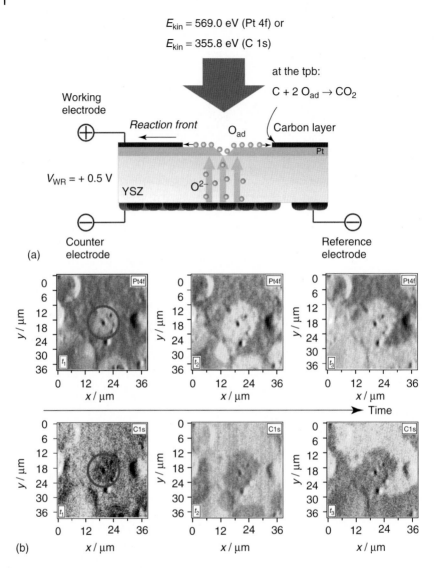

Figure 16.15 Electrochemical spill-over pumping and its observation by SPEM; (a) experiment and (b) SPEM images for the Pt 4f and C 1s signals at three different times. Lighter pixels correspond to higher X-ray photon intensities of the chosen element.

that released from synchrotron light sources. These light sources might serve as single shot illumination devices.

Another interesting development is time-of-flight photoelectron emission microscopy (TOF-PEEM) using a pulsed excitation source like a synchrotron (see

Section 16.6.2) or a pulsed laser and a TOF detector [53–56]. The advantage of this technique, first proposed by Spiecker *et al.* [56], is the simultaneous collection of spectroscopic information and non-energy filtered PEEM images. A TOF-PEEM is equipped with a drift tube behind the imaging optics, and the energy resolution is achieved by the serial registration of all photoelectrons with their different kinetic energies. After the MCP a delay line detector (DLD) is placed in front of the fluorescent screen. This DLD not only records the *X* and *Y* coordinates of the incoming electrons but also their arrival time. The timescale can be converted into an energy-scale (kinetic energy) by synchronizing the detector with the pulsed excitation source so that the DLD offers energy and space resolution. A time resolution of about 125 ps has already been achieved [54].

16.5
Conclusions and Perspectives

Today, a large variety of photoelectron-based analytical techniques is available. Among these are UPS and XPS/ESCA (and AES, using electrons as exciting species), widespread spectroscopic methods with large impact on materials characterization. However, with interest not only in stationary or static surface states but also in the kinetics of surface processes, photoelectron detection with energy, time, and spatial resolution is required. With PEM in its variants PEEM and SPEM we can achieve these requirements under UHV conditions.

UV-PEEM is a commercially available technique which offers high spatial and temporal resolution in the imaging of the work function, and which does not require access to large-scale facilities. Thus, chemical or electrochemical processes which are related to local changes in the work function can be imaged. Using energy-filtering, we can even obtain chemical information, however, in this case we need a much higher photoelectron yield and in general have to use a synchrotron photon source.

SPEM requires an intense photon source *a priori* which in the past was only available at synchrotrons. The spatial resolution which can be achieved today is already high, and the energy resolution is high enough for chemical analysis–therefore SPEM can also be addressed as µ-ESCA. The temporal resolution is worse than in the case of PEEM, and, therefore, fast surface processes cannot be imaged with spatial resolution.

Up to now, the most attractive field of application is the study of non-linear surface reactions which show concentration patterns in space and time. Recently, the study of the electrode structures of fuel cells or other electrochemical systems at elevated temperatures is emerging as a new field of application. In both cases, the requirement of UHV conditions is a serious disadvantage from the chemical point of view. However, recently there are attempts to build ambient-pressure synchrotron-based SPEMs which will overcome the limitation of investigations under UHV conditions and the so-called pressure gap.

16.6
Supplementary Material

16.6.1
PEEM – The Development

After Eugen Goldstein had obtained the first unfocused electron emission image of a metal coin in 1880 it took more than 50 years until the first systematically developed electron emission microscope was introduced in 1932. As a major step the necessary electron optics had to be developed. Here Ruska did pioneering work, and much later in 1986 he received the Nobel prize in physics for his "fundamental work in electron optics, and for the design of the first electron microscope".

The progress in photoelectron emission microscopy has always been connected with developments in other fields, for example, UHV technology, the improvement of photo-detectors, electron amplifiers, and not least synchrotron radiation. Modern PEEMs do not differ from the first prototypes in their essential components. For instance Brüche and Johannson already used a mercury discharge lamp to generate the photoelectrons. The investigated cathode material was zinc metal and the acceleration voltage was chosen between 10 kV and 30 kV. For focusing on a phosphorus screen they used one electrostatic lens. At the same time Knoll, Houtermans, and Schulze developed a first PEEM–with two magnetic lenses instead of one electrostatic lens. Both PEEMs achieved a magnification factor of 100–150.

Originally, the investigation of metal surfaces was in the spotlight of interest and soon the profit for metallurgy was recognized: PEEM images not only show the surface topography but also allow the observation of grains and grain boundaries in polycrystalline metals. Surface impurities can also be made visible. Early examples are the PEEM images published in 1934 by Pohl (Figure 16.16) [20]. They show a fingerprint on a scratched aluminum surface (Figure 16.16a) and a polycrystalline Pt metal surface (Figure 16.16b).

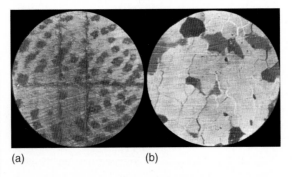

(a)   (b)

Figure 16.16 Early PEEM images of aluminum metal surfaces taken by Pohl: (a) fingerprint (seven-times magnified) and (b) polycrystal (25-times magnified) [20].

In the following years different "laboratory" emission microscopes were developed. The idea by Mecklenburg to produce a commercial PEEM was stopped due to World War II. From 1947 on Philips sold a thermionic emission microscope. The fast spread of the method is reflected by the "First International Conference on Electron Microscopy" in 1950 in Paris. At the beginning of the 1960s a PEEM originally developed by Möllenstedt und Dünker was sold by Trüb, Täuber, and soon by the Balzer company. The name "Metioskop" points out the main application area in metallurgy. A few non-commercial PEEMs were developed in parallel, for example, at the Institute for Applied Physics at the University of Tübingen (Möllenstedt) or in Halle (Bethge). In the 1970s Balzer sold more than 10 PEEMs (model "Metioskop KE3" build by Wegmann and Ruska), and the method was more and more accepted in metallurgy and physics – if we take the multitude of publications from this period and the "First International Conference on Emission Microscopy" in 1979 as a measure.

Meanwhile, the scanning electron microscope (SEM) had been developed as an alternative and powerful method for the imaging of electrically conducting surfaces and soon became commercially available. Today, SEM is the standard tool for the study of surface topography. With the fast rise of "surface science", the first PEEMs for UHV conditions were developed, and Engel thereafter built a PEEM with a resolution of 12 nm. In 1972 the first biological samples were investigated by Griffith and PEEM entered biology. In the course of time the PEEM design, especially the electron optics, was considerably improved. To reach the aim of a spatial resolution of 1 nm–2 nm a magnetic sector field in combination with a hyperbolic electron mirror (aberration correction) has been suggested and currently realized. spectro-microscope for all relevant techniques (SMART) at Bessy II, Berlin [57] and PEEM3 at advanced light source (ALS), Berkeley [58]). Today, different commercial PEEMs [16, 59–61] are available. A survey over PEEM/LEEM groups and additional information can be found under *http://www.leem-user.com/*. PEEM does not rank among the standard methods in physical chemistry and surface science, but it has become a powerful tool for the study of nonstationary surface processes.

A detailed historical review of the developments in electron microscopy until 1991 can be found in [62].

16.6.2
Synchrotron Radiation

Synchrotron radiation is generated whenever charged particles (usually electrons) move with almost the speed of light and are deflected by magnetic fields [63, 64]. This type of radiation is extremely intense and bright and is always released tangentially to the trajectory of the particles and focused in a narrow cone along the main flight direction (see Figure 16.17).

Synchrotron radiation offers a number of advantages compared with conventional laboratory radiation sources:

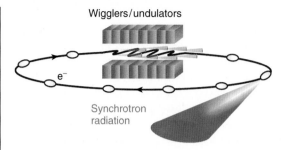

Wigglers/undulators

e⁻

Synchrotron
radiation

Figure 16.17 The origin of synchrotron radiation (shown schematically). The oval shapes indicate the traveling electron bunches in the ring.

- Synchrotron light is not monochromatic but shows a continuous spectrum from infrared to hard X-ray radiation; this allows us to extract the desired energy by the use of monochromators. This type of light can be extracted at so-called bending magnets.
- Synchrotrons are discontinuous light emitting sources. Small bunches of particles circulate in the synchrotron ring and emit very short pulses of light with a frequency in the GHz range whenever they pass a bending magnet or so-called "insertion device"
- Special magnetic insertion devices—so called undulators and wigglers—are mounted in the straight sections of third generation synchrotrons and enable a further increase in the light intensity. By these magnets (mounted in alternating geometrical orientations) the particles are forced on a kind of "slalom" (see Figure 16.17) and the emitted light superimposes.
- The divergence γ^{-1} is of the order $\frac{1}{\gamma} \cong \sqrt{1 - \left(\frac{v}{c}\right)^2} = \frac{m_0 c^2}{E}$ (where E is the beam energy, v the velocity, c the speed of light, and m_0 the rest mass of an electron) and is very small; after 10 m the beam is usually expanded by only a few millimeters
- The radiation is by nature polarized in a specific way, linearly or circularly, depending on the geometry of the light emitting device. This can be used favorably in the investigation of magnetic materials.

In synchrotron facilities (over 50 around the world at this time) this radiation is used in a wide range of spectroscopic and microscopic applications. The SPEM experiments described above were all performed at the ESCA microscopy beamline at the ELETTRA synchrotron in Trieste. A photon energy of $E \approx 650\,\text{eV}$ is used, and the photon beam is focused on the surface of the sample using a Fresnel zone plate followed by a so-called order sorting aperture (see Figure 16.4). The measured photon flux on the sample in the light spot of 100 nm × 100 nm is $10^9 - 10^{10}$ photons s^{-1}.

Further information can be found for example on the ELETTRA homepage (*http://www.elettra.trieste.it/*). An attractive survey of "The Physics of Light" is published on the websites of the "Deutsche Physikalische Gesellschaft" (*http://www.weltderphysik.de*).

16.6.3
Potentials and Work Function

Figure 16.18 visualizes the different potentials which are used in the description of surfaces and the photoelectric effect. The *inner* or *Galvani potential* corresponds to the electrical work $(e \cdot \phi)$ which is required to transfer a unit charge from infinity (E_∞) to the interior of a solid. This inner potential is usually decomposed into the *outer* or *Volta potential* ψ and the *surface potential* χ, by defining the so-called vacuum level (E_{vac}), which is located above the surface of the solid at a distance where image forces can be neglected.

The *electrochemical potential* of the electrons $\tilde{\mu}_e$ (as a molar quantity) consists of the chemical part (the chemical potential of the electrons μ_e, accounting for the ensemble properties of the electrons) and the electrical part $(F \cdot \Phi)$: $\tilde{\mu}_e = \mu_e - F\varphi$. It is identical to the *Fermi energy* (which is usually given in terms of a single electron). The *work function* Φ is defined as the energy which is required to extract an electron from the Fermi level (E_F) and to bring it to the vacuum level $(\Phi = E_{vac} - E_F)$, that is, the work function is related to the electrochemical potential or Fermi energy by $\Phi = -\frac{\tilde{\mu}_e}{N_A} - e\varphi + e\chi$ with N_A denoting Avogadro's number. For the sake of completeness we add the real potential α of an electron $(\alpha/N_A = -\Phi)$ as the absolute value of the work function.

Example: In our electrochemical example we image work function changes of both the metal electrode (platinum) and the solid electrolyte (the oxide "YSZ"). In general, a change in the work function can be caused by a change in either (i) the position of the vacuum level or (ii) the Fermi level. (i) A modified surface potential generates a shift in the vacuum level, as shown in Figure 16.18b,c. The surface

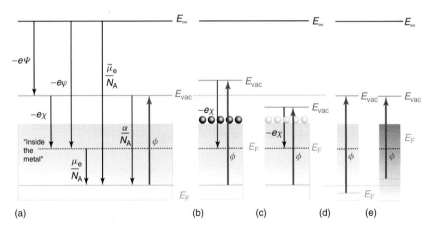

Figure 16.18 Work function and different potentials: (a) reference case of a "clean" surface, (b) and (c) work function changes by the adsorption of electronegative or electropositive species, (d) and (e) work function changes caused by a shift of the Fermi level (composition changes).

potential depends on the surface dipole field, which varies with the adsorption of a compound (e.g., oxygen) on the surface. The adsorption of an electronegative adsorbate like oxygen on platinum increases the surface dipole, and, thus, increases the work function which corresponds to Figure 16.18b – a darker PEEM image is the result. A reduced work function and a brighter PEEM image appear after adsorption of an alkali metal on a platinum surface (Figure 16.18c). (ii) The position of the Fermi level can be shifted by a change in the composition of the material. This is, for example, observed when YSZ is reduced [46] (Figure 16.18d and e).

Essentially, PEEM can be considered as a "work function microscopy" under UHV conditions. Both adsorption effects and chemical composition changes contribute to the work function contrast. In the case of compound surfaces this can lead to problems in the interpretation of PEEM images.

Acknowledgments

The authors thank M. Kiskinova and L. Gregoratti (ELETTRA, Trieste) for continuous support in SPEM studies. One of us (E.M.) is grateful to the FCI (Fonds der Chemischen Industrie) and the DFG (German Research Foundation) for a scholarship. J.J. and B.L. are grateful to the DFG for financial support of the project Ja 648/10-1.

References

1. Hüfner, S. (2003) *Photoelectron Spectroscopy – Principles and Applications*, 3rd edn, Springer-Verlag, Berlin.
2. Günther, S., Kaulich, B., Gregoratti, L., and Kiskinova, M. (2002) *Progress in Surface Science*, **70**, 187.
3. Kinoshita, T. (2002) *J. Electron Spectrosc. Relat. Phenom.*, **124**, 175–194.
4. Mundschau, M. (1991) *Ultramicroscopy*, **36**, 29–51.
5. Ertl, G. and Küpper, J. (1985) *Low Energy Electrons and Surface Chemistry*, Wiley-VCH Verlag GmbH, Weinheim, p. 35.
6. Briggs, D. and Seah, M.P. (1990) *Practical Surface Analysis, Auger and X-ray Photoelectron Spectroscopy*, Vol. **1**, John Wiley & Sons, Ltd, Chichester.
7. Cardona, M. and Ley. L. (eds) (1978) *Photoemission in Solids I*, Springer-Verlag, Berlin, Heidelberg, New York, p. 84.
8. Tanuma, S., Powell, C.J., and Penn, D.R. (1991) *Surf. Interface Anal.*, **17**, 927–939.
9. Jablonski, A. and Powell, C.J. (1999) *J. Electron Spectrosc. Relat. Phenom.*, **100**, 137–160.
10. Moulder, J.F., Stickle, W.F., Sobol, P.E., and Bomben, K.D. (1992) in *Handbook of Photoelectron Spectroscopy* (ed. J. Castaign), Perkin-Elmer Corporation, Physical Electronics Division, Eden Prairie.
11. Marsi, M., Casalis, L., Gregoratti, L., Günther, S., Kolmakov, A., Kovac, j., Lonza, D., and Kiskinova, M. (1997) *J. Electron Spectrosc.*, **84**, 73–83.
12. Casalis, L., Jark, W., Kiskinova, M., Lonza, D., Melpignano, P., Morris, D., Roesei, R., Savoia, A., Abrami, A., Fava, C., Furlan, P., Pugliese, R., Vivoda, D., Sandrin, G., Wei, F.-Q., Contarini, S., DeAngelis, L., Gariazzo, C., Nataletti, P., and Morrison, G.R. (1995) *Rev. Sci. Instrum.*, **66**, 4870–4875.
13. Kordesch, M.E. (1995) *The Handbook of Surface Imaging and Visualization*, CRC Press, pp. 581–596.

14. Bauer, E. (2001) *J. Phys.: Condens. Matter*, **13**, 11391–11404.

15. Rempfer, G.F. and Griffith, O.H. (1998) *Ultramicroscopy*, **27**, 273–300.

16. Elmitec (Manufacturer of LEEM and PEEM systems), *http://www.elmitec-gmbh.com* (latest access: 22.11.2011).

17. Houle, W.A., Engel, W., Willig, F., Rempfer, G.F., and Griffith, O.H. (1982) *Ultramicroscopy*, **7**, 371–380.

18. Engel, W., Kordesch, M.E., Rotermund, H.H., Kubala, S., and Oertzen, Av. (1991) *Ultramicroscopy*, **36**, 148–153.

19. Schmidt, T., Heun, S., Slezak, J., Diaz, J., Prince, K.C., Lilienkamp, G., and Bauer, E. (1998) *Surf. Rev. Lett.*, **5** (6), 1287–1296.

20. Pohl, J. (1934) *Z. Technol. Phys.*, **22**, 579–581.

21. Schwarzer, R.A. (1981) *Microsc. Acta*, **84** (1), 51.

22. Schönhense, G. (1999) *J. Phys. Condens. Matter*, **11**, 9517–9547.

23. Stöhr, J., Padmore, H.A., Anders, S., Stammler, T., and Scheinfein, M.R. (1998) *Surf. Rev. Lett.*, **5** (6), 1297–1308.

24. Gudat, W. and Kunz, C. (1972) *Phys. Rev. Lett.*, **29**, 169–172.

25. Stöhr, J. and Anders, S. (2000) *IBM J. Res. Dev.*, **44** (4), 535–551.

26. Stöhr, J. (1992) *Nexafs Spectroscopy*, Springer-Verlag, Berlin.

27. Imbihl, R. and Ertl, G. (1995) *Chem. Rev.*, **95**, 697–733.

28. Gottschalk, N., Mertens, F., Bär., M., Eiswirth, M., and Imbihl, R. (1994) *Phys. Rev. Lett.*, **73**, 3483–3486.

29. Mertens, F., Gottschalk, N., Bär, M., Eiswirth, M., and Mikhailov, A. (1995) *Phys. Rev. E*, **51**, R5193–R5196.

30. Mertens, F. and Imbihl, R. (1994) *Nature*, **370**, 124–126.

31. Thomas, J.M. and Thomas, W.J. (1997) *Principles and Practice of Heterogeneous Catalysis*, Wiley-VCH Verlag GmbH, Weinheim, pp. 576–590.

32. Eiswirth, M. and Ertl, G. (1994) in *Chemical Waves and Patterns* (eds R. Kapral and K. Showalter), Kluwer, Dordrcht.

33. Mikhailov, A.S. (1990) *Foundations of Synergetics I*, Springer-Verlag, Berlin, Heidelberg, New York.

34. Mertens, F., Schwegmann, S., and Imbihl, R. (1997) *J. Chem. Phys.*, **106** (10), 4319–4326.

35. Comelli, G., Dhanak, V.R., Kiskinova, M., Prince, K., and Rosei, R. (1998) *Surf. Sci. Rep.*, **32** (5), 165–231.

36. Günther, S., Esch, F., Gregoratti, L., Marsi, M., Kiskinova, M., Schubert, U.A., Grotz, P., Knözinger, H., Taglauer, E., Schütz, E., Schaak, A., and Imbihl, R. (2000) Spectromicroscopy of catalytic relevant processes with sub-micron resolution. AIP Conference Proceedings (X-Ray Microscopy), Vol. 507, pp. 219–224.

37. Schaak, A., Günther, S., Esch, F., Schütz, E., Hinz, M., Marsi, M., Kiskinova, M., and Imbihl, R. (1999) *Phys. Rev. Lett.*, **83** (9), 1882–1885.

38. Schmidt, T., Schaak, A., Günther, S., Ressel, B., Bauer, E., and Imbihl, R. (2000) *Chem. Phys. Lett.*, **318**, 549–554.

39. Makeev, A., Hinz, M., and Imbihl, R. (2001) *J. Chem. Phys.*, **114** (20), 9083–9098.

40. Vayenas, C.G., Bebelis, S., Pliangos, C., Brosda, S., and Tsiplakides, D. (2001) *Electrochemical Activation of Catalysis*, Kluwer Academic/Plenum Publishers, New York.

41. Toghan, A., Roesken, L.M. and Imbihl, R. (2010) *Phys. Chem. Chem. Phys.*, **12**, 9811–9815.

42. Luerßen, B., Mutoro, E., Fischer, H., Günther, S., Imbihl, R., and Janek, J. (2006) *Angew. Chem. Int. Ed.*, **45** (9), 1473–1476.

43. Luerßen, B., (2003) In situ-spektroskopische Untersuchungen an Pt/YSZ-Elektroden, Dr. rer. nat. thesis, Justus-Liebig-Universität Gießen, *http://geb.uni-giessen.de/geb/volltexte/2003/1207/*.

44. Poppe, J., Schaak, A., Janek, J., and Imbihl, R. (1998) *Ber. Bunsenges. Phys. Chem.*, **102**, 1019–1022.

45. Luerßen, B., Günther, S., Marbach, H., Kiskinova, M., Janek, J., and Imbihl, R. (2000) *Chem. Phys. Lett.*, **316**, 331–335.

46. Luerßen, B., Janek, J., Günther, S., Kiskinova, M., and Imbihl, R. (2002) *Phys. Chem. Chem. Phys.*, **4**, 2673–2679.

47. Janek, J., Luerssen, B., Mutoro, E., Fischer, H., and Günther, S. (2007) *Top. Catal.*, **44** (3), 399–407.

48. Rempfer, G.F. (1990) *J. Appl. Phys.*, **67**, 6027.

49. Ade, H., Yang, W., Englisch, S.L., Hartman, J., Davis, R.F., Nemanich, R.J., Litvinenko, V.N., Pinayev, I.V., Wu, Y., and Madey, J.M.J. (1998) *Surf. Rev. Lett.*, **5** (6), 1257–1268.

50. Samokhvalov, A., Garguilo, J., Wang, W.-C., Edwards, G.S., Nemanich, R.J., and Simon, J.D. (2004) *J. Phys. Chem. B*, **108** (42), 16334–16338.

51. Hamburger Synchrotronstrahlungslabor HASYLAB at Deutsches Elektronen-Synchrotron DESY, *http://hasylab.desy.de* (latest access: 22.11.2011).

52. Free Electron Lasers of Europe EuroFEL, *http://www.iruvx.eu/e20* (latest access: 22.11.2011).

53. Schönhense, G. (2004) *J. Electron. Spectrosc. Relat. Phenom.*, **137–140**, 769–783.

54. Oelsner, A., Krasyuk, A., Fecher, G.H., Schneider, C.M., and Schönhense, G. (2004) *J. Electron Spectrosc. Relat. Phenom.*, **137–140**, 757–761.

55. Khursheed, A. (2002) *Optik*, **113** (11), 505–509.

56. Spiecker, H., Schmidt, O., Ziethen, C., Menke, D., Kleineberg, U., Ahuja, R.C., Merkel, M., Heinzmann, U., and Schönhense, G. (1998) *Nucl. Instrum. Methods A*, **406**, 499.

57. XUV Light Sources BESSY and BESSY II, Helmholtz-Zentrum Berlin, *http://www.helmholtz-berlin.de/* (latest access: 22.11.2011).

58. Advanced Light Source, Lawrence Berkeley National Laboratory, *http://www-als.lbl.gov/* (latest access: 22.11.2011).

59. Focus GmbH (Instruments for Electron Spectroscopy and Surface Analytics), http://www.focus-gmbh.com/, Omicron NanoTechnology, *http://www.omicron.de/en/home* (latest access: 22.11.2011).

60. Staib Instruments, *http://www.staibinstruments.com*.

61. Jeol Ltd., *http://www.jeol.com* (latest access: 22.11.2011).

62. Griffith, O.H. and Engel, W. (1991) *Ultramicroscopy*, **36**, 1–28.

63. Wille, K. (1992) *Physik der Teilchenbeschleuniger und Synchrotron-strahlungsquellen*, Teubner Studienbücher Physik, Teubner-Verlag, Stuttgart.

64. Attwood, D. (1999) *Soft X-rays and Extreme Ultraviolet Radiation: Principles and Applications*, Cambridge University Press, Cambridge.

17

Secondary Ion Mass Spectrometry – a Powerful Tool for Studying Elemental Distributions over Various Length Scales

Roger A. De Souza and Manfred Martin

▪ Method Summary

Acronyms, Synonyms
- Secondary ion mass spectrometry (SIMS)
- Time-of-flight secondary ion mass spectrometry (TOF-SIMS).

Benefits (Information Available)
- The detection of all elements, from hydrogen to uranium, in vacuum compatible solids
- High sensitivity: a detection limit of milligrams per kilogram (ppm) is usual for most elements in most matrices; in some cases a detection limit of micrograms per kilogram (ppb) is possible
- Ability to differentiate between isotopes of the same element
- Logarithmic intensity scale: one can determine changes in concentrations over orders of magnitude
- Spatially resolved information: the depth resolution is of the order of nanometers, the lateral resolution of the order of micrometers.
- Surface sensitivity and chemical structure for measurements performed under static conditions.

Limitations (Information Not Available)
- Separate standard required for each element in each matrix (the secondary ion intensity depends strongly on various parameters, such as the element of interest, the matrix in which the element is located, etc.)
- Composition and structure of sample ahead of sputtering front altered by the primary ion beam
- Mass interferences (two or more secondary ions of same nominal mass) can complicate analysis
- No information available on the charge state of the element in the matrix
- No information available on atomic structure or physical topography of the surface/sample.

Methods in Physical Chemistry, First Edition. Edited by Rolf Schäfer and Peter C. Schmidt.
© 2012 Wiley-VCH Verlag GmbH & Co. KGaA. Published 2012 by Wiley-VCH Verlag GmbH & Co. KGaA.

17.1
Introduction

Secondary ion mass spectrometry (SIMS) is a sophisticated analytical technique for determining the elements present in materials, and especially on surfaces, with trace level sensitivity and high spatial resolution. The aim of this article is to introduce the SIMS technique to students of physical chemistry by focusing on the fundamentals, on the principles of operation and on selected case studies. It is, therefore, by no means a comprehensive overview, but rather a personal selection of the salient aspects. Experienced SIMS analysts will certainly find some issues given insufficient attention or none at all. For further reading we recommend the standard SIMS textbooks [1–4], and some extended reviews [5–8], to which we have referred in writing this article.

17.2
Basic Principles

17.2.1
The General Principle of Analysis

All methods for analyzing condensed matter can be considered in terms of the same general principle: question and answer. Whether we are interested in determining the elements present in a sample, their distribution, their crystallographic arrangement, or even ensemble properties, such as electronic or magnetic structure, the principle of analysis consists of questioning the sample with a suitable probe particle (e.g., ion, electron, photon, neutron, or muon) and receiving an answer in the form of a different particle (ion, electron, photon, etc.) or the same particle with modified properties. In order to obtain any sort of information in response to a probe particle we need the probe particle to have interacted in some way with a certain volume of material.

$$Question \xrightarrow{\text{Interaction}} Answer$$

This interaction may alter significantly the questioned volume of sample, and thus we must exercise due care when interpreting the answer received. In Table 17.1 we sort selected analytical techniques according to this scheme of question and answer, in order to place SIMS in the range of analytical techniques and to indicate that SIMS employs ions as the question and as the answer.

To illustrate this general principle, we show in Figure 17.1 the basis of SIMS and three other analysis techniques that use energetic ions as probe particles. SIMS, for instance, uses energetic ions (primary ions) to eject target material from a sample surface and mass analyses the ionized target fragments (secondary ions). The basic idea of Rutherford back-scattering (RBS) is to determine the type and distribution of elements, particularly heavy elements in a matrix of lighter elements, by analyzing the energy loss spectrum of MeV He ions that have been elastically scattered from the screened nuclei of the target atoms. Elastic recoil detection analysis (ERDA),

Table 17.1 Selected analytical techniques.

Question	Answer		
	Ion	Electron	Photon
Ion	SIMS, RBS, ERDA	–	PIGE
Electron	–	TEM, AES	IPES, EDX
Photon	–	PES (UPS, XPS)	XRD, IR, UV–VIS

the third method depicted, uses projectile ions heavier than the target species of interest to knock these lighter species out of the target. Thus, analysis with He projectile ions provides one of the best methods for analyzing ^{1}H and ^{2}H in solids. Finally, nuclear reaction analysis (NRA) refers to the suite of techniques, in which a nuclear reaction between a probe particle and a specific target nucleus is induced; one could describe it as analysis by atomic scale alchemy. The variant shown in the figure is known specifically as proton induced gamma-ray emission (PIGE): highly energetic protons surmount the coulomb barrier of the target nucleus, leading to the transmutation of the nucleus and the emission of a characteristic gamma-ray. PIGE is particularly well-suited to the analysis of light elements such as Li, Be, B, F, and Na.

17.2.2
The SIMS Phenomenon

17.2.2.1 The SIMS Principle
The SIMS process is illustrated schematically in more detail in Figure 17.2. Primary ions with an energy between 0.25 and 30 keV are fired at a solid; in common use

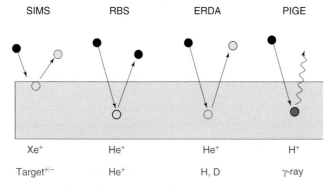

Figure 17.1 Four techniques of ion beam analysis schematically illustrated in terms of the general principle, question and answer: secondary ion mass spectrometry (SIMS), Rutherford backscattering (RBS), elastic recoil detection analysis (ERDA), and PIGE (proton induced gamma-ray emission).

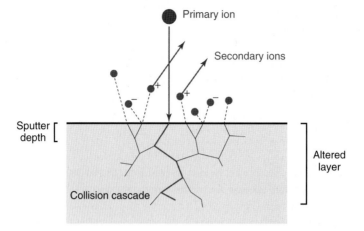

Figure 17.2 Schematic illustration of the principle of SIMS.

as primary ions in SIMS are Cs^+, O_2^+, Ar^+, Xe^+, and Ga^+, with the recent developments of Bi and Au cluster ions (Au_n^{z+}, Bi_n^{z+}) and even C_{60}^+. Upon entering the solid, the primary ions transfer their energy, in binary collisions, to the atoms of the target. These target atoms are displaced from their original sites and collide in turn with other target atoms, and so on, until the energy transferred is insufficient to cause atom displacement. In other words, a short-lived cascade of collisions takes place, as in, say, a game of snooker or billiards. Collision cascades that reach the surface may cause target material to be ejected from the top two to three monolayers in the form of atomic and molecular moieties – a process known as *sputtering* – if sufficient energy and momentum are imparted to surface species to overcome binding energies. Most of these sputtered particles are neutral,[1] but some are charged: these are the secondary ions.[2] Secondary ions are emitted in various directions from the sample surface with a range of energies that peaks around 10 eV and that is independent of the energy of the primary ions. Some secondary ions of given polarity can be collected by means of a suitable electrical potential and sorted in a mass spectrometer according to their mass to charge ratio (m/q).

17.2.2.2 SIMS Capabilities
SIMS is, therefore, essentially the mass spectrometric analysis of secondary ions that have been produced by ion beam erosion of a target. As a consequence the technique offers an outstanding combination of capabilities for the analysis of

1) It is possible to post-ionize the secondary neutral particles and analyze these species in a mass spectrometer. This technique is known as secondary neutral mass spectrometry (SNMS). For details see Refs. [1, 2].
2) The phenomenon of sputtering was first observed in 1853 by W.R. Grove as the occurrence of a metallic coating on the glass walls of a discharge tube [9]. To physical chemists, Grove is better known as the *inventor of the fuel cell*. The beginnings of SIMS can be attributed to J. J. Thompson, the discoverer of the electron, who, in 1910, found that most of the species ejected from the cathode of a discharge tube are neutral, but some are ions [10].

vacuum-compatible solids. It is capable of detecting all elements, from H onwards, and of differentiating between isotopes. It can routinely determine concentrations from the matrix level down to the parts per million level, and, in favorable cases, down to the parts per billion level. And it can provide this information as a function of spatial coordinate, routinely with sub-nanometer depth resolution or with a lateral resolution of a few hundred nanometers.

17.2.2.3 The Altered Layer

There are of course a number of drawbacks associated with SIMS, as with any analytical technique. The most trivial of these is that ion beam erosion of a surface makes it impossible to analyze the same place twice. Far more critical for the SIMS analyst is the fact that ion bombardment alters the composition and structure of the sample ahead of the sputtering front. As might be expected, this so-called altered layer will be highly enriched with implanted primary ions. This means that already implanted primary ions will be sputtered and give rise to secondary ions along with matrix material. Primary ion implantation is not only a drawback, however, since it can be put to good use (see Section 17.2.2.5). The composition of the altered layer is also homogenized on account of ion beam mixing, that is, the continual redistribution of surface and sub-surface species by the primary ions directly (recoil mixing), and within the collision cascades (cascade mixing) prior to their emission as secondary particles. In compound targets, preferential sputtering may occur, leading to the depletion of one or more components from the altered layer. Various other effects are also known [1–3]. The continual redistribution of target species results additionally in the altered layer being highly defective and, in many cases, even amorphous. Ion bombardment may also cause, under certain conditions, an initially flat surface to develop microtopographical features.

The main consequences of the altered layer are first, that the measured intensity variations with sputter time do not *necessarily* reflect changes in sample composition; and secondly, that the ultimate depth resolution is not a few monolayers, the maximum depth from which secondary ions are emitted, but is determined, due to redistribution ahead of the sputtering front, by the thickness of the altered layer. It is thus comparable to the penetration depth of the primary ions into the target and, hence, for the primary ions and primary ion energies commonly used in SIMS is of the order of nanometers to tens of nanometers. Topography development severely degrades the depth resolution beyond this ultimate limit.

17.2.2.4 Mass Interferences

Since the technique utilizes a mass spectrometer, the SIMS analyst must be aware of the possibility of mass interferences. This is the occurrence of two or more secondary ions with the same nominal m/q, for example, $^{56}Fe^+$ and $^{28}Si_2^+$; their appearance complicates the assignment of peaks in mass spectra and compromises detection limits for trace element analysis. Such interferences may be separated on account of the small differences in their masses (m and $m + \Delta m$), if the resolution of the mass spectrometer, $m/\Delta m$, is sufficient. Thus, for the example given above, a mass resolution of $55.934939/(2 \times 27.976927 - 55.934939) \approx 2.9 \times 10^3$ is required.

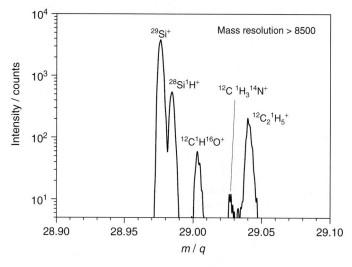

Figure 17.3 The region of a positive secondary ion mass spectrum around $m/q = 29$ obtained from the polished surface of a metallic sample with a TOF-SIMS machine under high resolution conditions.

If the mass spectrometer is unable to resolve the mass interferences, then detecting, say, trace amounts of iron in silicon will be problematic, since the tiny iron signal will be swamped by the intense signal from the silicon dimer. One alternative is to monitor ^{54}Fe, albeit with reduced sensitivity (^{54}Fe/^{56}Fe $= 0.064$). It is worth noting that on moving from elemental semiconductors to compounds, and especially oxides, the number of possibilities for mass interferences increases drastically. In Figure 17.3 we show a high mass resolution mass spectrum at $m/q = 29$. The high mass resolution of 8500 allows the separation of the various secondary ions and their unambiguous assignment. We point out that the number of constituent atoms in the secondary ions increases from left to right on account of the reduced nuclear binding energy. Two final examples we mention, in order to illustrate that mass interferences not only arise from the sample, are (i) phosphorus dopant analysis in silicon (^{31}P$^+$ and ^{30}Si^{1}H$^+$), as the hydrogen in the interfering ion may originate from the residual gas in the analysis chamber and (ii) the detection of gold in FeS$_2$ (pyrite) using a Cs$^+$ beam (Au$^-$ and ^{133}Cs^{32}S$_2{}^-$), as Cs$^+$ primary ions, which provide the highest sensitivity for gold, also give rise to a mass interference.

17.2.2.5 The SIMS Equation and Quantification

The main drawback of SIMS is arguably the strong dependence of the secondary ion intensity on a number of parameters, such as the element of interest A, the matrix M in which it is located, the background pressure in the analysis environment, the crystal orientation, the type of primary ions, their energy, and their angle of incidence with respect to the surface normal. The relative importance of these parameters also varies from case to case. As a consequence, quantification – the process of converting a secondary ion intensity into a concentration – is, at best, simple but time consuming.

One can relate the measured secondary ion intensity i_A^s to the atomic fraction of element A, x_A, via

$$i_A^s = I^P Y \alpha_A \eta_A \theta_A x_A \tag{17.1}$$

where I^P is the primary ion intensity (ions per second); Y, the sputter yield (removed target atoms per incident primary ion); α_A, the ionization probability (α_A^+ for positive secondary ions, α_A^- for negative secondary ions) (ions of type $A^{+/-}$ emitted per removed target atom); η_A, the combined transmission efficiency of the extraction optics, the mass spectrometer, and the detector (ions detected per ion emitted); and θ_A, the isotopic abundance. Obtaining x_A from a measured i_A^s consequently requires knowledge of the other quantities in Equation 17.1. Whereas I^P and Y can be measured separately, and values of θ_A are, of course, tabulated [11], this is not the case for $\alpha_A^{+/-}$ or η_A. Moreover, the ionization probabilities vary by orders of magnitude, depending on the secondary ion and the matrix under consideration, the type, energy, and angle of incidence of the primary ions, and so on. As a result, quantification has proceeded largely by analyzing, under identical measurement conditions, the unknown and a sample that (i) contains a known amount of the relevant element and (ii) is close to the unknown in matrix stoichiometry. As suitable standards are rarely if ever available, the most frequently used calibration samples are those made by ion implanting a known areal dose of the element of interest into a matrix-matched substrate. In view of the fact that $\alpha_A = f$ (Element, Matrix), it is clear that each dopant–matrix combination requires a separate standard.

The calibration procedure is most satisfactorily performed by comparing the ratio i_A^s/i_M^s measured for the unknown with that measured for the calibration standard (i_M^s is the secondary ion intensity of a matrix-related ion). This approach has the advantages that monitoring a matrix-related ion provides a means of confirming the stability of the instrument for the duration of the measurement, and that the ratio is insensitive to changes in measurement conditions (e.g., sputter rate) between analyzing the unknown and the standard. For semiconductor materials, such as silicon and gallium arsenide, Wilson, Stevie, and Magee have collated conversion factors that relate, for most elements, secondary ion intensity ratios to concentrations [3]. These conversion factors are known as relative sensitivity factors (RSFs)

$$c_A = RSF \frac{i_A^s}{i_M^s} \tag{17.2}$$

and depend on primary beam parameters and, via η, SIMS machine type.

We close this section with a few comments on the choice of primary ion and secondary ion parameters. With respect to secondary ion polarity, one can say, very generally, that elements that tend to form cations exhibit higher α_A^+ than α_A^-, and, hence, it is advantageous to detect these elements as positive secondary ions; for elements that tend to form anions, the opposite is true. In this regard, it is extremely important to note that the charge of the detected ion tells us nothing about the charge state of that species in the matrix, a prime example being hydrogen, which is best detected as H^-, even when it is present in the matrix as H^+.

We mentioned above that implanting primary ions into the target alters the composition but should not necessarily be considered a drawback. This is because employing suitable primary ions can enhance ionization probabilities by orders of magnitude. In comparison with noble gas primary ions, electronegative O_2^+ primary ions increase α_X^+ for species that tend to form cations, whereas electropositive Cs^+ primary ions increase α_X^- for species that tend to form anions. With regard to other primary ions, Ga^+ is employed when extremely high lateral resolution is required, as Ga^+ guns deliver finely focused beams; C_{60}^+ and Bi_n^{z+} are finding increasing use in the analysis of polymers and biomolecules, as they produce improved yields of higher molecular weight fragments. The energy of the primary ions can also be optimized for a given analysis. Higher energies mean higher sputter yields, and thus better sensitivity, but at the expense of depth resolution; higher energies also result in better focused ion beams. In summary, the choice of instrumental parameters is dictated by which factor – sensitivity, detection limit, depth resolution, lateral resolution – is the most critical factor in the analysis, with the optimal set of parameters often representing a compromise between several factors.

17.3
Experimental Methods

17.3.1
SIMS Machines and Modes of Operation

The basic components of all SIMS machines are shown in Figure 17.4: one or more ion guns to produce beams of primary ions; an extraction electrode to collect as many secondary ions of given polarity as possible; a mass spectrometer to sort the secondary ions according to their mass to charge ratio, and, finally, a detector. Most machines also have an electron-flood gun, in order to compensate the charge that builds up during ion bombardment of poorly conducting samples. All components are housed in an ultra-high vacuum chamber (see Section 17.6.1).

There are three main types of SIMS machine, which differ according to the type of mass spectrometer used to sort the secondary ions: quadrupole, double focusing magnetic sector, and time-of-flight. Each mass spectrometer has its own advantages and disadvantages in terms of mass resolution, transmission efficiency, detectable mass range, mass detection, complexity of operation, and cost. A summary of selected characteristics is given in Table 17.2. Detailed descriptions of the working principles of these mass spectrometers are beyond the scope of this article; the interested reader is referred to the extensive literature [1–4]. A simple discussion of the working principle of a time-of-flight secondary ion mass spectrometry (TOF-SIMS) machine is given in Section 17.6.2.

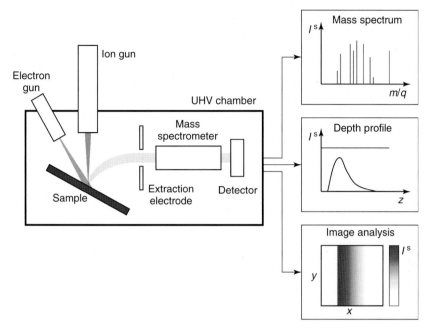

Figure 17.4 Schematic diagram of a SIMS machine, indicating the main components. Three modes of analysis are also depicted.

Table 17.2 Characteristics of the three main types of mass spectrometer used in SIMS machines: quadrupole, double focusing magnetic sector, and time-of-flight [2].

	Mass resolution	Mass range (amu)	Transmission efficiency	Mass detection	Sensitivity (relative to Q-MS)
Q-MS	$10^2 - 10^3$	$\leq 10^3$	0.01–0.1	Sequential	1
DFMS-MS	$> 10^4$	$> 10^4$	0.1–0.5	Sequential	10
TOF-MS	$10^3 - 10^4$	$10^3 - 10^4$	0.5–1	Parallel	10^4

There are basically two modes of SIMS operation: dynamics SIMS and static SIMS. As indicated in Figure 17.4, the information acquired from such measurements can be viewed as a mass spectrum, that is, secondary ion intensity against mass to charge ratio; as ion images, that is, the lateral variation of secondary ion intensities; and additionally for dynamic SIMS, as a depth profile, that is, secondary ion intensities as a function of depth.

Dynamic SIMS refers to the progressive erosion of a defined region of the sample surface by a high current beam of primary ions. As noted above, the primary application is the depth profile: by recording the intensities of various secondary ions as a function of sputter time, one can monitor changes in the chemical composition. Quadrupole and magnetic sector instruments achieve this by switching between secondary ions signals, since they cannot register more than one element at a time (sequential mass detection, see Table 17.2). TOF-SIMS instruments, on the other hand, acquire an entire mass spectrum for every point addressed on the sample surface, but do have to be operated in pulsed mode and require separate primary beams for analysis and sputtering.

Since primary beams usually exhibit Gaussian-type intensity profiles, the use of stationary beams would lead to secondary ions being collected from a variety of depths, and,consequently, to rapid loss of depth resolution. In order to avoid this crater-wall effect, the primary beam is raster scanned over an area of, say, 300 μm × 300 μm, but secondary ion intensities are only recorded from the central, flat area of the crater (e.g., 100 μm × 100 μm). This procedure is known as *gating*.

Static SIMS refers to the bombardment of a sample surface with a very low dose of primary ions ($<10^{12}$ ions cm^{-2}), such that, during the analysis, secondary ions are emitted from regions of the surface that, statistically speaking, have not yet been irradiated. Under such conditions, since the damage to the sample surface is minimized, ions are predominantly emitted from the uppermost monolayer, and, furthermore, the desorption of large fragments is promoted. A static SIMS measurement thus makes true *surface* analysis possible. An example is shown in Figure 17.5.

17.4
Case Studies

A glance through the proceedings of the biannual international SIMS conferences [12–14], for instance, provides a comprehensive survey of areas of science and technology to which SIMS is being applied: the semiconductor industry (fabrication control and troubleshooting), materials science (glass, paper, steel, coatings), life sciences (pharmaceuticals, hair, cells, plants), geo- and cosmo-chemistry (rocks, meteorites, pre-solar grains), and physical chemistry (catalysis, corrosion, mass transport). In these diverse areas one can find numerous examples that convincingly demonstrate the strengths of the SIMS technique for studying elemental distributions over various length scales. Thus, for reasons of cohesion we have decided to concentrate, with one notable exception, on systems we have investigated in the field of solid state physical chemistry. Unless otherwise stated, the measurements were made on a TOF-SIMS IV TOF-SIMS machine (IONTOF GmbH, Münster, Germany) at the Institute of Physical Chemistry, RWTH Aachen University.

17.4.1
Oxygen Tracer Diffusion in LiNbO₃: Depth Profiling

Tracer diffusion is an important method for studying transport processes in solids [15–20]. Tracers of component A, denoted A*, are, by definition, chemically

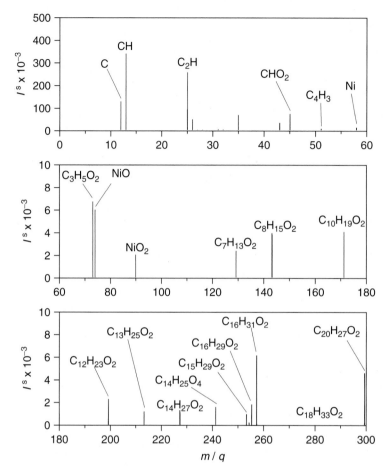

Figure 17.5 A negative secondary ion mass spectrum of a nickel surface acquired under static SIMS conditions. In addition to atomic and simple molecular ions, large molecular cluster ions are observed. The ion fragmentation patterns allow the identification of molecular species, in this case, fatty acids, such as stearic acid $C_{17}H_{35}COOH$.

identical to A and have a very low concentration. Since one can distinguish between A^* and A, the motion of the indistinguishable A particles can be followed with the help of the tracers, A^*. The analysis of tracer concentration profiles in a chemically homogeneous solid yields the tracer diffusion coefficient, D_A^* (or the mean square displacement of tracers). As ideal tracers one can use either radioactive isotopes or stable isotopes of an element. Since the radioactive isotopes of oxygen are not suitable for diffusion studies, their half-lives being too short, the investigation of oxygen diffusion in oxides with stable isotopes necessitates the use of an ion beam analysis technique, such as SIMS or NRA, because they are able to distinguish between the stable isotopes as a function of position within a sample.

The stable minor isotopes of oxygen are ^{17}O, with a natural isotopic abundance of 0.04%, and ^{18}O, with a natural isotopic abundance of 0.20% [11]. Generally, ^{18}O is employed as the labeled species, since it is less expensive than ^{17}O and since there is mass interference at $m/q = 17$ from the usually more intense ^{16}O^{1}H signal that arises from residual gas adsorption. Mass resolution of $m/\Delta m > 6000$ is required to separate ^{17}O clearly from ^{16}O^{1}H. The introduction of an oxygen isotope profile into an oxide is usually accomplished by annealing an equilibrated sample in a large volume of dry ^{18}O$_2$-enriched gas, and after a time t, quenching the sample to room temperature (see Refs [21–23] and references therein for a detailed description of this procedure). This approach also provides the surface exchange coefficient k_O^*, which characterizes the dynamic equilibrium between oxygen in the gas phase and that in the solid.

In Figure 17.6 we present raw SIMS data obtained by TOF-SIMS depth profiling [24] of an exchanged single crystal sample of LiNbO$_3$ [25]. Negative secondary ions were monitored because of the high yields of such species. In addition to the ^{16}O$^-$ and ^{18}O$^-$ secondary ion signals, molecular ions containing ^{16}O are also plotted, as they show two interesting features. First, we see that these molecular ion signals trivially exhibit identical profiles to the ^{16}O$^-$ signal. Second, we find that the ratio ^{93}Nb^{16}O$^-$: ^{7}Li^{16}O$^-$ of 35 : 1 clearly does not reflect the stoichiometry of the sample, even when corrected for ^{7}Li^{16}O$^-$ representing 92.4% [11] of the total Li^{16}O$^-$ emitted, because of the different ionization probabilities of these two species (as discussed in Section 17.2.2.5).

If mass fractionation effects may be neglected (this would lead to η in Equation 17.1 being different for ^{16}O$^-$ and ^{18}O$^-$), quantification does not present a problem: the use of the isotope fraction c^* in analysis means that the unknown terms in Equation 17.1 cancel, as the tracer is chemically identical to the host:

$$c^* = \frac{c_{18}}{c_{16} + c_{18}} = \frac{i_{18}^s}{i_{16}^s + i_{18}^s} \tag{17.3}$$

The raw ^{16}O$^-$ signal plotted in Figure 17.6 was corrected for deadtime effects, as detailed elsewhere [24]. The depth scale was calibrated by determining the crater depth after the SIMS measurement by means of interference microscopy and assuming a constant sputter rate. A 3D image of a SIMS crater obtained by interference microscopy is shown in Figure 17.7a. The crater depth is extracted from the image either by examining a cross-section (b) or the depth histogram (c).

The end result is plotted in Figure 17.6b: the ^{18}O isotope fraction normalized with respect to the background and gas phase isotope abundances, c_r^*, as a function of depth. Also shown is the fitted curve of the appropriate solution to the diffusion equation (see Section 17.6.3). It is seen that a very good fit to the experimental data is possible and this allows precise measurement of the transport parameters: $D_O^* = (4.7 \pm 0.2) \times 10^{-15}$ cm^2 s^{-1}, $k_O^* = (1.7 \pm 0.1) \times 10^{-9}$ cm s^{-1}.

Lastly, we emphasize that, by virtue of determining the isotope profile in the solid, one has incontrovertible evidence that diffusion in the solid state has taken place, if one can describe the profile with the relevant solution of the diffusion equation. Furthermore, one can unambiguously separate surface and bulk transport

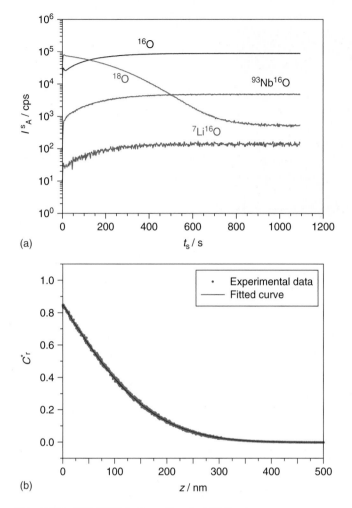

Figure 17.6 TOF-SIMS depth profile of a LiNbO$_3$ single crystal after $^{18}O/^{16}O$ exchange for $t = 23\,700$ s at $T = 1073$ K in $p_{O_2} = 220$ mbar (89% ^{18}O enrichment): (a) raw data showing the measured intensity versus sputter time t_S and (b) quantified profile together with fitted curve (see Section 17.6.3).

steps and one can find indications that fast diffusion along grain boundaries or dislocations has taken place [26] (see also Section 17.4.3).

17.4.2
Oxygen Diffusion in Fe-Doped SrTiO$_3$

Having introduced the oxygen isotope exchange technique in the previous section, we present here results on a second system, Fe-doped SrTiO$_3$, in order to discuss

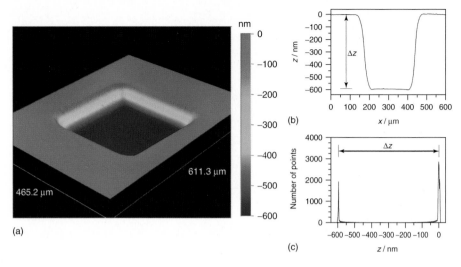

Figure 17.7 Depth determination of a SIMS crater in single crystal LiNbO$_3$ by interference microscopy: (a) 3D image, (b) cross-section, and (c) depth histogram. Analysis gives a depth of $\Delta z = (595 \pm 5)$ nm.

two important points. Specifically, we want to demonstrate, first, the large range of length scales that the technique can probe, and, second, an unconventional application of the technique.

In Figure 17.8a we show the normalized ^{18}O penetration profile obtained by TOF-SIMS imaging of the exposed cross-section of a single crystal sample of Fe-doped SrTiO$_3$. SIMS imaging was used, since the profile length of around 220 µm is beyond the limit of depth profiling; a short depth profile of the original surface was carried out in addition, so that the isotope fraction at the sample surface could be fixed. This short profile is shown in Figure 17.8b. We see in Figure 17.8a that imaging analysis is also capable of providing high quality data, even when the isotope enrichment in the solid is low, and that the data is described well by the relevant solution to the diffusion equation (see Section 17.6.3), yielding the transport parameters $D_O^* = (3.7 \pm 0.9) \times 10^{-10}$ cm^2 s^{-1} and $k_O^* = (1.05 \pm 0.07) \times 10^{-9}$ cm s^{-1}.

Returning now to Figure 17.8b, we reiterate that this depth profile corresponds to the first 20 nm of the 220 µm oxygen diffusion profile. More importantly we draw attention to a sharp, initial decrease in c_r^* at the surface, which extends circa 10 nm into the bulk. It is thus of the same order of magnitude as the depth resolution achievable in SIMS, and one must, therefore, question if this additional profile is properly resolved. In order to establish whether this was the case, we examined the sputter removal of species that arise from residual gas adsorption and from surface contaminants, that is, species that are expected to be found only in the first one or two monolayers. The normalized signal extracted from the raw data file for ^{16}O^{1}H$^-$ is also plotted in Figure 17.8b. As this signal decreases over a far

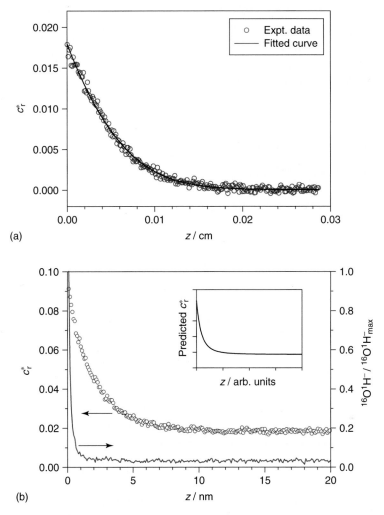

Figure 17.8 Normalized oxygen isotope fraction, c_r^*, in an Fe-doped $SrTiO_3$ single crystal. Sample annealed at $T = 973$ K in $pO_2 = 500$ mbar for $t = 86\,300$ s. (a) Imaging analysis of exposed cross-section. (b) Depth profile of the original surface (inset shows a profile calculated for tracer incorporation through a depletion space-charge layer); also shown is the sputter removal of a surface contaminant (see text). Note the different depth scales in (a) and (b).

smaller depth interval than does c_r^*, we can be reasonably confident that the sharp, initial decrease is not significantly affected by ion beam mixing. We attributed this additional profile to D_O^* being lower in the surface region on account of a space-charge layer depleted of oxygen vacancies being present [24, 27]. It is well documented that grain boundaries in Fe-doped $SrTiO_3$ have space-charge layers in the adjacent bulk regions [28–32]; this result indicates that the (001) surface of an acceptor-doped $SrTiO_3$ does too.

In summary, we have demonstrated that an oxygen isotope exchange experiment combined with a high depth resolution TOF-SIMS analysis can be used to probe a space-charge layer at the surface of a crystalline oxide under thermodynamically well-defined conditions. In addition to presenting this unconventional application, we have shown that SIMS is able to access a large range of length scales: by depth profiling we were able to resolve profiles of 10 nm extent, whilst at the other end of the scale we were able, with ion imaging, to measure profiles extending hundreds of micrometers.

17.4.3
Cation Diffusion in LSGM: 3D Analysis – Depth Profiling with Image Analysis

Perovskite-type oxides based on $LaGaO_3$ are of great technical interest because of their high oxygen-ion conductivity. Lanthanum gallates doped with strontium on the A site and magnesium on the B site, $La_{1-x}Sr_xGa_{1-y}Mg_yO_{3-(x+y)/2}$ (LSGM), attain higher oxygen-ion conductivities than yttria-doped zirconia (yttria-stabilized zirconia, YSZ) [33, 34]. Thus LSGM represents a promising alternative to YSZ as the electrolyte in solid oxide fuel cells (SOFCs) [35]. Cells using LSGM as the electrolyte are expected to operate at intermediate temperatures around 700 °C for more than 30 000 h, and very thin electrolyte layers (down to some microns) are desired to reduce the cell resistance. Although cation diffusion in simple perovskites is known to be very slow, there are several important processes that are determined by the slowest moving species, such as sintering or creep. If the cations exhibit different diffusivities, kinetic demixing of the electrolyte [36] can be an additional origin of long term degradation. It is, therefore, important to obtain data for cation diffusion in LSGM. By means of SIMS, cation impurity diffusion of Y, Fe, and Cr [37] and cation tracer-diffusion of La, Sr, and Mg[3] in LSGM have been investigated [38].

A typical mass spectrum acquired under dynamic TOF-SIMS conditions is shown in Figure 17.9. Various atomic and molecular secondary ions from the LSGM matrix are observed, some of which are not stable as solids, for example, Mg_2O or SrO_2. The spectrum also indicates the presence of impurities, for example, Na, Al, K, Ca, though it does not give their concentrations, or even their amounts with respect to one another.

A typical depth profile of ^{138}La, ^{84}Sr, and ^{25}Mg obtained by SIMS[4] is shown in Figure 17.10. In each profile one can distinguish two different contributions. At small penetration depths (up to 100 nm) the profile has a steep slope and corresponds to bulk diffusion. At larger penetration depths the slope decreases, and the profile is due to combined grain boundary and bulk diffusion according to Fisher's model [39] of simultaneous bulk and grain boundary diffusion in polycrystalline materials. The bulk diffusion coefficients of ^{138}La, ^{84}Sr, and ^{25}Mg are

---

3) It was not possible to measure Ga tracer diffusion coefficients, owing to the natural abundances of the two stable gallium isotopes, ^{69}Ga and ^{71}Ga, being 60 and 40%, respectively. The maximum enrichment of,

say, ^{71}Ga above this background level is therefore a factor of 2.5, which is insufficient for determining diffusion profiles.

4) Measured with a DFMS-SIMS, Cameca ims 5f.

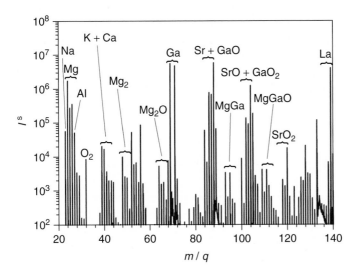

Figure 17.9 Selected region of a positive secondary ion mass spectrum of a $La_{0.1}Sr_{0.1}Ga_{0.1}Mg_{0.1}O_{2.9}$ ceramic obtained under dynamic SIMS conditions. A 25 keV Ga^+ beam was used to generate secondary ions for analysis and a 2 keV O_2^+ beam was used for sputtering.

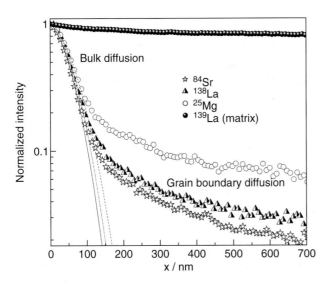

Figure 17.10 Typical diffusion profiles of ^{138}La, ^{84}Sr, and ^{25}Mg in polycrystalline $La_{0.9}Sr_{0.1}Ga_{0.9}Mg_{0.1}O_{2.9}$ ($T = 1673$ K, $t_{diff} = 1.61$ h). All profiles were corrected by the background signal of the corresponding isotope and the matrix-element ^{139}La and then normalized to 1. The lines show the fits of the thick-film solution to the bulk part of the profiles.

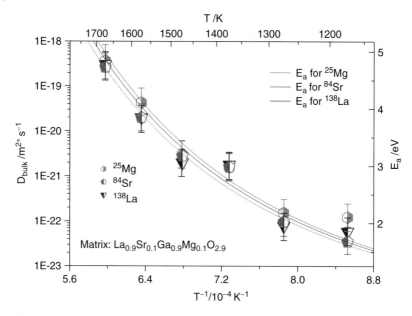

Figure 17.11 Cation tracer diffusion coefficients of La, Sr, and Mg in $La_{0.9}Sr_{0.1}Ga_{0.9}Mg_{0.1}O_{2.9}$. The activation energies obtained from the local slopes are also shown.

obtained by fitting the corresponding solution of the diffusion equation (thick-film solution [40]) to the bulk part of the experimental profiles (see Figure 17.10). The resulting cation diffusion coefficients are shown in Figure 17.11 as a function of inverse temperature.

As expected the cation diffusion coefficients are extremely small, for example, $D_{cation}^{bulk} \approx 10^{-18}$ cm^2 s^{-1} at $T = 1200$ K. This value is much smaller than the oxygen diffusion coefficient, $D_O \approx 10^{-6}$ cm^2 s^{-1}, demonstrating that SIMS is capable of measuring diffusion coefficients in a broad range of more than 12 orders of magnitude. The two surprising observations of a temperature-dependent activation energy and nearly identical diffusion coefficients of cations occupying A- and B-sites in the perovskite structure can be attributed to frozen-in defects and cluster-formation, as discussed in detail in Ref. [38].

The grain boundary diffusion coefficients that are obtained from the tails of the diffusion profiles are about 3–4 orders of magnitude higher than the bulk diffusion coefficients, both for self-diffusion (La, Sr, and Mg) and impurity diffusion (Y, Fe, and Cr). This difference in diffusion coefficients can be excellently visualized by combining depth profiling with imaging. Figure 17.12 shows the three-dimensional distribution of the impurity Fe after diffusion from the surface into the bulk (slow) and along grain boundaries (fast). From this three-dimensional distribution two-dimensional distributions can be easily extracted. The schematic concentration contours for diffusion from the surface of a polycrystalline material into the bulk and along grain boundaries are depicted in Figure 17.13a; the concentration

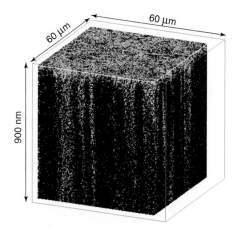

Figure 17.12 Three-dimensional Fe distribution in a poly-crystalline LSGM sample after diffusion from the sample surface into the bulk (slow) and along grain boundaries (fast).

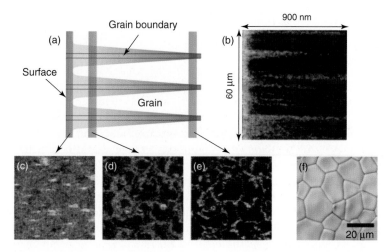

Figure 17.13 (a) Schematic concentration contour for diffusion from the surface of a polycrystalline material into the bulk and along grain boundaries. (b) Experimental concentration contour for Fe diffusion into polycrystalline LSGM as obtained by SIMS imaging. (c–e) Fe distribution in cross-sections parallel to the sample surface and at different depths. (f) Micrograph of the sample surface as obtained by SEM.

contours for Fe diffusion into polycrystalline LSGM, as obtained by SIMS imaging, are shown in Figure 17.13b, and beautifully confirm the schematic concentration contours of Fisher's model [39]. Figure 17.13c–e show the Fe distribution in three cross-sections parallel to the sample surface, which refer to three different depths. Near the surface, Figure 17.13c, a nearly homogeneous Fe distribution is found, whereas at larger depths, Figure 17.13e, Fe is confined to the grain boundaries (due to fast grain boundary diffusion), and as a result reveals the sample's grain structure. For comparison a micrograph of the sample surface obtained by scanning electron microscopy (SEM) is shown in Figure 17.13f.

17.4.4
Nitrogen in YSZ: Depth Profiles and Quantification

The incorporation of nitrogen into oxides is a subject of growing importance as oxynitrides have recently been studied as new pigments [41, 42], materials for UV optics [43], or as nitrogen ion conductors at high temperatures [44]. Concerning the last topic, it is known that nitrogen can partially replace oxygen in YSZ at elevated temperatures and low oxygen activities by occupying oxygen vacancies. However, due to the high stability of the nitrogen molecule this mode of incorporation is limited to very high temperatures, typically $T > 1800$ K. A different approach makes use of the electrochemical incorporation of nitrogen [45] at relatively low temperatures, around 1000 K, and is, in addition, spatially resolved by the use of microelectrodes. As will be shown subsequently, SIMS provides the answer to three important questions: (i) Is it possible to incorporate nitrogen electrochemically? (ii) Can we quantify the nitrogen content? (iii) Can we determine the lateral distribution of nitrogen near the electrode to directly confirm the electrochemical incorporation reaction?

The detection of nitrogen with SIMS presents an interesting case. Because nitrogen is an electronegative element, one would expect that N^- is the secondary ion to monitor. N^-, however, has a positive electron affinity, and is, therefore, unstable with respect to N and e^-. Monitoring N^+ is an alternative, but the yields of this ion are low. The procedure that has proved itself in semiconductor analysis is to detect a negative molecular ion comprising a matrix moiety and nitrogen. In the case of nitrogen in YSZ, there are a number of possibilities; the best results were obtained for $^{90}Zr^{14}N^-$ and $^{14}N^{16}O^-$. As shown in Figure 17.14a,b both these species are present in the region round the point electrode contact; away from the electrode both these signals are essentially zero (results not shown). Thus, this confirms that we can electrochemically incorporate nitrogen into YSZ. It is important to note that this does not in any way indicate that zirconium nitride or NO molecules are present in the matrix, only that zirconium, oxygen, and nitrogen are present. Also worthy of note is that the lateral resolution achievable in SIMS is defined by the diameter of the primary ion beam; for this measurement it was of the order of $5\,\mu m$. The ultimate limit is given by the diameter of the collision cascade, that is, secondary ions will be emitted up to several nanometers from the exact point of primary ion impact.

In order to quantify these results, we analyzed a YSZ single crystal that had been implanted with nitrogen. The TOF-SIMS depth profile is shown in Figure 17.14c; instead of $^{90}Zr^{16}O^-$, $^{96}Zr^{16}O^-$ was used to determine the RSF, because the $^{90}Zr^{16}O^-$ signal was saturated. From Equation 17.2 we have

$$c_N = RSF \frac{i^s_{^{90}Zr^{14}N}}{i^s_{^{96}Zr^{16}O}} \tag{17.4}$$

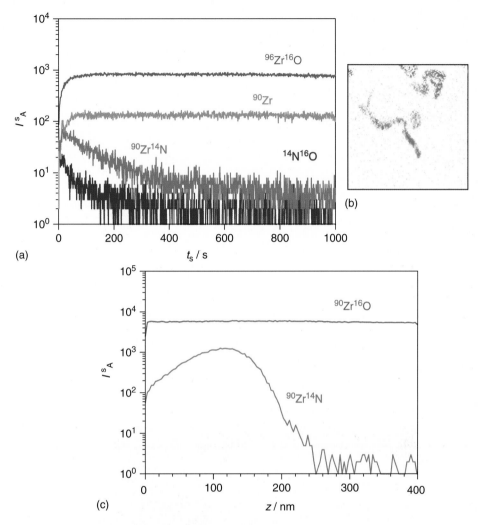

Figure 17.14 (a) Depth profile of the region around the point electrode contact. (b) Alternative presentation of the data shown in (a) as an ion image $\{^{90}Zr^{14}N/^{96}Zr^{16}O\}$. (c) Depth profile of an implanted YSZ sample. All data acquired as negative secondary ions.

Integrating both sides

$$\int c_N dz = \mathrm{RSF} \int \frac{i^s_{^{90}Zr^{14}N}}{i^s_{^{96}Zr^{16}O}} dz \tag{17.5}$$

we recognize that the integral on the left-hand side is simply the implanted dose, $\Phi = \int c_N dz$ (in this case, $\Phi = 10^{16}$ cm^{-2}), whilst the integral on the right-hand side can be obtained by numerical summation of the data shown in Figure 17.14c.

Hence we arrive at

$$RSF = \Phi / \int_{i^{s}_{96_{Zr}16_{O}}}^{i^{s}_{90_{Zr}14_{N}}} dz \tag{17.6}$$

Having obtained the RSF, we can now, with Equation 17.4, quantify the data in Figure 17.14a,b. We find that by means of electrochemistry we can incorporate at the surface $(8.0 \pm 0.5) \times 10^{20}$ cm^{-3} of nitrogen into zirconia [45].

17.4.5
Elemental Distribution in Plants: Imaging Analysis

The same questions that have concerned us so far with regard to inorganic solids are also important topics for biological systems [46, 47]: Which elements are present, and how are they distributed? How fast and along which paths does mass transport take place? The fact that SIMS is an ultra-high vacuum technique, though, would appear to exclude the possibility of analyzing biological samples. However, preparation techniques have been developed that aim to preserve the elemental composition and distribution in biological systems as close as possible to their native state, whilst satisfying instrumental requirements. These techniques are based on fixing the biological samples either chemically or cryogenically. For more details on preparation procedures, the interested reader is referred to Ref. [48].

The images presented in Figure 17.15 show the distributions of K, Mg, and Ca acquired by SIMS analysis of a twig cross-section.

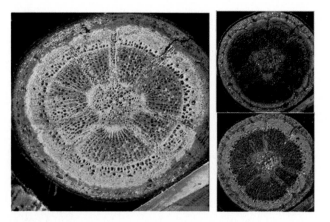

Figure 17.15 K, Mg, and Ca ion images of a cryosubstituted twig cross-section. The field of view in each case is 1.8 mm × 1.8 mm. Images obtained on a Fisons SIMSLAB 410 SIMS/SNMS machine with a beam of 22 keV Ga$^+$ primary ions (Ionoptika Ga-gun). Images courtesy of W.H. Schröder and U. Breuer.

17.5
Conclusions and Perspectives

In this chapter we have introduced the SIMS technique, discussed its advantages and disadvantages, and presented a number of case studies, mainly from our own work. We have demonstrated that SIMS currently plays an important role, especially in the field of solid state physical chemistry, in investigating the distribution of elements and their mass transport in solids. Specifically, we have shown that SIMS is capable of providing high quality data on the rate of mass transport in homogeneous solids over various length scales, ranging from some nanometers to hundreds of micrometers; we have also shown that SIMS is capable of resolving diffusion processes in polycrystalline materials; future generations of SIMS machines with (hopefully) simultaneous lateral resolution and depth resolution in the nanometer range will prove invaluable in the study of individual grains and grain boundaries. We have also drawn attention to the use of diffusion experiments to elucidate inhomogeneous defect distributions at oxide homo- and hetero-interfaces. All evidence suggests that the application of SIMS within this field [49], and in many others, for example, biophysical chemistry, is set to grow.

17.6
Supplementary Material

17.6.1
UHV

There are two reasons why SIMS measurements are generally performed under ultra-high vacuum (UHV) conditions, that is, in the range 10^{-7}–10^{-10} Pa [50, 51]. The first reason is to avoid scattering of the primary and secondary ion beams by the residual gas molecules in the analysis chamber. In a well-prepared system, the vacuum composition will be dominated by H_2 and CO. The second reason is to maximize the time taken for residual gas molecules to form a monolayer on the surface of the sample under investigation; any species deposited onto the sample surface during analysis will be sputtered along with sample material, and this leads to high background signals (e.g., for H, C, and O) and more complicated mass spectra.

To begin with, we note that the number of molecules per unit volume N in a vacuum chamber is still significant. Using

$$N = \frac{nN_A}{V} = \frac{p}{kT}$$

with, for example, $T = 300\,\text{K}$ and $p = 10^{-10}$ Pa, we obtain $N = 2.4 \times 10^{10}\,\text{m}^{-3}$.

In order to determine the pressure below which scattering of the ion beams by the residual gas particles is no longer significant, we need the average distance a gas molecule travels between collisions with other gas molecules, that is, the mean free path λ. Using Maxwell's kinetic theory of gases λ can be expressed as

$$\lambda = \frac{1}{N\sigma\sqrt{2}} = \frac{kT}{p\sigma\sqrt{2}}$$

where σ is the collision cross-section. Assuming $\sigma = \pi d^2$, and taking for the purpose of illustration CO as the molecule (of diameter $d = 0.3$ nm), we find that a pressure of 10^{-3} Pa guarantees a mean free path (≈ 10 m) far larger than the dimensions of a SIMS machine. In other words, a residual gas molecule is far more likely to collide with the walls of the analysis chamber rather than with another molecule.

Let us consider an atomically clean surface in a vacuum. The time for this surface to be covered with a monolayer of adsorbed gas molecules, t_{ML}, is governed by the flux of residual gas molecules impinging on the surface Z_w, the probability of an impinging molecule sticking to the surface, otherwise known as the *sticking coefficient* S, and the areal density of adsorption sites N_{ads}.

$$t_{ML} = \frac{N_{ads}}{SZ_w} = \frac{N_{ads}}{S} \frac{\sqrt{2\pi mkT}}{p}$$

(m is the mass of the molecule.) If we assume that the sticking coefficient is unity and that N_{ads} is of the order of 10^{19} m^{-2}, then we require pressures below 10^{-7} Pa so that the time taken to contaminate the surface with a monolayer is greater than 1 h.

17.6.2
Basic Principle of TOF-SIMS[1]

In TOF-SIMS we require all secondary ions to start their flight at the same time. This is achieved by using a short pulse of primary ions. For analysis at high mass resolution the pulse length is below 1 ns. When one such pulse strikes the sample surface, it causes a pulse of secondary ions to be emitted. By means of the extraction potential U_{ex}, most of the secondary ions of given polarity can be collected and accelerated into the field-free flight tube. As a result of the acceleration stage, these secondary ions now all have (approximately) the same kinetic energy.

$$\tfrac{1}{2}mv^2 = U_{ex}q$$

and thus travel with varying velocities along the flight tube to the detector, with the lightest ions arriving first, followed by increasingly heavier species. The flight time of a specific secondary ion is given by (l_f is the length of the flight tube. Figure 17.16).

$$t = \frac{l_f}{v} = \sqrt{\frac{l_f^2}{2U_{ex}} \frac{m}{q}}$$

It is therefore only dependent on fixed instrumental parameters, l_f and U_{ex}, and the mass-to-charge ratio of that secondary ion.

---

[1] See also Section 2.6.3, page 58

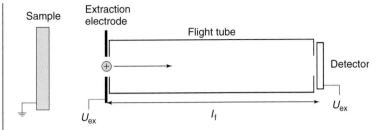

Figure 17.16 Schematic time-of-flight mass spectrometer.

For the case shown U_{ex} is negative, so that positive secondary ions are extracted for mass filtering.

As mentioned in Section 17.2.2.1, secondary ions are emitted with a range of energies. For this reason, TOF-SIMS instruments are not based on such simple linear mass spectrometers, but utilize some sort of energy focusing, in order to compensate for the different flight times of secondary ions with identical m/q but slightly different kinetic energies [2, 4].

17.6.3
The Mathematics of Oxygen Diffusion and Surface Exchange

The determination of the tracer diffusion coefficient of oxygen is commonly accomplished by using the stable oxygen isotope ^{18}O. The sample is annealed in an isotope-enriched atmosphere and the diffusion profile is determined subsequently by SIMS. Mathematically, we have to solve the diffusion equation for the concentration $c_{18}(x,t)$ of the isotope ^{18}O. However, instead of the isotope concentration, c_{18}, we can use the isotope fraction $c^* = c_{18}/(c_{18} + c_{16})$, the quantity which is directly measured by SIMS (see Section 17.4.1), since the sample remains chemically homogeneous during the isotope exchange, $c_{18}(x, t) + c_{16}(x, t) = $ const. The diffusion equation can therefore be written as

$$\frac{\partial c^*}{\partial t} = D_O^* \frac{\partial^2 c^*}{\partial x^2}$$

where D_O^* is the oxygen tracer diffusion coefficient. To solve this differential equation we need an initial condition and two boundary conditions. The initial condition is

$$c^*(x \geq 0, t = 0) = c_{bg}^*$$

where c_{bg}^* is the isotope fraction in the solid before the isotope anneal. If diffusion may be assumed to take place in a semi-infinite solid, then the first boundary condition corresponds to the situation for which, at large penetration depths, the isotope fraction always corresponds to the initial value, c_{bg}^*.

$$c^*(x = \infty, t) = c_{bg}^*$$

For the second boundary condition, at $x = 0$, we assume continuity of the isotope flux crossing the surface and entering the bulk of the solid. The first flux is generally assumed to be proportional to the difference between the time-invariant isotope fraction in the gas phase, c_g^*, and the time-dependent fraction at the surface of the solid, $c_s^*(t) = c^*(x = 0, t)$; the second flux is simply given by Fick's first law.

$$k_O^* \cdot \left\{ c_g^* - c_s^*(t) \right\} = -D_O^* \cdot \left. \frac{\partial c^*}{\partial x} \right|_{x=0}$$

Here k_O^* is the tracer surface exchange coefficient. The solution of the diffusion equation subject to these initial and boundary conditions is [40].

$$c^*(x, t) = c_{bg}^* + (c_g^* - c_{bg}^*) \cdot \left\{ \operatorname{erfc}\left(\frac{x}{\sqrt{4D_O^* t}} \right) - \exp\left(\frac{k_O^* x}{D_O^*} + \frac{k_O^{*2} t}{D_O^*} \right) \right.$$

$$\left. \cdot \operatorname{erfc}\left(\frac{x}{\sqrt{4D_O^* t}} + k_O^* \sqrt{\frac{t}{D_O^*}} \right) \right\}$$

The parameters D_O^* and k_O^* are then obtained by fitting the above equation to the experimental profile. A detailed analysis of the conditions where D_O^* and k_O^* can be determined unambiguously can be found in Refs [22, 23]. We note that the ^{18}O surface fraction.

$$c_s^*(t) = c_{bg}^* + (c_g^* - c_{bg}^*) \cdot \left\{ 1 - \exp\left(\frac{k_O^{*2} t}{D_O^*} \right) \cdot \operatorname{erfc}\left(k_O^* \sqrt{\frac{t}{D_O^*}} \right) \right\}$$

depends only on the parameter $\alpha = k_O^{*2} t / D_O^*$. Thus, $\alpha \ll 1$ (small annealing times, or slow oxygen surface exchange or fast oxygen bulk diffusion) results in an ^{18}O surface fraction, $c^*(x = 0, t > 0)$, that is smaller than the gas phase fraction, c_g^*. On the other hand, $\alpha \geq 10$ leads to an ^{18}O surface fraction that is identical to the gas phase fraction, c_g^*. This behavior is illustrated in Figure 17.17 where two diffusion profiles are shown for identical diffusion coefficients but different surface exchange coefficients. Finally, we note that for fast surface exchange,

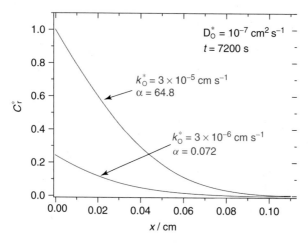

Figure 17.17 Normalized isotope fraction $c_r^*(x, t) = (c^*(x, t) - c_{bg}^*)/(c_g^* - c_{bg}^*)$.

$\alpha \gg 1$, the above solution transforms to the *constant-source solution* [40] $c^*(x,t) = c_{bg}^* + (c_g^* - c_{bg}^*) \cdot \mathrm{erfc}\left(x/\sqrt{4D_O^* t}\right)$.

Acknowledgments

We are extremely grateful to Dr. W. Schröder and Dr. U. Breuer (Forschungszentrum Jülich) for generously allowing the use of unpublished images.

References

1. Trends, A., Benninghoven, F., Rüdenauer, G., and Werner, H.W. (eds) (1987) *Secondary Ion Mass Spectrometry: Basic Concepts, Instrumental Aspects, Applications*, John Wiley & Sons, Inc., New York.

2. Vickerman, J.C., Brown, A., and Reed, N.M. (eds) (1989) *Secondary Ion Mass Spectrometry: Principles and Applications*, Clarendon Press, Oxford.

3. Wilson, R.G., Stevie, F.A., and Magee, C.W. (eds) (1989) *Secondary Ion Mass Spectrometry: A Practical Handbook for Depth Profiling and Bulk Impurity Analysis*, John Wiley & Sons, Inc., New York.

4. Vickerman, J.C. and Briggs, D. (eds) (2001) *ToF-SIMS: Surface Analysis by Mass Spectrometry*, IM Publications and Surface Spectra.

5. Zalm, P.C. (2000) *Microchim. Acta*, **132**, 243.

6. Hofmann, S. (1998) *Rep. Prog. Phys.*, **61**, 827.

7. Stephan, T. (2001) *Planet. Space Sci.*, **49**, 859.

8. McPhail, D.S. (2006) *J. Mater. Sci.*, **43**, 873.

9. Grove, W.R. (1852) *Philos. Trans. R. Soc. London*, **142**, 87.

10. Thompson, J.J. (1910) *Philos. Mag*, **20**, 752.

11. Rosman, K.J.R. and Taylor, P.D.P. (1998) *Pure Appl. Chem.*, **70**, 217.

12. Benninghoven, A., Bertrand, P., Migeon, H.N., and Werner, H.W. (eds) (2000) *SIMSXII, Proceedings of the 12th International Conference on Secondary Ion Mass Spectrometry*, Elsevier.

13. Benninghoven, A. (2003) SIMSXIII, Proceedings of the 13th International Conference on Secondary Ion Mass Spectrometry and Related Topics, *Appl. Surf. Sci.*, **203**, 203.

14. Benninghoven, A., Hunter, J.L. Jr., Schueler, B.W., Smith, H.E., and Werner, H.W. (eds) (2004) SIMS XIV, Proceedings of the 14th International Conference on Secondary Ion Mass Spectrometry and Related Topics, *Appl. Surf. Sci.*, **231–232**, 1.

15. Jost, W. and Hauffe, K. (1957) *Diffusion*, Steinkopff, Darmstadt.

16. Howard, R.E. and Lidiard, A.B. (1964) *Rep. Prog. Phys.*, **27**, 161.

17. Manning, J.R. (1968) *Diffusion Kinetics for Atoms in Crystals*, Van Nostrand, Princeton, NJ.

18. Le Claire, A.D. (1970) in *Physical Chemistry – an Advanced Treatise*, vol. X (eds H. Eyring, D. Henderson, and W. Jost), Academic Press, New York, p. 261.

19. Philibert, J. (1985) *Diffusion et Transport de Matière dans les Solides*, Les Ulis, Editions de Physique, Les Ulis.

20. Martin, M. (2005) in *Diffusion in Condensed Matter, Methods, Materials, Models* (eds P. Heitjans and J. Kärger), Springer, p. 209.

21. Kilner, J.A., Steele, B.C.H., and Ilkov, L. (1984) *Solid State Ionics*, **12**, 89.

22. Fielitz, P. and Borchardt, G. (2001) *Solid State Ionics*, **144**, 71.

23. De Souza, R.A. and Chater, R.J. (2005) *Solid State Ionics*, **176**, 1915.

24. De Souza, R.A., Zehnpfenning, J., Martin, M., and Maier, J. (2005) *Solid State Ionics*, **176**, 1465.

25. Masoud, M., Fielitz, P., De Souza, R.A., Heitjans, P., Borchardt, G., and Martin, M. (2008) *Solid State Sci.*, **10**, 746.

26. Kaur, I., Mishin, Y., and Gust, W. (1995) *Fundamentals of Grain and Interphase Boundary Diffusion*, 3rd edn, John Wiley & Sons, Ltd, Chicester.

27. De Souza, R.A. and Martin, M. (2008) *Phys. Chem. Chem. Phys.*, **10**, 2356.

28. Denk, I., Claus, J., and Maier, J. (1997) *J. Electrochem. Soc.*, **144**, 3526.

29. Guo, X., Fleig, J., and Maier, J. (2001) *J. Electrochem. Soc.*, **148**, J50.

30. De Souza, R.A., Fleig, J., Maier, J., Kienzle, O., Zhang, Z., Sigle, W., and Rühle, M. (2003) *J. Am. Ceram. Soc.*, **86**, 922.

31. De Souza, R.A., Fleig, J., Maier, J., Zhang, Z., Sigle, W., and Rühle, M. (2005) *J. Appl. Phys.*, **97**, 053502.

32. De Souza, R.A. (2009) *Phys. Chem. Chem. Phys.*, **11**, 9939.

33. Ishihara, T., Matsuda, H., and Takita, Y. (1994) *J. Am. Chem. Soc.*, **116**, 3801.

34. Feng, M. and Goodenough, J.B. (1994) *Eur. J. Solid State Inorg. Chem.*, **31**, 663.

35. Feng, M., Goodenough, J.B., Huang, K., and Milliken, C. (1996) *J. Power Sources*, **63**, 47.

36. Martin, M. (2000) *Solid State Ionics*, **136–137**, 331.

37. Schulz, O., Flege, S., and Martin, M. (2003) in *The Electrochemical Society Proceedings Series: SOFC-VIII* (eds S.C. Singhal and M. Dokiya), PV 2003-07, p. 304.

38. Schulz, O., Martin, M., Argirusis, C., and Borchardt, G. (2003) *Phys. Chem. Chem. Phys.*, **5**, 2308.

39. Fischer, J.C. (1951) *J. Appl. Phys.*, **22**, 74.

40. Crank, J. (1975) *The Mathematics of Diffusion*, Oxford University Press, Oxford.

41. Jansen, M. and Letschert, H.P. (2000) *Nature*, **404**, 980.

42. Guenther, E. and Jansen, M. (2001) *Mater. Res. Bull.*, **36**, 1399.

43. Zu, P., Tang, Z.K., Wong, G.K.L., Kawasaki, M., Ohtomo, A., Koinuma, H., and Segawa, Y. (1997) *Solid State Commun.*, **103**, 459.

44. Wendel, J., Lerch, M., and Laqua, W. (1999) *J. Solid State Chem.*, **142**, 163.

45. Valov, I., Korte, C., De Souza, R.A., Martin, M., and Janek, J. (2006) *Electrochem. Solid-State Lett.*, **9**, F23.

46. Metzner, R., Schneider, H.U., Breuer, U., and Schroeder, W.H. (2008) *Plant Physiol.*, **147**, 1774.

47. Metzner, R., Thorpe, M.R., Breuer, U., Blümler, P., Schurr, U., Schneider, H.U., and Schroeder, W.H. (2010) *Plant Cell Environ.*, **33**, 1393.

48. Chandra, S., Sod, E.W., Ausserer, W.A., and Morrison, G.H. (1992) *Pure Appl. Chem.*, **64**, 245.

49. De Souza, R.A. and Martin, M. (2009) *Mater. Res. Soc. Bull.*, **34**, 907.

50. O'Hanlon, J.F. (2003) *A User's Guide to Vacuum Technology*, John Wiley & Sons, Inc., Hoboken, NJ.

51. Wedler, G. (1987) *Lehrbuch der Physikalischen Chemie*, Wiley-VCH Verlag GmbH, Weinheim.

18

Application of the Quartz Microbalance in Electrochemistry

Karl Doblhofer and Konrad G. Weil[1)]

◾ **Method Summary**

Acronyms, Synonyms
- quartz (crystal) microbalance (QCM, QMB)
- electrochemical quartz crystal microbalance (EQCM).

Benefits (Information Available)
- the QCM detects mass changes of deposits in a straightforward, reproducible manner
- nanogram resolution allows the detection of atomic or molecular monolayers
- it operates with the quartz electrodes in contact with vacuum, with gases, and with viscous liquids
- it has developed into a valuable versatile tool for *in situ* studies of mass changes resulting from electrochemical processes (electrodeposition, corrosion etc.).

Limitations (Information Not Available)
- the analysis is facile only with plane, rigid deposits
- the QCM signal does not contain information on the chemical nature of the deposit.

18.1
Introduction

The pioneering work of Sauerbrey, who showed that the resonance frequency of a quartz plate depends linearly on mass that is present on one side of the plate led to numerous applications of quartz devices for mass determination [1]. These applications are mainly in the field of film formation by evaporation,

---

1) We communicate with great sorrow that
Professor Weil passed away in May 2009.

Methods in Physical Chemistry, First Edition. Edited by Rolf Schäfer and Peter C. Schmidt.
© 2012 Wiley-VCH Verlag GmbH & Co. KGaA. Published 2012 by Wiley-VCH Verlag GmbH & Co. KGaA.

sputtering, and chemical reactions. Later Nomura and Okuhara [2], and a research group at Almaden Research Center of IBM [3] demonstrated that mass change determination with oscillating quartzes is also possible when the quartz plate is in contact on one side with a viscous liquid. This opened the way to the application of such a balance in electrochemistry, pioneered by Bruckenstein and Swathijan [4] and the group at IBM [3].

Simultaneous determination of mass changes and charges proved to be a versatile tool for the elucidation of complex electrode reactions [5]. Another important benefit of the quartz microbalance is that the device is inexpensive and is easy to use. The main investment is a frequency counter. Every electronics shop is able to fabricate the HF circuit for the excitation of the quartz oscillations and the electrochemical cell is a simple task for the glassblower. Our opinion is that, because of these benefits, the electrochemical quartz microbalance should become an important tool in laboratory classes in universities as well as for research groups faced with electrochemical problems.

We do not attempt to acknowledge the many important contributions from many laboratories all over the world. The examples given in the following are from our own work. This is because we are familiar with them, we know from own experience that the relevant experiments can be performed reliably and, last but not least, they were easily accessible to us. If somebody feels unduly neglected in this paper we already apologize.

18.2
Basic Principles

18.2.1
Piezoelectricity of Quartz in Static Electric Fields

P. and J. Curie found, in 1880, that crystals of certain substances, such as tourmaline or quartz, develop positive and negative surface charges on several crystal faces when the crystal is subjected to mechanical stress (pressure, tension, distortion) [6]. This phenomenon is known as *piezoelectricity*.

Consider a unit cell of a trigonal quartz crystal consisting of three SiO_2-molecules [7]. If pressure is applied in the direction of the crystallographic x-axis (the "polar axis"), the center of positive charge (originating from charged silicon atoms) is displaced further than the center of negative charge (from charged oxygen atoms). Thus, an electric dipole forms in the unit cell. The sum of such dipoles causes an electric field to build up across the quartz crystal, and this is measurable as a potential difference between metal layers deposited on the two opposite crystal surfaces. For a detailed description of the piezoelectric effect on quartz the elastic anisotropy of the quartz crystal and the vector property of the piezoelectric effect have to be taken into account [8].

Piezoelectricity is characterized by a one-to-one correspondence of direct and reverse effect. Thus, the application of an electric field in a direction defined in the $X-Y-Z$-space of the quartz crystal axes leads reproducibly to one of many well

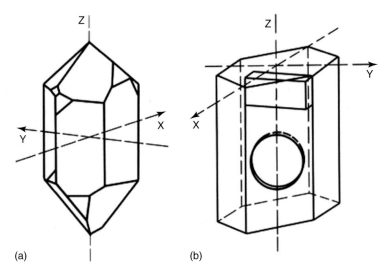

Figure 18.1 (a) A quartz crystal and (b) illustration of cuts of the quartz producing piezoelectric plates with exact orientation with respect to the crystallographic axes X, Y, and Z.

defined shape changes. This phenomenon has become the basis for numerous valuable technical devices, such as frequency standards and mass sensors. For particular applications, the quartz crystal is cut at defined angles with respect to the crystallographic axes such that the voltage applied to two deposited metal electrodes generates the desired mode of deformation (Figure 18.1).

18.2.2
Piezoelectricity of Quartz in Oscillating Electric Fields

Consider a quartz disk or plate obtained by cutting the quartz crystal as illustrated in Figure 18.1. Typically, thin gold layers are deposited onto the two opposite plane faces. Application of an AC voltage to these electrodes will generate a rhythmic deformation. The exact character of the deformation will depend on the orientation of the electrode planes in the crystal's X–Y–Z-space. If the frequency of the ac voltage corresponds to resonance of this mechanical oscillation, the acoustic amplitude of the quartz will have a maximum. Correspondingly, the AC current and the electric admittance have a maximum.

For mass sensor application it is customary to use the "AT cut" of the quartz crystal that oscillates in the thickness shear mode. Figure 18.2 illustrates the standing transverse wave of the fundamental resonance of the AT quartz. The nodal plane of the oscillation is in the middle of the plate, at $y = 0$. The crystal planes move parallel to one another in the x-direction, without deformation. The maximum deflection (ζ in Figure 18.2) occurs at the antinodes, $y = \pm d/2$.

According to the above considerations, the plate thickness d corresponds exactly to one half wavelength of the shear wave if the quartz oscillates at its fundamental

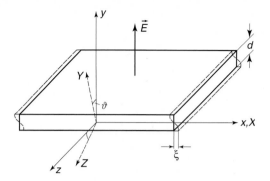

Figure 18.2 Shear mode deformation of the AT cut quartz under electric AC polarization. X, Y, and Z are the crystallographic axes of the quartz, as in Figure 18.1; $\vec{E}$ is the electric field vector; see Ref. [1].

resonance frequency, f_0. Thus, this resonance frequency is defined by

$$f_0 = \frac{v_{tr}}{2d} = \frac{1}{2d}\sqrt{\frac{\mu_Q}{\rho_Q}} \tag{18.1}$$

where v_{tr} is the velocity of the wave propagation in the Y-direction (see Figure 18.2), $\mu_Q = 2.947 \times 10^{10}$ N m^{-2} is the relevant piezoelectrically stiffened shear modulus of the quartz, and $\rho_Q = 2651$ kg m^{-3} is the density of quartz. With a value of $v_{tr} = 3334$ m s^{-1} [9] the resonance frequency of a plate is 1.67 mm MHz d^{-1}.

The AT cut quartzes are obtained by cutting the crystal at angles between $\vartheta = 35°10'$ and $35°15'$ with respect to the crystallographic Z-axis, parallel to the X-axis. An important reason for using this cut is the temperature stability of the resonance frequency. For instance, the resonance frequency of a 10 MHz quartz cut at $\vartheta = 35°15'$ will vary by -2.5 Hz K^{-1} at ambient temperature, as shown in Figure 18.3.

18.2.3
The Mass Load Dependence of the Resonance Frequency

Consider an AT cut plate large enough that the boundaries will not affect the thickness shear mode oscillation. The surface of the plate will constitute an antinode, as illustrated in Figure 18.2. This means that the quartz layers near the surface influence the resonance frequency only by their rigid mass, but not by their elastic properties. A thin homogeneous layer of any rigid material deposited onto the surface of the quartz plate will consequently influence the resonance frequency of the plate in the same way as would a quartz layer of the same mass.

The change in the resonance frequency, Δf, caused by the deposition of a rigid thin layer of mass Δm is therefore defined by

$$\frac{\Delta f}{f_0} = -\frac{\Delta d}{d} = -\frac{\Delta m_Q}{\rho_Q A d} = -\frac{\Delta m}{\rho_Q A d} \tag{18.2}$$

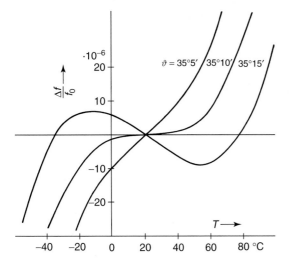

Figure 18.3 The temperature dependence of the resonance frequency $\Delta f/f_0$ of AT quartz resonators cut at the indicated angles ϑ (see Ref. [1]).

where the subscript Q relates to quartz and A is the surface area. Combined with Equation 18.1 one obtains the well known "Sauerbrey equation" [1]

$$\Delta f = -\frac{f_0}{d\rho_Q}\left(\frac{\Delta m}{A}\right) = -\frac{v_{tr}}{2d^2\rho_Q}\left(\frac{\Delta m}{A}\right)$$

$$= -\frac{2f_0^2}{\sqrt{\mu_Q\rho_Q}}\left(\frac{\Delta m}{A}\right) = -S\left(\frac{\Delta m}{A}\right) \tag{18.3}$$

Note that the mass sensitivity constant S depends on the resonance frequency squared. For example, for a 14 MHz quartz (thickness $d = 0.12$ mm) operating at the fundamental resonance frequency, the sensitivity constant $S = 4.43 \times 10^7$ Hz kg^{-1} m^2, corresponding to 0.443 Hz ng^{-1} cm^2. For a 5 MHz quartz $S = 0.0565$ Hz ng^{-1} cm^2. For more details see Section 18.6.1.

18.2.4
Measurements of the Resonance Frequency

The electrical behavior of the considered quartz resonator can be adequately described by the equivalent circuit represented in Figure 18.4. It has two parallel branches [10, 11]. The dynamic branch is the series combination of an inductance L_0, a resistance R_0, and a capacitance C_0. The capacitance of the static branch, C^*, is defined by the concepts of a parallel plate capacitor with the metal films as the electrodes and the quartz as the dielectric medium. The values of the circuit elements can be calculated from the mechanical, dielectric, piezoelectric, and dimensional parameters of the quartz [8, 11] (see Section 18.6.2). They may also be determined experimentally.

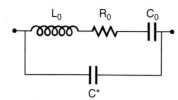

Figure 18.4 Equivalent circuit representing the quartz oscillator.

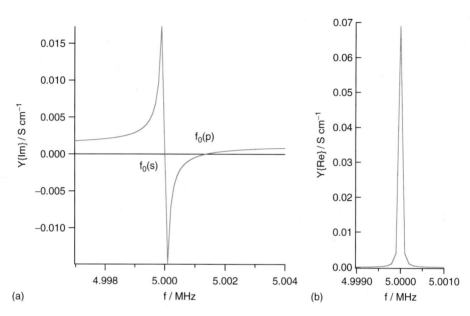

(a)

(b)

Figure 18.5 Calculated curves of the admittance Y near resonance of a 5 MHz quartz oscillator: (a) the imaginary and (b) the real part. The values of R_0, L_0, and C_0 used for the simulation are given in Section 18.6.2. The quartz oscillates in vacuum, $C^* = 40$ pF.

The electric admittance Y of this equivalent circuit is given by

$$Y(\omega) = \frac{1}{R_0 + j\left(\omega L_0 - \frac{1}{\omega C_0}\right)} + j\omega C^* \qquad (18.4)$$

where ω is $2\pi f$ (f is the AC frequency), and $j = \sqrt{-1}$. With typical values for a 5 MHz quartz plate the imaginary part of the admittance near resonance is shown in Figure 18.5a (for details see Section 18.6). There are two frequencies of resonance designated as f_s and f_p. Crystal-controlled oscillators operating at either frequency f_s or f_p may be built [11, 12]. The frequency f_s corresponds to the resonance of the dynamic arm. This is the frequency that is usually generated and measured with a frequency counter. The real part of the electric admittance, $Y\{Re\}$ shows a sharp maximum at f_s, see Figure 18.5b.

Changes in the resonance frequency of about 1 Hz are conveniently measurable. Thus, deposits of mass corresponding to atomic or molecular monolayers can be determined. For instance, a monolayer of water molecules (1.3×10^{15} cm^{-2}) would weigh about 40 ng cm^{-2}. Devices designed to measure mass changes in this way are known as quartz crystal microbalances (QCMs).

18.2.5
Quartz Oscillators in Viscous Media

In electrochemical applications of QCMs, one of the two thin metal layers deposited on the quartz plate is used simultaneously as an electrode for the quartz oscillator and as the working electrode in the electrochemical system.

The adhesive molecular forces between the surface of the metal film on an AT quartz and the contacting liquid cause the electrolyte at the interface to follow the oscillation of the quartz. A damped sinusoidal shear wave is generated that extends into the liquid phase. It is apparent that solution dragged with the quartz surface increases the vibrating mass, that is, the resonance frequency in solution will decrease upon immersion of the quartz in the liquid. Furthermore, the viscous liquid will cause the damping resistor R_0 to increase substantially. On the other hand, the liquid has normally negligible elasticity, leaving C_0 unaffected [13].

The coupling of the elastic shear wave in the crystal to the viscous shear wave in the liquid electrolyte was first described in a straightforward quantitative manner by Kanazawa and Gordon in 1985 [14]. Figure 18.6 shows the shear velocity profile of an aqueous solution in contact with the surface of a 5 MHz quartz oscillator. The characteristic decay length δ of the shear wave in the liquid is:

$$\delta = \sqrt{\frac{\eta_L}{\pi \rho_L f}} \tag{18.5}$$

where η_L is the viscosity of the liquid and ρ_L is its density. With the values for water at 20 °C ($\eta_L = 0.0010$ N s m^{-2} and $\rho_L = 997$ kg m^{-3}) one obtains the decay length $\delta = 0.23$ µm. Note that the wavelength of the shear wave is much larger than the decay length, with the consequence that the sinusoidal character of the damped shear wave is not apparent.

The shift of resonance frequency of the quartz oscillator caused by the solution contact was found to be [14]

$$\Delta f = -f_0^{3/2} \sqrt{\frac{\rho_L \eta_L}{\pi \rho_Q \mu_Q}} \tag{18.6}$$

Accordingly, the resonance frequency of a 5 MHz quartz is expected to decrease by $\Delta f = -712$ Hz as one quartz surface contacts the aqueous solution. In practice, the oscillation frequency is quite sensitive to stresses caused by mounting and to changes in the hydrostatic pressure of the fluid. This is why frequency changes resulting from changes in ρ_L and η_L are usually reported relative to the pure solvent rather than to vacuum or air [15].

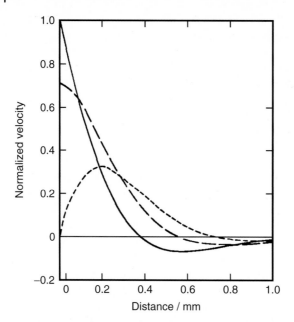

Figure 18.6 Shear velocity profiles of a liquid with the properties of water in contact with a quartz surface oscillating in the shear mode. (From Kanazawa and Gordon [14].)

Despite the coupling of the quartz to a liquid, a rigid mass deposited on the quartz resonator under electrochemical quartz crystal microbalance (EQCM) conditions will yield the linear frequency response expected from Equation 18.3, as long as the interface remains smooth. The effect of surface roughness will be discussed below.

18.3
Experimental Methods

18.3.1
Metalization of the Quartz Surfaces

In order to be able to excite the shear vibrations of a quartz crystal plate both opposite faces of the plate must be covered with a conductive layer. In most cases these layers will consist of gold. Gold films can be deposited onto the crystal either by evaporation or by sputtering. The adherence of gold on quartz is not very strong, therefore a thin layer (~20 nm) of chromium is deposited first, followed by the deposition of gold up to a thickness of ~200 nm. The circular gold layers on both opposite faces will each have a small flag at which thin wires are fixed that lead to the high frequency source. Figure 18.7 shows this arrangement schematically. The reason for the unequal diameters of the layers will be discussed in Section 18.6.

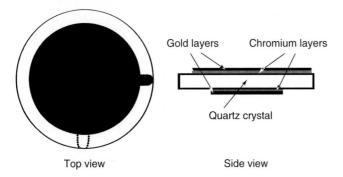

Figure 18.7 The metalized quartz plate.

18.3.2
The Electrochemical Cell

The metalized quartz crystal must be connected to the electrochemical cell in such a way that only one side of it is in contact with the electrolyte, while the other side will be in contact with the outer atmosphere. Crystal holders for this purpose are available commercially. The most important requirement for these holders is that they fit the crystal tightly in order to prevent the liquid dropping out of the cell but without preventing the crystal from vibrating freely. We want to describe a very simple construction which worked satisfactorily in many of the investigations reported later. It can be seen in Figure 18.8.

The electrochemical cell has on one of its sides a circular hole. An O-ring is glued onto the cell around the hole. A Teflon stopper with a diameter slightly larger than this O-ring contains another O-ring. The crystal plate will sit just between the two O-rings. An elastic rubber band around the whole cell will press the stopper against the cell. By careful selection of the rubber band one can ascertain that this combination just prevents the solution from dripping out of the cell but permits the crystal to vibrate freely. This arrangement will work fine as long as the wires that connect the metal layers with the outer circuit are sufficiently thin. The accepted name for this device, including the external HF circuit is electrochemical quartz crystal microbalance (EQCM).

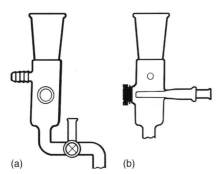

(a) (b)

Figure 18.8 Sketch of the electrolyte vessel.

18.3.3
Calibration of the Microbalance

A simple way to calibrate the EQCM is the comparison of the observed frequency change during an electrochemical deposition process with the mass change calculated from Faraday's law. One should choose a process that is known to proceed uniformly and with a current efficiency of 1. This the case for copper deposition from a bath containing 50 g ethanol, 50 g H_2SO_4 (conc.), 80 g copper sulfate pentahydrate, and 1000 g double-distilled water [16].

Figure 18.9 shows the frequency change when copper deposition occurs with a current density of 1 mA cm^{-2}.

The relation between the current density and the rate of frequency change can be obtained by inserting Faraday's law

$$\Delta m/\Delta t = -iM/zF \tag{18.7}$$

(remember: cathodic currents are, by definition, negative) with $\Delta m/\Delta t$ rate of mass change per unit area, i = current density, M = molar mass of the deposited metal, and F, the Faraday constant, into the Sauerbrey equation, leading to

$$\Delta f/\Delta t = -iM \, S/zF \tag{18.8}$$

With $\Delta f/\Delta t$ = rate of frequency change, i = current density, M = molar mass of the deposited metal, z = valence of the metal cation, F = Faraday's constant, and S = sensitivity constant. The appropriate units for the sensitivity constant are Hz cm^2 ng^{-1}. For a 5 MHz crystal the experimental results from Figure 18.9 lead to $S = 0.054$ Hz cm^2 ng^{-1}. Hence, a copper deposition current of 5 μA cm^{-2} leads to the easily measurable rate of frequency change of about -4.8 Hz min^{-1}.

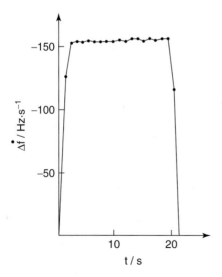

Figure 18.9 Rate of frequency change during copper deposition.

18.4
Case Studies

18.4.1
The Effect of Surface Roughness

In the theory outlined in Section 18.2 it is assumed that the interface with the electrolyte is planar and remains planar during the process under study. In many cases of metal corrosion or metal deposition this is not the case. Nonuniform corrosion or deposition will lead to an increase in surface roughness. Therefore, the effect of surface roughness on the response of the EQCM has to be studied carefully. In order to simplify the discussion we will use a very crude model for a rough surface, depicted in Figure 18.10a. The important parameters of this model are the depth of the individual cavities and their diameter. These quantities have to be compared with the thickness of the adhering viscous layer, which was discussed previously. In order to prepare a rough surface we used a method that is widely applied to the preparation of sufficiently rough surfaces for surface enhanced Raman spectroscopy (SERS), see, for instance, [17]. The method consists of the subsequent oxidation and reduction of a silver electrode in the presence of chloride ions. The obtained roughness will increase with the number of steps. Then we have the series:

$$\text{first anodic step: } Ag + Cl^- \longrightarrow (AgCl)_{layer} + e^-$$
$$\text{first cathodic step: } (AgCl)_{layer} + e^- \longrightarrow (Ag)_{layer} + Cl^-$$

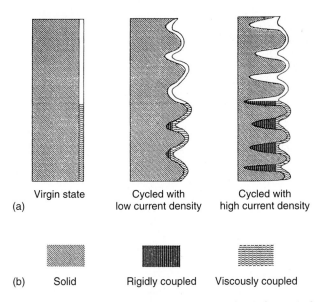

Virgin state   Cycled with   Cycled with
(a)   low current density   high current density

(b)   Solid   Rigidly coupled   Viscously coupled

Figure 18.10 Schematic cross-section of a silver film before and after roughening cycles.

Table 18.1 EQCM results obtained with a silver electrode oxidized and reduced in a chloride electrolyte at different current densities (i).

$i/\text{mA cm}^{-2}$	0.2	0.4	1.0	2.0
$(\Delta f/q_{ox})/\text{Hz mC}^{-1}$	77.6	75.8	74.4	73.4
$(\Delta f/q_{red})/\text{Hz mC}^{-1}$	69.6	63.1	54.9	49.7

second anodic step: $(\text{Ag})_{\text{layer}} + \text{Cl}^- \longrightarrow (\text{AgCl})_{\text{layer}} + \text{e}^-$

second cathodic step: $(\text{AgCl})_{\text{layer}} + \text{e}^- \longrightarrow (\text{Ag})_{\text{layer}} + \text{Cl}^-$

The subscript "layer" indicates that the respective atom or molecule sits on top of the bulk silver. In all these steps the charge is the same with opposite signs. If there was no roughening of the surface, we would expect that $\Delta f/q_{ox} = \Delta f/q_{red}$ for all four steps, independent of the applied current density. This is not the case. In Table 18.1 we give some values for $\Delta f/q_{ox}$ and $\Delta f/q_{red}$ for the first two steps with different current densities.

From this table we see that $(\Delta f/q_{ox})/(\Delta f/q_{red})$ is always >1. The deviation increases markedly with increasing current density. That means we lose less mass during a reduction step than we gained though the preceding oxidation step. A simple explanation of this effect is that during the oxidation/reduction steps cavities and pores are formed that are filled with electrolyte. Hence, the reduction step consists of a partial replacement of chloride by the liquid and *vice versa*. A careful analysis of the variation of different impedance elements of the quartz impedance during the electrochemical steps, described in Ref. [18], leads to results that are schematically depicted in Figure 18.10b.

At low current densities roughening leads predominantly to an increase in the interface area. Therefore, the vibrating mass and the viscous loss will increase likewise. At higher current densities the cavities become deeper and then they will contain liquid that is rigidly coupled to the quartz. It can be shown [18] that rigidly coupled liquid contributes to the imaginary part of the quartz impedance only, while viscously coupled liquid contributes equally to the real and to the imaginary parts.

18.4.2
Corrosion Studies

18.4.2.1 General Considerations
Corrosion is always correlated with a mass change of the object under investigation. The quartz microbalance tells us the rate of mass change. In order to obtain the corrosion rate from these measurements one must know the stoichiometry of the corrosion reaction. This is especially simple, when the corrosion product is soluble.

The corrosion reaction in this case can often be written as

$$Me + zH^+ \longrightarrow Me^{z+} + (z/2)H_2$$

or

$$Me + (z/4)O_2 + (z/2)H_2O \longrightarrow Me^{z+} + zOH^-$$

In both cases 1 mol metal atoms per formula unit is removed from the electrode.
More complicated is the situation, when a solid product is formed, for example,

$$Zn + 2H_2O \longrightarrow Zn(OH)_2 + H_2$$

Here the electrode mass increases during the corrosion time.

$$-dn_{Zn}/dt = dn_{Zn(OH)_2}/dt = \underline{{}^{1\!/\!2}dn_{OH}/dt}$$

The underlined term in the equation leads to a mass change, the rate of which can be measured with the balance and which is proportional to the corrosion rate.

A severe limitation of the application of the EQCM in corrosion studies stems from the requirement that the material to be investigated must be a thin film which is rigidly fixed at the surface of the quartz. Hence, only materials that can be electrodeposited, sputtered, vacuum deposited, or chemically deposited can be investigated.

18.4.2.2 Uniform Corrosion

During uniform corrosion an initially flat surface remains flat; the corrosion rate is constant at any point of the corroding surface. This type of corrosion can easily be studied with help of the quartz microbalance. When the corrosion products are soluble in the corrosive medium, the rate of frequency change can, with the help of the Sauerbrey equation, be converted to the rate of mass loss, that is, the corrosion rate.

As an example we report the dissolution of copper in a solution of pH 3 in the presence of oxygen. Here we measure a rate of frequency change of 15 Hz per 600 s = 0.025 Hz s^{-1}. With the Sauerbrey equation this leads to a corrosion rate of 0.00325 ng cm^{-2} and with the Faraday equation we finally obtain an equivalent current density of 1.4 µA cm^{-2}. It should be stressed that this very small rate was measured within 600 s.

18.4.2.3 Composites

Protective films can be produced by co-deposition of tin and zinc. Both metals do not show mutual solubility, hence, the deposit consists of a mixture of pure tin and zinc crystals. At pH 3 zinc should corrode selectively. If an anodic scan with a scan rate of 10 mV s^{-1} is applied to an electrode consisting of a thin layer of the tin/zinc composite Sn$_{70}$Zn$_{30}$ one sees a flat current maximum between the open circuit potential (OCP) and the beginning of the tin oxidation. Integration of the current–time curve leads to the result that during this period the total amount of zinc in the layer was oxidized. When the composite was deposited onto an EQCM the mass loss during the anodic scan can be monitored. This mass loss is equal to the amount of dissolved zinc minus the amount of electrolyte in the resulting

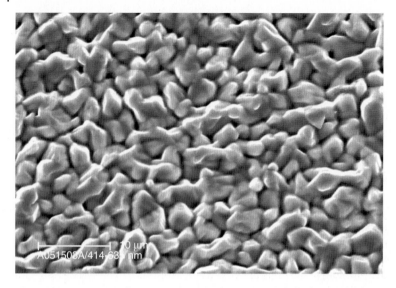

Figure 18.11   Scanning electron micrograph of a tin/zinc film after partial oxidation.

cavities. This experiment [19] confirms the assumption of selective oxidation of zinc from the composite. Figure 18.11 shows a micrograph of the residual tin structure.

18.4.2.4   Inhibitors

Inhibitors are substances which reduce the corrosion rate of a material when present in small amounts in the corroding medium. The EQCM is particularly suitable to study the effect because of the fast acquisition of the necessary data. This is demonstrated in Figure 18.12 for tolyltetrazole (TTA) as an inhibitor for copper corrosion. Copper was exposed to a sodium sulfate solution + sulfuric

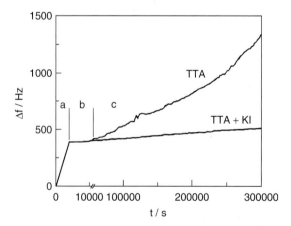

Figure 18.12   The effect of TTA and of additional potassium iodide (KI) on the corrosion rate of copper.

acid to adjust the pH to 2.96. During the first period (a in Figure 18.12) copper was corroding at a rather high and constant rate, indicated by the initial slope of the frequency versus time curve. At the beginning of period b, the corrosive solution was replaced by one that also contained $0.01 \, mol \, l^{-1}$ of the inhibitor TTA. The corrosion rate immediately changes to nearly zero. After replacing the solution with the inhibitor-free solution (period c) the corrosion rate stays low for an extended period. This effect is even more pronounced when $0.01 \, mol \, l^{-1}$ potassium iodide is added to the inhibitor solution [20].

18.4.3
Metal Deposition

18.4.3.1 Galvanic Deposition
Galvanic metal deposition is a widely used technique in industry. Examples are copper deposition in the fabrication of integrated circuits and chromium deposition for corrosion protection or decorative purposes.

We have assumed in the previous section that copper deposition from a solution containing Cu^{2+} ion occurs with a current efficiency of 1. This is the case for the deposition of sufficiently noble metals like gold, silver, and copper. For less noble metals we will always have the competing reaction of hydrogen evolution. In this case the observed mass change during deposition, calculated with Sauerbrey's equation will be smaller than that calculated from Faraday's law. As an example, we will show some results for nickel deposition. If nickel is deposited from a bath containing $0.1 \, mol \, l^{-1}$ nickel ions at pH 1.2 onto Ag, Ni, or Au with current densities of 2.4 or $4.5 \, mA \, cm^{-2}$ the deposition rate becomes constant and independent of the substrate after a few seconds. The current efficiency η, obtained from $(\Delta f_{stationary}/dt)/i$, is shown in Table 18.2 [21].

It is interesting to see, that the current efficiency increases with increasing current density. Furthermore, as to be expected, the efficiencies are independent of the substrate because after a short deposition time they are completely covered with nickel.

Table 18.2 Current efficiency (η) of cathodic nickel deposition on the indicated metals and current densities (i).

Substrate	i (mA cm^{-2})	$(\Delta f_{stationary}/dt)$ i^{-1} (Hz mA^{-1} s^{-1})	η
Ag	2.4	3.0	0.17
Ni	2.4	3.1	0.18
Au	2.4	3.0	0.17
Ag	4.5	7.6	0.43
Ni	4.5	7.8	0.44
Au	4.5	8.4	0.47

18.4.3.2 Electroless Deposition

Electroless metal deposition means the reduction of metal ions by a reducing agent that is present in the solution.

$$\text{Metal} + \text{Red} \longrightarrow \text{Metal} + \text{Ox}$$

This technique is especially advantageous when a layer of uniform thickness is desired on a material of complicated shape or when many similar objects are to be treated.

The most prominent examples for such reactions are copper deposition with formaldehyde in alkaline solution and the reduction of nickel with sodium hyphosphate. In the latter case not only nickel ions are reduced, but also some hypophosphate is reduced to phosphorus. The product is a mixture of glassy nickel–phosphorus and hexagonal nickel, supersaturated with phosphorus. We will return to this reaction in the next section. In order to stabilize copper ions in an alkaline medium a complex former, in this case ethylendiamonotetraacetate (EDTA), must be present.

In order to obtain a metal deposit only on a certain surface, it is necessary that the reaction occurs with negligible rate in the homogeneous phase and on the remaining other surfaces. This is achieved if the reduction reaction is heterogeneously catalyzed by the metal's surface. Hence, electroless deposition must be an autocatalytic process. If one wants to metalize an isolating surface like a polymer surface very small particles of a metallic catalyst like palladium must be deposited first. The electroless deposition can easily be monitored with an EQCM. If, for instance, during the deposition a copper surface is the working electrode in an electrochemical cell, polarization leads to an external current. This current is the algebraic sum of the cathodic copper deposition current and the anodic formaldehyde oxidation current. The EQCM permits one to monitor the cathodic current as a function of the potential because it is the only reaction that leads to a mass change of the electrode. The anodic current then is the difference between the total current and the cathodic current. Schumacher et al. [22] detected, by use of the EQCM, that an intermediate of the formaldehyde oxidation catalyzes the copper reduction from the Cu-EDTA-complexes. It can be demonstrated by cathodic polarization that the copper reduction current is zero when no formaldehyde is present and increases with increasing formaldehyde concentration in spite of the fact that the necessary electrons are supplied by the outer circuit [23], see Figure 18.13. Also, when, by anodic polarization, the copper reduction current vanishes the formaldehyde oxidation rate becomes zero. Copper reduction forms the active sites for the formaldehyde oxidation [24].

18.4.3.3 Anodic Stripping

In this section we want to demonstrate that a combination of current versus potential curves with mass loss versus potential curves can be used to determine the composition of two-component mixtures. As an example, we will discuss results obtained with galvanic deposits from nickel and H_3PO_3-containing solutions. Similarly as with electroless deposition from nickel and H_3PO_2-containing solutions

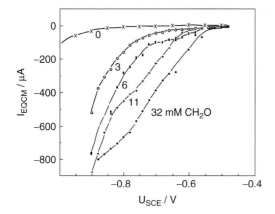

Figure 18.13 Dependence of the copper reduction rate on the concentration of formaldehyde. I_{EQCM} is the current, calculated from the rate of frequency change.

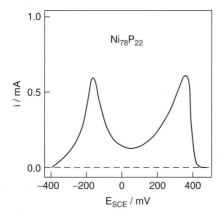

Figure 18.14 Voltammogram of a Ni_xP_y film during anodic stripping.

the deposit contains nickel and phosphorus. With films of a thickness in the range 500–700 nm the film can be totally oxidized during a single anodic sweep. Unexpectedly the voltammogram shows two well separated peaks [25], shown in Figure 18.14 for a sample of the overall composition $Ni_{78}P_{22}$. During the scan one can continually record the mass loss. There are two potential regions where the mass loss occurs and they coincide with the regions where the oxidation occurs. It can be shown that the oxidation leads to nickel in the oxidation state +2 and phosphorus in the oxidation state +5. Let us denote all quantities related to the first peak and the second peak, respectively, by the upper indices 1 and 2. Then we have for the total mass loss and charge during one of the peaks

$$\Delta m^{1,2} = n_{NI}^{1,2} M_{Ni} + n_P^{1,2} M_P; \quad \Delta Q^{1,2}/F = 2n_{Ni}^{1,2} + 5n_P^{1,2}$$

($n_i^{1,2}$ are the amounts of nickel and phosphorus oxidized during peak 1 and 2).

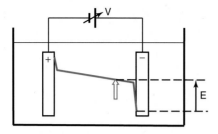

Figure 18.15 Potential distribution in an electrochemical cell during electrolysis (schematic). The unfilled arrow indicates the position of the Luggin capillary.

For each peak these are two equations for two unknowns which can easily be solved. We obtain for peak 1, $Ni_{88}P_{12}$, and for peak 2, $Ni_{64}P_{36}$. The lower limit of the phosphorus content of glassy NiP phases is known to be slightly below 30. Hence, similarly as with electroless deposition the product is a mixture of two phases: glassy NiP and supersaturated fcc nickel crystals.

18.4.4
Oscillating Electrochemical Reactions

Oscillating electrochemical reactions can occur, when the current–voltage characteristics show a region of negative differential resistance. Under potentiostatic conditions one has to place the Luggin capillary at a short distance away from the electrode surface, now electrode potential and current can oscillate, see Figure 18.15. We will demonstrate this for the example of the reduction of H_2O_2 at a silver electrode.

In Figure 18.16 cyclic voltammograms are shown for the first and a later cycle, starting at the indicated potential. Silver oxidation takes place when the potential

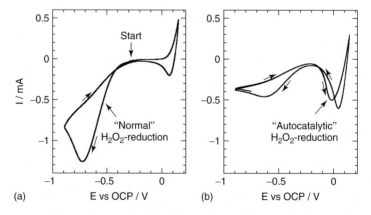

Figure 18.16 Cyclic voltammogram of a silver electrode in H_2O_2-containing solution. (a) 1st cycle and (b) later cycle. OCP: open circuit potential of a virgin silver electrode. Details in Ref. [27].

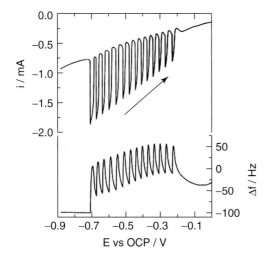

Figure 18.17 Simultaneous recordings of the current (upper trace) and the response of the EQMB (lower trace) during oscillations of the H_2O_2 reaction, details in Ref. [27].

is more positive than the OCP of a virgin silver electrode. The small hump during the cathodic scan, at slightly negative potentials, is due to the reduction of silver ions that are still close enough to the electrode surface. At more negative potential the H_2O_2 reduction starts, leading to a sharp increase in the negative current. In the later scans a new hump in the cathodic current appears close to the OCP. It was suggested by Flätgen *et al.* [26] that this hump indicates a new reaction path for H_2O_2 reduction, catalyzed by an adsorbed species which is formed during the reduction at more negative potentials. In the regime where the negative current increases with decreasing potential we have a negative differential resistance. Here oscillations occur. It is noteworthy that during these oscillations the electrode potential moves periodically into the regime of silver oxidation. During these periods the electrode mass should decrease. Figure 18.17 shows that indeed the EQCM response demonstrates synchronous mass oscillation with the observed current oscillation [27].

18.4.5
Self-Assembled Monolayers

A subject of great present interest is the modification of surfaces with organic thin films or adsorbed organic monolayers. The EQCM is a valuable tool for studying the formation of such layers on electrodes. As an example, the "self-assembly" of thiol monolayers on gold electrodes will be discussed. This study was conducted simultaneously with the EQCM and with the electrochemical impedance technique [28]. Inspection of the results represented in Figure 18.18 shows systematic differences between the EQCM and the impedance results. Initially, the adsorption

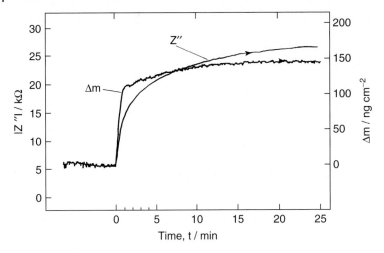

Figure 18.18 The mass change, Δm, determined with the EQCM, and the change in the imaginary part of the electrochemical impedance, $Z'' = Z\{\text{Im}\}$, both observed simultaneously after the injection (at $t = 0$) of n-octadecanethiol into the solution of 0.1 M NaClO$_4$ in ethanol.

progresses readily, as indicated by the sharp rise in Δm observed with the EQCM. However, the dielectric properties of the deposit are relatively poor. This suggests that initially the film is disordered. It contains solvent and ions, which add to the mass, but render the effective dielectric thickness relatively small.

At the later times of Figure 18.18, both the Z'' and the Δf curves approach constant values. However, both values are significantly smaller than expected from straightforward considerations. A perfect self-assembled monolayer formed from n-C$_{18}$H$_{37}$SH consists of 9.3×10^{-10} mol cm^{-2}, corresponding to a mass of 265 ng cm^{-2}. After 25 min of "self-assembly" in the considered alcoholic electrolyte the mass detected by the frequency shift is only 55% of this value. The corresponding impedance measurement yields a similar result. Using the concept of a parallel plate capacitor with the octadecanethiol layer as the dielectric, a value of the capacitance of $C_F = 0.85 \, \mu\text{F cm}^{-2}$ would be expected. This corresponds to $Z'' = 46$ KΩ at 10 Hz and an electrode area of 0.33 cm^2. The observed value is again significantly smaller.

These results suggest that also at later times in the alcoholic electrolyte the self-assembled monolayer remains solvated. The aliphatic chains do not form a completely rigid layer, and because of the solvent in the layer the ions are not excluded, lowering the dielectric barrier. This view is supported by measurements conducted with the modified electrodes transferred from the alcoholic electrolyte into an aqueous electrolyte that does not contain the thiol. After some time in the aqueous phase the impedance was found to increase towards the value expected for the compact film.

18.5
Conclusions and Perspectives

The heart of a quartz microbalance is shown to consist of a piezoelectric quartz crystal with metal films deposited as electrodes on two opposite plane faces. Application of an alternating voltage to the electrodes leads to mechanical oscillation of the crystal. Its resonance frequency can be derived from the frequency dependence of the electric impedance. With a 14 MHz quartz, a conveniently measurable change in the resonance frequency by 1 Hz corresponds to a variation of the mass of the vibrating quartz/electrode system by about $2\,\mathrm{ng\,cm^{-2}}$. This means that the deposition or removal of atomic or molecular monolayers at the metal electrode is readily detectable.

The influence of a viscous medium in contact with the oscillating quartz is discussed, and a device is described which permits the application of a QCM in electrochemical research. A number of examples are given. They include: the current efficiency of galvanic metal deposition, the rate of uniform corrosion, the effect of corrosion inhibitors, the determination of alloy composition in thin layers, the mechanism of electroless metal deposition, the elucidation of the mechanism of an autocatalytic reduction reaction, and the self-assembly of an organic monolayer. All examples are taken from the literature.

This work is aimed at graduate students, at organizers of electrochemical laboratory classes, and at all scientists from fundamental and applied research who want an uncomplicated, versatile tool for the study of complex electrode reactions.

18.6
Supplementary Material

18.6.1
Some Experimental Details

In Section 18.1 we explained that the temperature dependence of the resonance frequency of an AT cut crystal is small. Nevertheless, temperature control is important for many experiments. That is because the properties of the adhering liquid layer depend strongly on the temperature. Remember: the shift of resonance frequency of the quartz oscillator caused by the solution is (see Section 18.2.5),

$$\Delta f = -f_0^{3/2} \sqrt{\frac{\rho_L \eta_L}{\pi \rho_Q \mu_Q}}$$

The temperature dependences of both the density and the viscosity of water near room temperature are

$$d\rho_L/dT = -0.21\,\mathrm{kg\,m^{-3}K^{-1}},\ d\eta_L/dT = -2 \times 10^{-5}\,\mathrm{N\,s\,m^{-2}\,K^{-1}}$$

Assuming constant quartz parameters, the resonance frequency of the EQCM will vary by $-7.2\,\mathrm{Hz\,K^{-1}}$. Clearly, if one intends to measure frequency changes of $\Delta f = 1$ Hz, the solution temperature must be kept constant to better than 0.1 K.

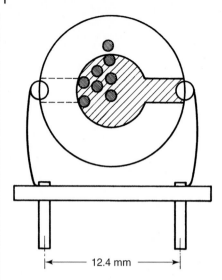

Figure 18.19 Experimental set-up for studying the mass sensitivity of an AT shear mode oscillator as function of the distance of an added mass from the center of the circular electrode.

The sensitivity constant of the EQMB must be determined for each experiment separately. One of the reasons is that the oscillation frequency is quite sensitive to stresses caused by mounting and to changes in the hydrostatic pressure of the fluid. Furthermore, the experimental sensitivity constant is an average over the total surface. This was investigated by Sauerbrey [1] who designed an experiment, illustrated in Figure 18.19.

Spots of metal films of small diameter and known mass were deposited systematically on various positions of the electrode and on the neighboring quartz surface. The resulting frequency changes were recorded in dependence on the distance of the spot from the electrode center. The mass sensitivity was calculated for each spot. The dependence of the mass sensitivity constant ($Hz\,g^{-1}\,cm^{-2}$) on the distance of the deposited spot from the electrode center is represented in Figure 18.20. In order to avoid complications in the determination of the sensitivity constants one uses crystals where the metal film on the electrolyte side is smaller than on the other side, see Figure 18.7.

Finally, because the driver circuits that excite the crystal oscillations have to be grounded at one of their outlets it is important that the potentiostat that is used for the electrochemical experiments can be connected to a grounded electrode. That is, for instance, possible for Wenking Potentiostats from Bank Electronics,Germany, and for the potentiostats from Gamry, Pennsylvania, USA.

For most applications of the ECQM the rate of frequency change is rather small ($<$ a few $Hz\,s^{-1}$). Hence the time resolution of the quartz frequency measurement can be rather poor. In these cases one uses the "reciprocal" technique' for the frequency determination. The number of zero transitions of the signal during a

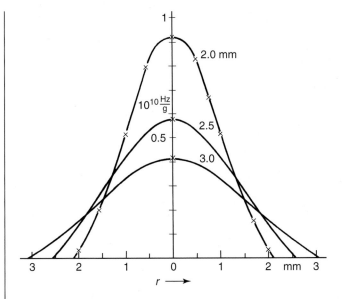

Figure 18.20 Differential sensitivity plots for mass deposition on the surface of three circular 14 MHz AT quartzes operating in the thickness shear mode. The values of the radii, R_e, of the circular gold electrodes are indicated, r is the distance from the electrode center.

well defined time, the gate time, is measured. Electrochemical oscillations, on the other hand, frequently occur within the 10^1–10^{-2} Hz regime. Here we need a time resolution better than about 10 ms. This can be obtained with the so-called "continuous technique." This method is described in Ref. [5].

18.6.2
The Impedance Analysis

The electric energy supplied to the quartz oscillator forces the quartz to do mechanical work. It is, therefore, clear that the electric behavior of the quartz oscillator is closely related to the material constants and the dimensions of the quartz plate. For the parameters R_0, L_0, C_0, and C^* of the equivalent circuit (Figure 18.4) one finds at the fundamental resonance frequency (see Arnau and Soares [11]):

$$R_0 = \frac{\pi^2 \eta_Q d}{8e_{26}^2 A} \qquad L_0 = \frac{\rho_Q d^3}{8e_{26}^2 A} \qquad C_0 = \frac{8e_{26}^2 A}{\pi^2 d \mu_Q} \qquad C^* = \frac{\varepsilon_{22} A}{d} \qquad (18.9)$$

Here, A is the active area of the quartz, that is, the area of the metal electrodes and d is the plate thickness. The following values apply for an AT-cut quartz [11]:

$$\varepsilon_{22} = 3.982 \times 10^{-11} \text{ A}^2\text{s}^4 \text{ kg}^{-1}\text{m}^{-3} \text{ (permittivity)}$$

$$\eta_Q = 9.27 \times 10^{-3} \text{ Pa s (effective viscosity)}$$

$\eta_Q = 2.947 \times 10^{10}$ N m^{-2} (piezoelectrically stiffened shear modulus)

$e_{26} = 9.657 \times 10^{-2}$ A s m^{-2} (piezoelectric constant)

$\rho_Q = 2651$ kg m^{-3} (density).

A 5 MHz AT quartz has a thickness of $d = 3.34 \times 10^{-4}$ m (see Equation 18.1). With electrodes of diameter 6 mm ($A = 2.83 \times 10^{-5}$ m^2) one obtains $R_0 = 14.5\ \Omega$, $L_0 = 0.0468$ H, $C_0 = 2.17 \times 10^{-14}$ F, and $C^* = 3.37 \times 10^{-12}$ F. In practical work one has to consider that the capacitances of the connecting wires and of the measuring instrument contribute to C^*. To present a realistic picture, the admittance of the 5 MHz quartz represented in Figure 18.5a,b was calculated with a value of $C^* = 40$ pF instead of the oscillator's intrinsic C^* value.

As shown in Figure 18.5, one is normally interested in the behavior of the oscillator over a narrow frequency range near resonance. Consider the role of the parallel capacitance C^* on the $Y\{\mathrm{Im}\} - f$ curve in this frequency range. According to Equation 18.4, C^* contributes to the $Y\{\mathrm{Im}\} - f$ curve a nearly constant value of $Y\{\mathrm{Im}\} = \omega C^*$, that is, C^* causes the curve to shift almost parallel to more positive values. In Figure 18.21 the $Y\{\mathrm{Im}\} - f$ curve (a) is calculated for the quartz in vacuum, as in Figure 18.5a, but with $C^* = 0$. The comparison between the $Y\{\mathrm{Im}\} - f$ curves of Figures 18.5a and 18.21a illustrates that it is the upward shift caused by C^* that leads to the second frequency at which $Y\{\mathrm{Im}\} = 0$, defining the second resonance frequency, f_p.

Consider now the EQCM. The oscillating quartz operates with one side in contact with a liquid electrolyte. To analyze this situation, note that for a Newtonian liquid the mechanical shear impedance is a complex quantity defined by

$$Z_L^{\mathrm{mechan}} = \sqrt{\eta_L \rho_L \pi f}(1 + \mathrm{j}) \tag{18.10}$$

The subscript L refers to the contacting liquid phase. Remarkably, the real and the imaginary component of this shear wave impedance both have the same value. This implies that the corresponding electric impedance corresponds to a resistor, R_L and an inductor, L_L, with the following property:

$$R_L = \omega L_L \tag{18.11}$$

Unlike true resistors and inductors, the values of R_L and L_L are not constants, but both depend on the square root of the frequency. According to Lucklum *et al.* [13], the electric impedance of R_L is defined by:

$$R_L = \frac{d}{8Ae_{26}^2}\sqrt{\frac{\rho_L \eta_L \mu_Q}{\rho_Q}}\sqrt{\frac{\pi}{f}} \tag{18.12}$$

Knowing R_L, one obtains the corresponding values of L_L with Equation 18.11 The electric impedance caused by the liquid, $Z_L = R_L + \mathrm{j}\omega L_L$, can be added to the dynamic arm of the equivalent circuit (Figure 18.4), and the impedance/admittance of the quartz oscillator in contact with a Newtonian liquid can be calculated. In Figure 18.21 imaginary admittance results for the 5 MHz quartz considered before, but now in contact with water, are represented as curves (b), (c), and (d). For

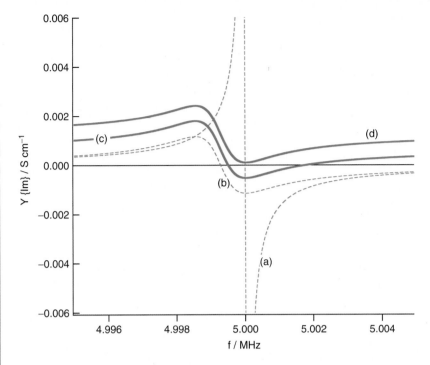

Figure 18.21 Imaginary part of the admittance calculated for a 5 MHz quartz oscillator with a circular metal electrode of diameter 6 mm: (a) The quartz operates in vacuum, with the assumption that the parallel capacitance $C^* = 0$; see Figure 18.5a for the peak values; (b) calculated curve for the same quartz in contact with water, with $C^* = 0$; (c) the same 5 MHz quartz in contact with water, with $C^* = 20$ pF; and (d) the same 5 MHz quartz in contact with water, with $C^* = 40$ pF.

the calculation of curve (b) the parallel capacitance C^* is again set to zero. The comparison with curve (a) demonstrates the dramatic damping effect of the liquid: At resonance $R_0 = 418\ \Omega$, while in vacuum $R_0 = 14.5\ \Omega$! As $C^* = 0$, the $Y\{Im\} - f$ curve(b) shows one resonance frequency, which is shifted to a smaller value, as discussed by Kanazawa and Gordon [14], see Equation 18.6

Curves (c) and (d) demonstrate the dramatic effect of adding (practically unavoidably) parallel capacitances. Consider first curve (c) with $C^* = 20$ pF. As discussed above, the $Y\{Im\} - f$ curve is shifted essentially parallel by ωC^*. It crosses the zero line twice, yielding series and parallel resonances similar to Figure 18.5a. This EQCM system can be operated with a standard driver.

On the other hand, the admittance curve (d) obtained with $C^* = 40$ pF is shifted to such high values that there is no longer intersection with the zero line. This means, a driver circuit designed for resonance at zero phase shift ($Y\{Im\} = 0$) will not work. In this situation one might make an effort to reduce the stray capacitances, and to get as close as possible to the intrinsic C^* value of the quartz oscillator. Alternatively, a modified driver circuit [12] might be employed. Of course, these considerations demonstrate the advantage of using an impedance analyzer, by which any response of the quartz oscillator can be recorded and analyzed conveniently.

Acknowledgments

Sabine Wasle prepared the figures and polished the final version of the manuscript. Her kind and very effective support is greatly acknowledged.

References

1. Sauerbrey, I.G. (1959) *Z. Phys.*, **155**, 206.
2. (a) Nomura, I.T. and Okuhara, M. (1982) *Anal. Chim. Acta*, **142**, 281; (b) Nomura, T. and Nagamune, T. (1983) *Anal. Chim. Acta*, **155**, 231.
3. (a) Schumacher, R., Borges, G., and Kanazawa, K.K. (1985) *Surf. Sci.*, **163**, L621; (b) Kanazawa, K.K. and Gordon, J.G. II (1985) *Anal. Chem.*, **57**, 1770.
4. (a) Bruckenstein, S. and Swathijan, S. (1985) *Electrochim. Acta*, **30**, 851 ; (b) Bruckenstein, S. and Shay, M. (1985) *Electrochim. Acta*, **30**, 1295.
5. Eickes, C., Rosenmund, J., Wasle, S., Doblhofer, K., Wang, K., and Weil, K.G. (2000) *Electrochim. Acta*, **45**, 3623–3628.
6. (a) Curie, P. and Curie, J. (1880) *C. R. Acad. Sci.*, **91**, 294; (b) Curie, P. (1880) *Oeuvres*, Paris.
7. Mierdel, G. (1972) *Elektrophysik*, 2nd edn, VEB Verlag Technik, Auflage, Berlin.
8. Scheibe, A. (1938) *Piezoelektrizität des Quarzes*, Theodor Steinkopff, Dresden und Leipzig.
9. Heising, R.A. (1946) *Quartz Crystals for Electrical Circuits*, Van Norstrand, New York.
10. von Dyke, K.S. (1925) *Phys. Rev.*, **25**, 895.
11. Arnau, A. and Soares, D. (2008) Fundamentals on piezoelectricity, in *Piezoelectric Transducers and Applications*, Chapter 1, 2nd edn (ed. A. Arnau), Springer-Verlag, Berlin, Heidelberg, pp. 1–38.
12. Soares, D.M., Kautek, W., Fruböse, C., and Doblhofer, K. (1994) *Ber. Bunsen-Ges. Phys. Chem.*, **98**, 219.
13. Lucklum, R., Soares, D., and Kanazawa, K. (2008) Models for resonant sensors, in *Piezoelectric Transducers and Applications*, Chapter 3, 2nd edn (ed. A. Arnau), Springer-Verlag, Berlin, Heidelberg, pp. 63–96.
14. Kanazawa, K.K. and Gordon, J.G. II (1985) *Anal. Chim. Acta*, **175**, 99.
15. Beck, R., Pittermann, U., and Weil, K.G. (1988) *Ber. Bunsen-Ges. Phys. Chem.*, **92**, 1363.
16. Müller, E. (1953) *Praktikum der Elektrochemie*, Steinkopf-Verlag, Berlin.
17. Pettinger, B., Wenning, U., and Kolb, D.M. (1978) *Ber. Bunsen-Ges. Phys. Chem.*, **82**, 1326.
18. Beck, R., Pittermann, U., and Weil, K.G. (1992) *J. Electrochem. Soc.*, **139**, 453.
19. Wang, K., Pickering, H.W., and Weil, K.G. (2001) *Electrochim. Acta*, **46**, 3835–3840.
20. Jope, D., Sell, J., Pickering, H.W., and Weil, K.G. (1995) *J. Electrochem. Soc.*, **142**, 2170.
21. Benje, M., Eiermann, M., Pittermann, U., and Weil, K.G. (1986) *Ber. Bunsen-Ges. Phys. Chem.*, **90**, 435.
22. Schumacher, R., Pesek, J.J., and Melroy, O.R. (1985) *J. Phys. Chem.*, **89**, 4338.
23. Wiese, H. and Weil, K.G. (1987) *Ber. Bunsen-Ges. Phys. Chem.*, **91**, 619.
24. Bittner, A., Wanner, W., and Weil, K.G. (1992) *Ber. Bunsen-Ges. Phys. Chem.*, **96**, 647–655.

25. Benje, M., Hofmann, U.,
    Pittermann, U., and Weil, K.G. (1988)
    Ber. Bunsen-Ges. Phys. Chem., **92**, 1257.
26. Flätgen, G., Wasle, S., Lübke, M.,
    Eickes, C., Radhakrishnan, G.,
    Doblhofer, K., and Ertl, G. (1999) *Elec-
    trochim. Acta*, **44**, 4499.
27. Eickes, C., Weil, K.G., and Doblhofer, K.
    (2000) *Phys. Chem. Chem. Phys.*, **2**,
    5691–5697.
28. Fruböse, C. and Doblhofer, K. (1995)
    J. Chem. Soc., Faraday Trans., **91**,
    1949.

19

The Scanning Tunneling Microscope in Electrochemistry: an Atomistic View of Electrochemistry

Dieter M. Kolb[1] and Marie Anne Schneeweiss

■ **Method Summary**

Acronyms, Synonyms
- Atomic force microscope (AFM)
- Scanning tunneling microscope (STM)
- Ultra-high vacuum (UHV)
- Low energy electron diffraction (LEED)
- Inner Helmholtz plane (IHP)
- Outer Helmholtz plane (OHP)
- Potential of zero charge (PZC).

Benefits (Information Available)
- Imaging of electrode surfaces in real space and under operating conditions (so-called *in situ* technique).
- Imaging surfaces with atomic resolution, when single crystal surfaces are used.

Limitations (Information Not Available)
- No chemical information available
- Relatively slow technique because of sequential imaging
- Imaged area is small (typically in the μm-range)
- Limitations by surface roughness; works best for flat single-crystal surfaces
- Close proximity of STM tip can influence reactions by mass transport limitation.

19.1
Introduction

The ongoing fascination of scanning tunneling microscopy (STM) is that it seemingly enables us to "see" atoms, something that might have been the "impossible dream" of generations of scientists. The technique utilizes a sharp needle which

---

[1] We communicate with great sorrow that Professor Kolb passed away in October 2011.

Methods in Physical Chemistry, First Edition. Edited by Rolf Schäfer and Peter C. Schmidt.
© 2012 Wiley-VCH Verlag GmbH & Co. KGaA. Published 2012 by Wiley-VCH Verlag GmbH & Co. KGaA.

traces the surface of a conductive substrate to eventually provide three-dimensional topographical maps of the smallest, nanometer-sized surface areas. STM was developed by Binnig and Rohrer [1–3] who first demonstrated its capability of atomic-scale resolution in the ultra-high vacuum (UHV). Once STM had bridged the gap from UHV to liquids and electrochemistry [4], it opened up a whole new field of applications. High resolution is not limited to the atomic structure of crystalline metal electrodes – organized monolayers of anions and molecules adsorbed on the metal surface can be imaged equally easily, as well as features on a slightly larger scale such as crystallites or substrate defects like grain and domain boundaries.

In fact, as the image is acquired in real space, that is, represents directly the local surface topography (as opposed to the diffraction techniques which provide information in k-space and are averaged over a large area), it will accurately depict localized structures, such as monoatomic high steps or screw dislocations on an atomically flat terrace. Such defects are crucial in determining the reactivity of a given surface. It allows the pinpointing of reaction sites, for example, nucleation sites for metal deposition.

Sequential images, snapshots of a certain area, acquired during an ongoing surface reaction, will provide an illustration of the process in real time, with changes in topography evident from one image to the next. Electrochemistry has the huge advantage that the rate of almost any given process can be accurately controlled and matched to the "recording" speed of an STM. Deposition rates and other processes can be carefully adjusted by choosing the right electrode potential or current value (in the case of galvanostatic processes).

Important questions in electrochemistry have always centered on both the structure of the electrode/electrolyte interface, and the rate of processes taking place at this interface. In this chapter we will first present a short overview of the STM technique as well as the basics of electrochemistry. The main body of this chapter will deal with typical applications, such as metal deposition, specific anion adsorption, organic monolayers, and electrochemical nanostructuring.

19.2
Basic Principles

19.2.1
The Scanning Tunneling Microscope

Gerd Binnig and Heinrich Rohrer were awarded the Nobel Prize in Physics in 1986 for the design of the STM. The other half of the prize that year went to Ernst Ruska "for his fundamental work in electron optics, and for the design of the first electron microscope," which highlights the equally revolutionary impact of both techniques at the time of their invention, both milestones not just for physics but for chemistry, biology, materials sciences, and life sciences, to name but a few (see Section 19.6.1).

Ever since that time, the technique has evolved into a well established method in surface science, particularly in those research areas that deal with single-crystal metal substrates (less so for semi-conductive substrates). STM quickly moved away from a UHV-only approach and was found to work just as well in liquid and, ultimately, in electrochemical surroundings, thus making it even more relevant for many areas. The technique can provide images of monocrystalline metal surfaces with atomic-scale resolution, as well as resolving the structure of chemi- or physic-sorbed monolayers of molecules and ions. Obvious advantages of this technique are imaging in real space and real time, while limitations include the inability for chemical identification.

The principle of STM is based on the interaction between the electron wavefunctions of a sharp metal tip and a conductive sample surface. If a voltage is applied between the two solids and they are brought into sufficiently close proximity (some angstroms), there is a finite probability that electrons will cross the gap, and a so-called tunnel current will flow, its size depending sensitively on the tip–sample separation. (The quantum mechanical basis of this phenomenon is sketched in Section 19.6.2.) An image is acquired by scanning the tip over the surface, while employing a feedback circuit to trace the outline of the sample topography.

A large number of scanning probe techniques have been developed in the past 20 years, all of them utilizing sharp probes to ensure high lateral resolution, and using the local interaction of probe and sample as a feedback parameter. The most significant of these is the atomic force microscope (AFM) [5], which has in fact long surpassed the STM in terms of popularity, being a routine technique these days in academia as well as in industry, with a comprehensive selection of commercial systems available. In the case of AFM, the interactions in question are in essence Van-der-Waals forces between a silicon tip (on the end of a cantilever) and the substrate. It can be applied to almost any surface regardless of conductivity, however, the routinely achievable resolution is slightly lower than with the STM.

Figure 19.1 illustrates the working principle of STM: The metal tip is fixed to a so-called piezo scanner, a device that allows motion of the tip in all three directions, lateral (x,y) and vertical (z). In the simplest case the scanner consists of three piezo elements, arranged at $90°$ to each other (see Figure 19.1); more sophisticated versions employ tube scanners. Under an applied voltage the piezo ceramics will change in length, the maximum change depending on the length of the piezo element (typically around 0.1% of the total length). The step resolution is often better than a fraction of an angstrom – in fact, in many cases maximum resolution depends neither on piezo nor on electronic resolution, but mainly on the size and shape of the tip apex and overall noise.

Then the tip is scanned over the surface in a predetermined grid (common grid sizes are 256×256 or 512×512 pixels). In order to collect information about the surface morphology, either the distance between tip and sample (equivalent to a constant tunnel current, *constant current mode*) or the absolute tip position can be held constant (*constant height mode*).

In the frequently used *constant current mode*, the current is kept constant through a feedback loop, with a selected current as the set-point. The voltage on the z-piezo

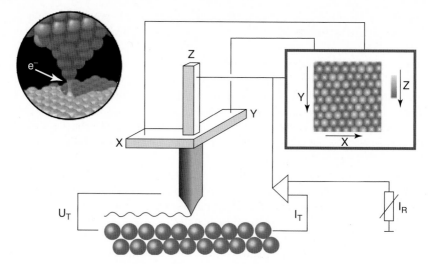

Figure 19.1 Schematic diagram of an STM. A fine metal tip is scanned across the surface with the help of the piezo elements x and y, while the tip–sample distance is held constant via a feed-back circuit that keeps the tunnel current I_T constant. Accordingly, the surface topography is reflected by the tip movement in the z direction, which is monitored by the voltage applied to the z-piezo. The height information is implemented into a gray-scale image, with higher parts of the surface presented in lighter and deeper lying parts in darker gray. The inset shows a model of tip and sample surface, illustrating that the tunnel current will originate exclusively from the most protruding part of the tip. From Ref. [6].

is adjusted, and it is this change in voltage required to keep the tip at a constant distance to the surface, that directly provides topographical and height information from the surface. This information is then displayed in three-dimensional (gray scale) images, as demonstrated below.

Strictly speaking, the STM image reflects a plane of constant tunneling probability, which in many cases, especially in the case of pure metals, is close enough to substrate topography (although seemingly random changes in image contrast, as well as imaging artifacts, do happen, especially when organic species are involved, and it is important to bear the tunneling mechanism in mind when interpreting images).

STM has obvious limitations: no direct chemical information is available regarding the imaged species, which in mixed monolayers often makes identification of the species involved difficult. A classic example would be an adlayer consisting of metal atoms and coadsorbed anions. In order to identify species correctly other methods have to be employed in addition.

A number of points specific to STM in electrochemical surroundings will be elaborated on later. The following section will give a short introduction to the basic principles and terminology of electrochemistry.

19.2.2
Some Basics of Electrochemistry

Electrochemistry is generally concerned with chemical reactions which are associated with a transfer of electrical charge. A considerable number of these processes take place at the solid/liquid interface. The solid phase is typically an electron conductor, while the liquid phase, the electrolyte, is an ion conductor. The interface, therefore, presents an abrupt change in the character of electric conductivity. In the absence of an electrochemical reaction a charge transfer through the interface is not possible. In that case, the interface will behave like a plate capacitor, which can be charged or discharged by applying an external voltage.

An electrochemical reaction, on the other hand, will act like a switch between electron- and ion-conductor and will facilitate a current flow through the electrode/electrolyte interface. Two different types of charge transfer reactions can be distinguished. *Electron transfer* describes a reaction where an electron is exchanged between the electrode and the solvated redox partners in the electrolyte (e.g., $Fe^{3+} + e^- \rightleftarrows Fe^{2+}$), whereas in the case of an *ion transfer* reaction the ion is discharged at the electrode surface (e.g., $Ag^+ + e^- \rightleftarrows Ag$) and subsequently remains there. Metal deposition reactions are prime examples of the latter type.

Another term for the aforementioned plate capacitor at the interface and an extremely important concept in interfacial electrochemistry is the electric *double layer* (Figure 19.2). In this picture, one "plate" or "layer" consists of the metal surface proper with the respective excess charge, the other is a plane of solvated ions, located as close to the metal surface as possible under the influence of the electric field. The latter is known as the *outer Helmholtz plane* (OHP) [7]. The gap between the planes is estimated to be of the order of 3 Å (roughly the radius of the solvated ions). The measured double layer capacitance depends on the electrode material, the electrolyte, and the applied voltage and can assume values of between 20 and $100\,\mu F\,cm^{-2}$ for aqueous solutions. With the above-mentioned 3 Å plate distance, a relative dielectric constant of 5–10 is estimated. This value can be explained by water molecules which have lost a considerable portion of their rotational degrees of freedom.

The high value of the double layer capacitance and the tiny gap between the planes of the capacitor lead to interesting consequences: With an externally applied voltage of 1 V, extremely high electric fields (some $10^7\,V\,cm^{-1}$) and unusually high excess charges (circa 0.1–0.2 electrons per surface atom) can be easily generated at the electrode surface.

As mentioned before, the excess charge on the electrolyte side is formed by solvated ions, provided that the ions have a strongly bound solvation shell. This applies to practically all cations (with Cs^+ being the only exception) and for a small number of anions, for example, fluoride or perchlorate. Such ions that are held in close proximity to the electrode by purely electrostatic forces are referred to as "*non-specifically adsorbed.*" Their cores formally define the OHP. Many anions, however, have only weakly bound solvation shells, which will be partly stripped off as the anion comes into contact with the electrode surface. These ions form a

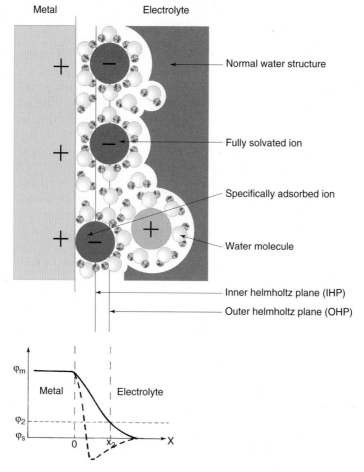

Figure 19.2 Schematic diagram of the metal/electrolyte interface, showing fully and partially solvated ions. The idea behind this model is that of a plate condenser. One plate is the metal proper with its surface excess charge; the other plate is built up by solvated ions at closest approach, held in place by purely electrostatic forces (outer Helmholtz plane). Ions with weakly bound solvation shells (mostly anions) usually lose part of their solvation shell and form

a chemical bond with the surface (so-called specific adsorption). Because the chemical interaction between such ions and the electrode surface causes more charge to be accumulated at the surface than required by electrostatics, counter charge is incorporated in the double layer for charge compensation. The potential drop across the interface in the case of non-specific (—) and specific (---) ion adsorption is also sketched in the figure.

chemical bond with the metal substrate and their cores form the so-called _inner Helmholtz plane_ (IHP). This process is known as _specific adsorption_ [8].

The anion's charge can be reduced considerably in direct contact with the metal surface, due to partial electron transfer. As a consequence of the reduced electrostatic repulsion, the coverage can be far higher than in the case of non-specific adsorption. Specifically adsorbed anions frequently form regular patterns and structures, which are readily imaged with the STM (see below).

Some words of explanation regarding the definition of the *electrode potential*: The potential difference between solid and liquid phase cannot be measured in absolute terms, because ultimately the measurement always occurs between two metallic wires. The difference is, therefore, determined with respect to a so-called reference electrode, with the electrochemical potential of the electrons in the reference electrode arbitrarily set to zero. It is useful to bear in mind that the electrode potential directly correlates with the electrochemical potential of the electrons in the metal, $\tilde{\mu}_{e^-}$, and corresponds to the Fermi energy E_F: $\tilde{\mu}_{e^-} = F(E_F - K)$ [9]. The constant K corresponds to the energy that is necessary to transfer an electron from the Fermi level of the normal hydrogen electrode to the vacuum, and, therefore, defines an absolute electrode potential. The experimentally determined values of K, which are reported in the literature, still carry a relatively large uncertainty; they vary between 4.4 and 4.7 eV [9, 10]. The above-said also demonstrates the close relationship between electrode potential and work function: A potential change of 1 V in the positive (negative) direction causes a change in the work function of exactly 1 eV toward higher (lower) values. Generally speaking, any change in electrode potential represents directly an equivalent change in the electrode's Fermi level relative to the Fermi level in solution (which is fixed with respect to that of the reference electrode).

19.3
Experimental Methods

The only fundamental prerequisite for imaging electrochemical processes with the STM is that tunneling and imaging are not hampered by the presence of the electrolyte. Although first purported to be a UHV technique, it was not long before the first reports of STM in liquid were seen [4] and the potential of the method for electrochemistry became clear [11, 12].

The following lists the experimental considerations to be taken into account for electrochemical STM experiments using commercial or home-built systems [6]. Some commercial systems provide the entire periphery, while others require modification by the user.

- An electrochemical cell normally consists of three electrodes: working, counter, and reference electrode. In addition to that, the STM tip will act as a fourth electrode, because it is in the electrolyte and there is a (tunnel) voltage applied to it. Hence, it will have a certain potential versus the reference electrode. This means that electrochemical processes take place at the tip and create a faradaic current, which will obliterate the tunneling current. This must be avoided, essentially by taking care of two issues. First, the tip potential must be freely adjustable in order to select a potential near the so-called rest potential, where, by definition, there should be no significant amount of current. This is achieved by controlling all potentials via a bipotentiostat, which allows selection of the potential of both the working electrode and the tip, independently against the reference electrode. In addition, STM tips for electrochemical studies must be coated for their largest part with an insulating substance (most commonly Apiezon wax [12], glass or polymers, nail varnish, and electrophoretic paint [13]). This reduces the surface area of the tip that is still in contact with solution to an absolute minimum, namely to the tunneling portion, and, in conjunction with choosing the right tip potential, will make the faradaic current negligible (typically <100 pA, while tunnel currents are normally in the 1–10 nA range).

- STM tips normally consist of tungsten, or more commonly a Pt/Ir (90 : 10) alloy, and are etched electrochemically from (in our case) a 0.25 mm wire [14].
- The electrochemical cell used for the experiments has to be miniaturized conforming to limitations dictated by the geometry of the STM set-up. The working electrode, in the simplest case a gold film on mica or glass, has to be located at the bottom of the cell, and it often *forms* the bottom of the cell. Other options include the use of commercial single crystal disks or so-called Clavilier beads (little single-crystal spheres made by carefully melting one end of a metal wire [15]), which both require an appropriate hollow in the base of the cell to nestle in. Counter and reference electrodes are miniaturized according to the small size of the cell [16]. This is easily achieved through using simple wires as reference electrodes (for example, a copper wire with Cu ions present in the electrolyte). Another option is to prepare microreference electrodes based on glass capillaries. The counter electrode is usually a platinum wire.

A further requirement for an electrochemical STM is the use of single crystalline surfaces. Strictly speaking, measurements can be carried out on polycrystalline substrates; however, the usual demand for "atomic resolution" as part of the STM attraction almost categorically requires that a structurally well-defined, that is, atomically flat surface, be present for the substrate. In addition, a single crystalline surface guarantees that the STM measurements, which typically cover areas of 100 nm × 100 nm or less, are fairly representative of the total surface. In many cases the crystallographic orientation has been shown to have a marked influence on reactivity [17]; therefore, studies often compare reactions and processes on different crystal faces. For the preparation of single crystal surfaces for electrochemical studies, see Ref. [18]. In Figure 19.3 are shown high-resolution STM images of all three low-index faces of silver in an aqueous solution under potential control [19].

19.4
Case Studies

In the following we shall demonstrate with a few selected examples the great potential of STM for developing a microscopic picture of the electrochemical interface and of processes occurring therein. The examples deal with structure studies of metal surfaces in contact with an aqueous solution, particularly with the reconstruction of gold surfaces, and with ordered adlayers of anions from the supporting electrolyte and of organic molecules. Then metal deposition, as an

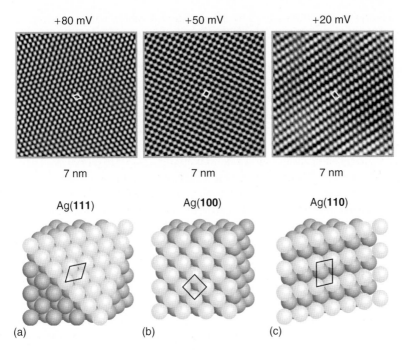

Figure 19.3 STM images of all three low-index faces of a silver single-crystal electrode in 0.05 M H_2SO_4 at the indicated potentials. The unit cell of the surface is shown in each image. Also shown are hard-sphere models of the three surfaces: (a) Ag(111), (b) Ag(100), and (c) Ag(110). From Ref. [19].

electrochemically rather simple yet technologically very important process, will be viewed by STM, and, last but not, least, it is demonstrated that STM tips are not only good for imaging surfaces in great detail, they can also be used as a tool for manipulating surfaces on a nanometer length scale.

19.4.1
The Reconstruction of Gold Surfaces

One of the main reasons for utilizing single crystal surfaces in electrochemical studies is the assumption that the arrangement of atoms in the surface is identical to that in the well-known volume lattice. This implies that if a crystal is cleaved to create the surface in question, the surface atoms maintain the same lattice positions they had before. This assumption is in many cases justified and true. However, there are some very relevant cases where the surface atoms experience a shift due to the change in the forces surrounding them, and reside in lattice positions significantly different from those in the bulk of the crystal [20]. Such lateral rearrangements, which are referred to as *surface reconstruction*, are caused by the accompanying decrease in surface energy.

Interestingly, particularly for electrochemical considerations, all three low-index faces of a gold single crystal are reconstructed when freshly prepared. As adsorbed species will generally lift the reconstruction, this applies mainly for the clean surface. *Lifting* of reconstruction refers to the surface atoms moving to resume the lattice structure of the bulk crystal.

Because surface atoms generally tend to arrange in a densely packed structure, a freshly annealed Au(100) electrode will exhibit an hexagonally close-packed structure, which is why this structure is often referred to as *Au(100)-hex* [21]. Figure 19.4a shows an STM image of such a reconstructed surface [22]. The hexagonal arrangement of the surface atoms is clearly visible. Another characteristic feature of this surface is the line structure, which is caused by the mismatch of the hexagonal surface layer and underlying square lattice of the volume crystal layers. Some surface atoms will end up directly on top of the atoms of the second layer ("on-top sites") and as such will be higher and, therefore, lighter in the STM image, whereas others will sit in the hollows formed by atoms of the second layer ("fourfold hollow sites"), and will, therefore, be lower and appear as darker on the image.

In the electrolyte, this reconstructed Au(100) surface is only stable at potentials sufficiently negative of the potential of zero charge (pzc), where solvated (i.e., non-specifically adsorbed) cations form the double layer on the electrolyte side. As soon as anions are specifically adsorbed (at the respective potentials), the reconstruction will be lifted and a so-called (hex) $\longrightarrow$ (1×1) transition will take place. Figure 19.4b shows an STM image of the unreconstructed surface, which clearly reveals the quadratic arrangement of the surface atoms, which one would expect from a mere termination of the crystal structure.

Under UHV conditions, the reconstruction of the Au(100) surface requires thermal activation. Therefore the clean unreconstructed (1×1) surface is, at room temperature, metastable. In an electrolyte, however, the (hex) structure will be restored rapidly if negative potentials are applied. This causes interesting

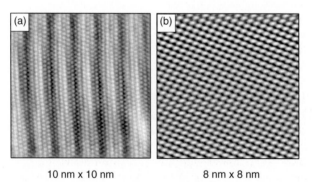

10 nm x 10 nm   8 nm x 8 nm

Figure 19.4 STM images of (a) reconstructed and (b) unreconstructed Au(100) in 0.1 M H_2SO_4. The electrode potentials are -0.2 V versus SCE (a) and $+0.5$ V versus SCE (b). From Ref. [22].

consequences for any electrochemical studies of Au(100) as well as the other two low-index faces of gold: The surface structure becomes a function of electrode potential! The square arrangement with fourfold symmetry that would be expected for the Au(100) surface due to its crystallographic orientation is only found at potentials positive of the pzc. At negative potentials the surface atoms rearrange to form a close-packed hexagonal structure with (111) character [23].

19.4.2
Ordered Adlayers of Ions and Organic Molecules

Anions of the supporting electrolyte have a strong tendency toward specific adsorption, that is, they lose part of their solvation shell and undergo a direct chemical bond with the substrate. As a consequence the ionic character of such ions is strongly reduced, which allows a significantly higher coverage of specifically adsorbing ions. Whereas for non-specifically adsorbing ions (i.e., purely electrostatic interaction with the substrate), the maximum coverage hardly exceeds 0.1 of a full monolayer, values up to 0.5, and more are found for specific adsorption (e.g., Cl^- or Br^- on Au(111)). In the latter case, ordered adsorption is often observed, that is, the adions form a regular pattern on the surface, the structure of which being often determined by two counteracting forces: the electrostatic repulsion due to the negative charge on the adions and the tendency of the adspecies to either nest in energetically favorable hollow sites of the substrate or to form a hydrogen-bridge bonded network with directional bonds. Sulphate adsorption on Au(111), which leads to the well-known $(\sqrt{3} x \sqrt{7})$ R19.1° superstructure, is a good example of the latter case [24]. In Figure 19.5 the STM image of a $PdCl_4^{2-}$ adlayer on Au(110) is shown [25a], which clearly reveals the alignment of the adsorbate along the [110] direction of the substrate (see model in Figure 19.5). There are two different adspecies seen on the surface: one that is imaged as little "windmills" with four bright spots at the corners of a square, and one that, in essence, consists of two bright spots. Considering the fact that $PdCl_4^{2-}$ is a square-planar complex, it appears reasonable to assign the "windmills" to that species, implying that not only the complex as such, but also the atomic constituents of it are imaged; at least the four corner (chlorine) atoms. With respect to the second species, one has to keep in mind that chloropalladate is not a very stable complex, and complexes with a lower number of ligands will also exist in solution, albeit at much lower concentrations [25b]. It seems not unreasonable to assign the second species to $PdCl_2$ being stabilized by the interaction with the substrate and being nested in the troughs of the (110) surface.

Adlayers from organic (i.e., neutral) molecules are another area that has been intensively studied by STM, as such layers again often form regular patterns. In these cases the structure is again the result of two forces: intermolecular and interaction with the substrate [26]. For thiols on gold, both types of interaction are such that these molecules form spontaneously densely-packed and ordered monolayers, so-called self-assembled monolayers (SAMs) [27]. These again are ideally suited for structure studies with an STM. In Figure 19.6 the STM image

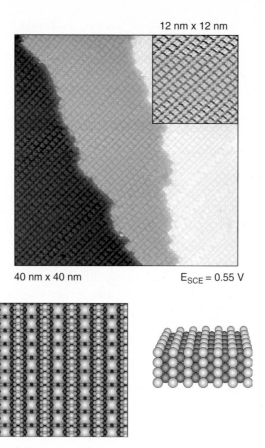

12 nm x 12 nm

40 nm x 40 nm $E_{SCE} = 0.55$ V

Figure 19.5 STM image of an Au(110) electrode in 0.1 M H_2SO_4 + 0.1 mM $PdCl_2$ + 0.6 mM HCl, showing the ordered adsorption of tetrachloropalladate ions (little square features) and a second species, presumably $PdCl_2$. A model of the Au(110) surface and the possible arrangement of the chloro palladate complexes in the adlayer is also shown. From Ref. [16].

of an ethane thiol SAM on Au(100) is shown. As seen by the square arrangement of the bright spots, the adlayer symmetry is dictated by the substrate surface. However, the adlayer structure is by no means in registry with the substrate, the exact form still being under investigation [28].

19.4.3
Metal Deposition

The study of metal deposition from aqueous solutions has a long tradition in electrochemistry. This reflects not only the tremendous importance of this process for the metal winning, refining, and plating industries, but also the great interest of scientists in electrocrystallization phenomena, for which metal deposition provides an ideal case [29]. When a metal is immersed in a solution containing its ion (e.g., a

10 x 10 nm²

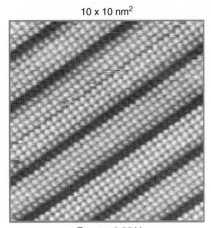

$E_{SCE} = +0.55\,V$

Figure 19.6 STM image of an ethanethiol self-assembled monolayer (SAM) on Au(100) in 0.1 M H_2SO_4 at +0.55 V versus SCE. The thiol molecules, presumably imaged via their sulfur atom, are not in registry with the gold substrate, which may be the cause of the apparent height variation within the adlayer. From Ref. [28].

sheet of Cu metal in a 1 M $CuSO_4$ solution), an equilibrium potential is established, at which the rates of deposition and dissolution are equal, that is, no net electric current is flowing. When applying an external potential to the electrode which is negative of this equilibrium potential E_r (the so-called Nernst potential), metal ions will be discharged and deposited onto the surface, while at potentials positive of E_r metal will be oxidatively dissolved; the process will continue in both cases until the resulting metal ion concentration matches the new equilibrium potential.

Metal plating frequently involves deposition onto a foreign metal substrate, which is particularly suited to study the initial states of this process. Metal deposition onto a foreign metal substrate (e.g., Cu onto Au) often starts with the formation of a monolayer at underpotentials, that is, at potentials positive to the Nernst potential for the respective bulk phase [30]. This, at first sight surprising observation (deposition should not be possible at potentials positive of E_r) can be easily understood by the fact that in such a case the metal–substrate interaction exceeds in strength that between the deposited metal atoms (e.g., Cu–Au as compared to Cu–Cu). This so-called underpotential deposition (upd) is most conveniently studied by cyclic current–potential curves, where the current due to deposition (during the potential scan in the negative direction) or dissolution (scan in the positive direction) of the monolayer represents, in essence, the first derivative of the corresponding adsorption isotherm. From such curves it has already been deduced that formation of the first monolayer should involve steps with ordered adsorption [31, 32]. Indeed, low energy electron diffraction (LEED) and reflection high energy electron diffraction (RHEED) studies with immersed electrodes showed superstructures in the diffraction patterns which proved the existence of ordered adlayers in upd well before the use of STM in electrochemistry [33, 34].

Metal deposits in submonolayer and monolayer amounts on foreign metal substrates obtained by upd have become a favorable playground for STM, as their manifold structures and structure transitions with changing coverage can be studied in detail and with relative ease by this technique. One example of a high-quality STM image with atomic resolution is given in Figure 19.7, which shows the adlayer of underpotentially deposited Ag on Au(111) in sulfuric acid solution at $\theta = 0.37$ [35]. This adlayer structure has been formed out of a $(\sqrt{3} \times \sqrt{3})$ R30° structure, which exists at slightly more positive electrode potentials (i.e., lower coverage), by a uniaxial compression of the former by about 10%. The resulting lattice misfit between this distorted hexagon structure and the underlying Au(111) substrate results in a height modulation that is clearly reflected in the stripe pattern of Figure 19.7. This image once more conveys the fascination of STM by being able to see individual atoms on surfaces, with an instrument that almost fits into one's jacket pocket (without controller!).

When we apply an electrode potential negative of the equilibrium potential E_r, bulk deposition takes place, the rate of which strongly depends on the so-called overpotential, that is, the difference between the actual and the equilibrium potential. Because metal deposition is a nucleation-and-growth process, surface defects such as steps or screw dislocations are known to play a crucial role as nucleation centers. Their density and local arrangement have a strong influence on the morphology of the deposit, particularly during the initial stages of growth. Because of the ability of STM to image surfaces in *real* space, this technique is especially suited to the study of nucleation at defects and the initial growth of metal

13 x 13 nm²   +400 mV

Figure 19.7 STM image of Ag atoms, regularly adsorbed on a Au(111) electrode in a distorted-hexagon structure. The image was recorded in 0.05 M H_2SO_4 + 1 mM Ag_2SO_4 at +0.4 V versus Ag/Ag$^+$. From Ref. [35].

clusters. Figure 19.8 shows two STM images of an Au(111) surface before and during bulk Cu deposition [34]. The bare gold surface has atomically flat terraces separated by three monoatomic high steps. After a potential step to negative values, deposition of bulk Cu is clearly seen to occur almost exclusively at the monoatomic high steps, the growing clusters virtually decorating the surface defects. It is only after some time that Cu clusters starts to grow on the terraces too.

19.4.4
Nanostructuring of Surfaces with an STM

The past decade has witnessed a second career of the STM: A tool for positioning single atoms and molecules on surfaces with an hitherto unprecedented precision. This was achieved by employing the tip–substrate interaction at close distance (say, of the order of an atomic diameter) to manipulate individual atoms or molecules with the tip, and directing them – one after the other – to predetermined positions.

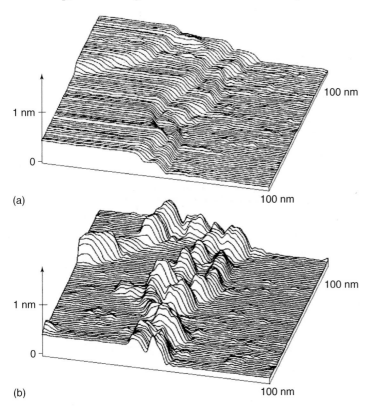

Figure 19.8 STM line-scan images of Au(111) in 5 mM $H_2SO_4 + 0.05$ mM $CuSO_4$ before (a) and during (b) copper electrodeposition. Image (b) demonstrates that nucleation and growth of Cu occurs preferentially at the steps of the substrate surface.

Impressive examples of this kind of nanostructuring of surfaces have been given by several groups working under UHV conditions and mainly at low temperatures [36–38]. Don Eigler's famous quantum corral, a circle of 48 Fe atoms arranged with an STM tip on a Cu(111) surface, is most likely the most frequently cited example of this kind [39].

Soon after surface scientists demonstrated how to use (or misuse) the STM for nanostructuring rather than mere imaging of surfaces, electrochemists started similar work. Their tip-generated entities were clearly larger than single atoms, because constructions from atoms would not have been stable enough to survive an electrochemical environment at room temperature. The most common approach to an electrochemical nanostructuring of surfaces was to create surface defects with the tip, which then acted as nucleation centers for the metal deposition at preselected positions [40]. The defects were often produced by a mechanical contact between tip and sample (so-called tip crash). A more recently developed strategy for positioning small metal clusters on electrode surfaces at will involved a two-step process, in which the metal was first deposited from solution onto the tip, followed by a burst-like dissolution and redeposition onto the sample right underneath the tip [41]. Such a procedure left the surface undamaged, but the cluster size was rather large, for example, in the tens of nanometers range. In the following we will briefly demonstrate a method, based on the so-called *jump-to-contact* between tip and sample [42], which also leaves the surface undamaged, but generates, quickly and reproducibly, metal clusters containing of the order of 100 atoms only.

The proposed mechanism for the tip-induced cluster formation is sketched in Figure 19.9 [43], highlighting the two essentials of this method: First, metal has to be deposited onto the tip, which is achieved simply by choosing a tip potential which is negative enough for this purpose. Secondly, the metal-loaded tip has to approach the surface for a short period of time during which the jump-to-contact occurs. The resulting connective neck breaks upon the subsequent retreat of the

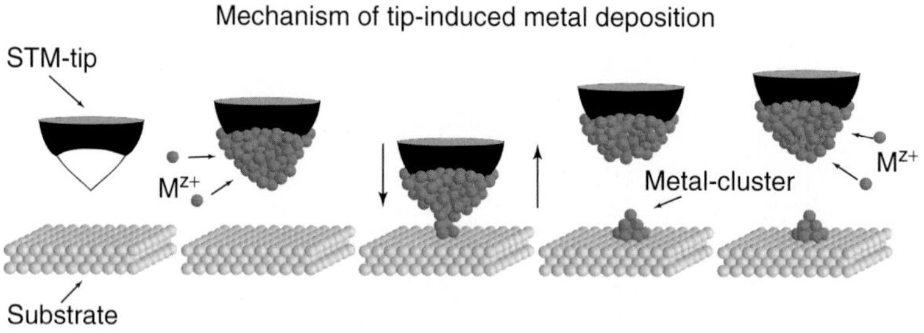

Figure 19.9 Schematic representation of the mechanism that leads to tip-generated metal clusters. First, the metal has to be deposited from solution onto the tip. Secondly, by applying an external voltage pulse to the z-piezo, the tip is made to approach the substrate for a short time for a so-called jump-to-contact to occur, where a metal bridge is formed between tip and substrate. Upon retreat of the tip the connecting bridge breaks, leaving a metal cluster on the substrate. The on-going metal deposition at the tip supplies enough material for cluster generation at a high rate (typically 10–100 clusters per second). From Ref. [43].

tip, leaving a small metal cluster on the surface. The direction of the material transfer depends on the cohesive energy of both sides, the situation depicted in Figure 19.9 being that for, for example, Cu and Au(111) substrate. (When nickel has been deposited onto the tip and the substrate is Au(111), gold will jump to the tip, leaving holes in the substrate.) Because of the negative tip potential, the tip is constantly "reloaded" with metal from solution and then is ready for another cluster formation.

The jump-to-contact requires a tip approach to about 0.3 nm tunnel gap, which is usually achieved by applying an external voltage pulse to the z-piezo. Actually, in our case, all three spatial coordinates of the tip can be externally controlled by a microprocessor, which makes the nanodecoration of an electrode surface with metal clusters a fully-automated process, allowing even complex patterns to be made quickly and reproducibly [44–46]. An example of tip-induced cluster formation is given in Figure 19.10 [44], again referring to Cu on Au(111), as this system has turned out to work best. It demonstrates the ability for complex surface patterning by showing a circle of 12 Cu clusters on Au(111), which is 40 nm in diameter. All Cu clusters are 0.8 nm high. It is important to point out that the electrode potential of the Au(111) during cluster fabrication and imaging has always been +10 mV vs. Cu/Cu^{++}. At such a value, the Au(111) surface is covered by a monolayer of Cu (because of upd), but bulk Cu deposition from solution into Au(111) is not possible. The cluster material originates exclusively from the Cu-covered tip.

Small clusters of catalytically active material are important objects for mechanistic studies in electrocatalysis. The dependence of the catalytic properties on cluster size is of particular interest. In order to determine the electrochemical properties of the tip-generated clusters, many tens of thousands are required for the electrocatalytic

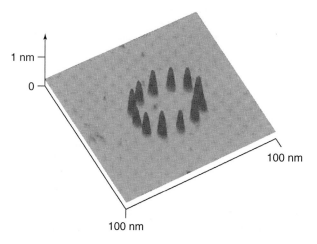

1 nm

0

100 nm

100 nm

Figure 19.10 STM image of 12 Cu clusters, arranged in a circle of 40 nm diameter on a Au(111) electrode. The clusters, which were generated by the tip of an STM, are about 0.8 nm high. From Ref. [44].

1500 x 1500 nm^2 $\qquad\qquad$ E$_{SCE}$ = 660 mV

Figure 19.11 STM image of 10 arrays of 2500 Pd clusters each, on Au(111) in 0.1 M H$_2$SO$_4$ + 1 mM PdSO$_4$. The clusters with an average height of 1 nm were generated by 4 ms pulses at a rate of 5 Hz. From Ref. [47].

reactions to be detectable and discriminated against any parasitic currents at the (electrocatalytically inactive) substrate, the area of which is simply colossal compared to that of the clusters. Figure 19.11 shows 10 fields of Pd clusters on a Au(111) electrode with 2500 clusters each [47]. By employing detector electrodes with sub-micrometer dimensions, measurement of the cluster reactivity becomes feasible (Laubender *et al.* in preparation).

19.5
Conclusion and Perspectives

STM certainly has contributed most significantly to the understanding in the field. Many reactions on the electrode surface, such as the initial stages of metal deposition, have been made visible for the first time. The technique has helped electrochemists to determine adlayer structures, and new insights have been gained into the dynamics surrounding the reconstruction of single crystal surfaces. In this article we have presented an overview of the various questions that can be addressed in electrochemistry with STM. As with any other scientific method, complementary techniques will have to be employed to support and corroborate information acquired by STM. However, STM data have proven to be fundamental to supplying key information and have formed the basis of many new insights, and, hence, STM is nowadays widely recognized as a central and indispensable technique in electrochemistry.

19.6
Supplementary Material

19.6.1
A Personal Recollection of the Time, the STM Came into Being (see also Ref. [48])

In the early 1980s I (the author Dieter M. Kolb) was in Madrid at a surface science conference, where I talked about our electrochemical work. There was a young man, whose name I had never heard before, but whose lecture emptied all parallel sessions, including mine. He described a microscope capable of imaging surfaces in real space with atomic scale resolution. It was the STM and the young man was Gerd Binnig. Those results were absolutely fascinating, and yet I remember the very frankly expressed disbelief of highly regarded surface physicists from the field-ion and -electron microscopy community, who centered the discussion around the STM tip. In order to image with atomic scale resolution, one would need a tip with a single atom as an apex, which would be almost impossible to obtain in such a simple manner like the one described by Binnig, and on a regular basis. But of course, it was not realized at that time that one does not have to worry about this single atom at the tip: Mother nature has given it for free to the brave. The natural roughness and the exponential dependence of the tunnel current ensure that imaging is done by the most protruding apex only.

Some years later, when the STM was firmly established among the UHV techniques, but still before 1986, Binnig, on the occasion of a visit to Berlin, came over to the Fritz-Haber-Institut with his host, Karl-Heinz Rieder, to discuss with me the possible use of STM in electrochemistry. Already at that time Binnig very clearly realized that this method would not only work in UHV, but also under ambient pressure and even in an electrolyte. And indeed, soon after, the first *in situ* measurements were reported by the Hansma group in Santa Barbara, albeit the experiments were done without potential control. And still somewhat later, at the Fritz-Haber-Institut in Berlin, upon initiation by H. Gerischer, two groups, J. Behm and I, joined forces to develop and build an *in situ* STM which could be operated under potential control. At that time Behm had a manually very gifted PhD student, J. Wiechers, who succeeded in building an STM, in designing an electrochemical cell for it, and – the most crucial point – found an insulation material for the tip.

In the late 1980s, we were able to image *in situ* monoatomic high steps, which led us to express the hope that one day in the not-too-distant future we might even be able to image individual atoms. On several occasions it was pointed out to us that this dream would never become true, simply because the electronic corrugation of the metal surface at the position of the STM tip is practically nil. And yet, in 1989, J. Wintterlin at the Fritz-Haber-Institut succeeded in producing an STM image of Al(111) inUHV, where the individual surface atoms could be seen. This was again a major breakthrough, although more in our minds than in experimental details. A few years later, electrochemists were also able to image individual surface atoms. Sometimes I have the feeling that progress in science is hampered more often by mental limits rather than by experimental problems. Once we knew we should be able to see atoms, we saw atoms.

19.6.2
Electron Tunneling

An electron of energy E incident on a barrier of finite height $V > E$ has an oscillating wavefunction outside the barrier, but shows an exponential decay inside. If the barrier is thin enough, the wavefunction is non-zero at the other side, and so oscillations can begin again. Such a penetration through a classically forbidden region is termed *tunneling*. In mathematical terms, this behavior can be described by a wavefunction with an imaginary wave vector $\kappa = \frac{i}{\hbar}\sqrt{2m(V-E)}$, and hence, e^{ikx} changes to e^{-kx}, k being real in both cases.

In Figure 19.12 an energy diagram is sketched for the contact between STM tip and sample. Applying a voltage U_T (tunnel voltage) between both solids causes a difference in the Fermi energies of exactly $e \cdot U_T$. In the figure the situation of a negative potential for the tip is shown and, hence, electrons would tunnel from the tip to the sample. Φ_{tip} and Φ_{sample} are the work functions of the tip and sample, respectively. E_F is the Fermi energy and E_{vac} the vaccum level.

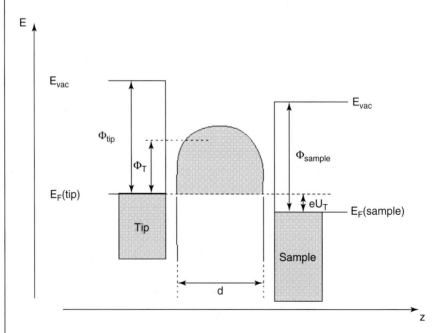

Figure 19.12 Energy diagram for the tip-sample tunneling, Φ_T: mean barrier height; Φ_{tip}, Φ_{sample}: work function of tip and sample, respectively; E_F: Fermi energy; U_T: tunnel voltage.

A quantum mechanical treatment of the tunnel process [49, 50] leads to the following expression for the tunnel current I_T:

$$I_T \sim U_T \exp(-A \cdot \sqrt{\Phi_T} d)$$

$A = 1.025\,\text{eV}^{-1/2}\,\text{Å}^{-1}$, d is, in practice, the tip–sample distance. Typical numbers for Φ_T are 3–4 eV for a vacuum gap, but only 1–2 eV for tunneling at metal/aqueous solution interfaces. From a more detailed treatment of the transfer formalism it becomes evident that I_T is proportional to the local density of states (LDOSs) for the electrons in both solids. Hence, the STM image of a surface represents lines of constant LDOS, which for pure metal surfaces reflects the surface topography in a very direct way.

References

1. Binnig, G. and Rohrer, H. (1982) *Helv. Phys. Acta*, **55**, 726.
2. Binnig, G., Rohrer, H., Gerber, C., and Weibel, E. (1982) *Phys. Rev. Lett.*, **49**, 57.
3. Binnig, G. and Rohrer, H. (1987) *Rev. Mod. Phys.*, **59**, 615.
4. Sonnenfeld, R. and Hansma, P.K. (1986) *Science*, **232**, 211.
5. Binnig, G., Quate, C.F., and Gerber, C. (1986) *Phys. Rev. Lett.*, **56**, 930.
6. Schneeweiss, M.A. and Kolb, D.M. (2000) *Chem. Unserer Zeit*, **34**, 72.
7. Bockris, J.O'M., Devanathan, M.A.V., and Müller, K. (1963) *Proc. R. Soc. London, Series A*, **274**, 55.
8. Habib, M.A. and Bockris, J.O'M. (1980) in *Comprehensive Treatise of Electrochemistry*, Chapter 4, vol. 1 (ed. J.O'M. Bockris), Plenum, New York.
9. Gerischer, H., Kolb, D.M., and Sass, J.K. (1978) *Adv. Phys.*, **27**, 437.
10. Trasatti, S. (1992) in *Electrified Interfaces in Physics, Chemistry and Biology* (ed. R. Guidelli), Kluwer, Dordrecht, p. 229.
11. Lustenberger, P., Rohrer, H., Christoph, R., and Siegenthaler, H. (1988) *J. Electroanal. Chem.*, **243**, 225.
12. Wiechers, J., Twomey, T., Kolb, D.M., and Behm, R.J. (1988) *J. Electroanal. Chem.*, **248**, 451.
13. Bach, C.E., Nichols, R.J., Beckmann, W., Meyer, H., Schulte, A., Besenhard, J.O., and Jannakoudakis, P.D. (1993) *J. Electrochem. Soc.*, **140**, 1281.
14. Wintterlin, J. (1989) Struktur und Reaktivität einer Metalloberfläche – eine Untersuchung mit dem Rastertunnelmikroskop am System Al(111)/Sauerstoff. Dissertation. Freie Universität Berlin.
15. Hoyer, R. (2004) Struktur und Reaktivität bimetallischer Einkristallelektroden der Platin-Metalle. Dissertation. Universität Ulm.
16. Kleinert, M. (2001) Strukturuntersuchungen mit dem Rastertunnelmikroskop an Gold- und Platineinkristall-Elektroden. Dissertation. Universität Ulm.
17. Adžić, R.R., Tripković, A.V., and O'Grady, W.E. (1982) *Nature*, **296**, 137.
18. Kibler, L.A. and Kolb, D.M. (2003) in Handbook of Fuel Cell – Fundamentals, Technology and Applications, vol. 2 (eds W. Vielstich, H.A. Gasteiger and A. Lamm), Wiley, Chichester, pp. 266–278.
19. Dietterle, M. (1996) Untersuchungen zur elektrolytischen Cu-Abscheidung und zur Dynamik von Stufenkanten auf niedrigindizierten Ag-Elektroden: eine *in situ* STM Studie. Dissertation. Universität Ulm.
20. Ertl, G. (1985) *Surf. Sci.*, **152/153**, 328.
21. Van Hove, M.A., Koestner, R.J., Stair, P.C., Bibérian, J.P., Kesmodel, L.L., Bartoš, I., and Somorjai, G.A. (1981) *Surf. Sci.*, **103**, 189 and 218.
22. Dakkouri, A.S. and Kolb, D.M. (1999) in *Interfacial Electrochemistry* (ed. A.

Wieckowski), Marcel Dekker, New York, p. 151.

23. Schneider, J. and Kolb, D.M. (1988) *Surf. Sci.*, **193**, 579.

24. Cuesta, A., Kleinert, M., and Kolb, D.M. (2000) *Phys. Chem. Chem. Phys.*, **2**, 5684.

25. (a) Kibler, L.A., Kleinert, M., Lazarescu, V., and Kolb, D.M. (2002) *Surf. Sci.*, **498**, 175; (b) Kibler, L.A., Kleinert, M., Randler, R., and Kolb, D.M. (1999) *Surf. Sci.*, **443**, 19.

26. Dretschkow, Th. and Wandlowski, Th. (2001) in *Solid/Liquid Interface Properties and Processes – A Surface Science Approach* (ed. K. Wandelt), Springer, Berlin.

27. Schreiber, F. (2000) *Progr. Surf. Sci.*, **65**, 151.

28. Schweizer, M., Hagenström, H., and Kolb, D.M. (2001) *Surf. Sci.*, **490**, L627.

29. Budevski, E., Staikov, G., and Lorenz, W.J. (1996) *Electrochemical Phase Formation and Growth*, Wiley-VCH Verlag GmbH, Weinheim.

30. Kolb, D.M. (1978) in *Advances in Electrochemistry and Electrochemical Engineering*, vol. 11 (eds H. Gerischer and C.W. Tobias), John Wiley & Sons, Inc., New York, p. 125.

31. Lorenz, W.J., Herrmann, H.D., Wüthrich, N., and Hilbert, F. (1974) *J. Electrochem. Soc.*, **121**, 1167.

32. Schultze, J.W. and Dickertmann, D. (1976) *Surf. Sci*, **54**, 489.

33. Nakai, Y., Zei, M.S., Kolb, D.M., and Lehmpfuhl, G. (1984) *Ber. Bunsen-Ges. Phys. Chem.*, **88**, 340.

34. Kolb, D.M. (2002) in *Advances in Electrochemical Science and Engineering*, vol. 7 (eds R.C. Alkire and D.M. Kolb), Wiley-VCH Verlag GmbH, Weinheim, p. 107.

35. Esplandiu, M.J., Schneeweiss, M.A., and Kolb, D.M. (1999) *Phys. Chem. Chem. Phys.*, **1**, 4847.

36. Meyer, G., Zöphel, S., and Rieder, K.H. (1996) *Appl. Phys. A*, **63**, 557.

37. Lyo, I.-W. and Avouris, P. (1991) *Science*, **253**, 173.

38. Cuberes, M.T., Schlittler, R.R., and Gimzewski, J.K. (1997) *Surf. Sci.*, **371**, L231.

39. Crommie, M.F., Lutz, C.P., and Eigler, D.M. (1992) *Science*, **262**, 218.

40. Nyffenegger, R.M. and Penner, R.M. (1997) *Chem. Rev.*, **97**, 1195.

41. Schindler, W., Hofmann, D., and Kirschner, J. (2000) *J. Appl. Phys.*, **87**, 7007.

42. Landman, U., Luedtke, W.D., Burnham, N.A., and Colton, R.J. (1990) *Science*, **248**, 454.

43. Ullmann, R., Will, T., and Kolb, D.M. (1995) *Ber. Bunsen-Ges. Phys. Chem.*, **99**, 1414.

44. Kolb, D.M., Ullmann, R., and Will, T. (1997) *Science*, **275**, 1097.

45. Kolb, D.M., Ullmann, R., and Ziegler, J.C. (1998) *Electrochim. Acta*, **43**, 2751.

46. Kolb, D.M., Engelmann, G.E., and Ziegler, J.C. (2000) *Solid State Ionics*, **131**, 69.

47. Kolb, D.M. and Simeone, F.C. (2005) *Electrochim. Acta*, **50**, 2989.

48. Kolb, D.M. (2002) in *Historical Perspectives on the Evolution of Electrochemical Tools*, Chapter 4 (eds J. Leddy, V. Birss, and P. Vanýsek), ECS Publication, SV 2002-29.

49. Simmons, J.G. (1963) *J. Appl. Phys.*, **34**, 1793, 2581.

50. Tersoff, J. and Hamann, D.R. (1983) *Phys. Rev.*, **B50**, 1998.

20
Low-Energy Electron Diffraction: Crystallography of Surfaces and Interfaces

Georg Held

■ Method Summary

Acronyms, Synonyms
- Low energy electron diffraction (LEED)
- Intensity versus voltage curves (IV curves)
- Intensity versus energy curves (I(E) curves).

Benefits (Information Available)
- technique is surface-sensitive.
- periodicity of surface layers (superstructure) – LEED pattern.
- degree of surface order (e.g., phase transitions, island size) – spot profiles of LEED pattern.
- positions of atoms (± 1–10 pm) in the layers near the surface (<1 nm) – LEED-IV structure determination.

Limitations (Information Not Available)
- not element-specific.
- no information about bulk structure (>1 nm below surface).
- Requires long-range order (limited information about amorphous or random surface structures).

20.1
Introduction

When Clinton Davisson and Lester Germer conducted the very first low-energy electron diffraction (LEED) experiments in April 1925 at Bell Labs in New York it hit them – quite literally – like a lightning stroke: "At that time we were continuing an investigation ... of the distribution in-angle of electrons scattered by a target of ordinary nickel. During the course of this work a liquid-air bottle exploded at a

Methods in Physical Chemistry, First Edition. Edited by Rolf Schäfer and Peter C. Schmidt.
© 2012 Wiley-VCH Verlag GmbH & Co. KGaA. Published 2012 by Wiley-VCH Verlag GmbH & Co. KGaA.

time when the target was at high temperature; the experimental tube was broken, and the target heavily oxidized by the inrushing air. The oxide was eventually reduced and a layer of the target removed by vaporization but only after prolonged heating at various high temperatures in hydrogen and in vacuum. When the experiments were continued it was found that the distribution-in-angle of the scattered electrons had been completely changed" [1]. They added, "We must admit that the results obtained in these experiments have proved to be quite at variance with our expectations."

The prolonged heating treatment had transformed the crystallites of the polycrystalline nickel sample into millimeter size crystals and the intensity distribution of elastically back-scattered electrons now showed sharp maxima instead of the smooth angular distribution before the accident. Davisson and Germer soon realized that these were interference patterns and, thus, the first experimental proof of the wave nature of electrons, which had been postulated only a few years before, in 1923, by Louis De Broglie. He had suggested that electrons have a wave length, which is proportional to the inverse of their momentum $m_e v$:

$$\lambda_e = h/(m_e v) = (1.50\,\text{eV}/E_{kin})^{1/2} \text{ (in nm)} \tag{20.1}$$

and a wave vector of length

$$k_e = 2\pi/\lambda_e = (2\pi/h) \cdot m_e v \tag{20.2}$$

which is proportional to the momentum of the electron (h is Planck's constant, m_e the electron mass, v the velocity, and E_{kin} the kinetic energy of the electron). For low kinetic energies between a few tens and a few hundreds of electron volts (eV) the wavelength is of the order of 0.1 nm, that is, comparable to typical interatomic distances in crystals and molecules, and it was soon realized that the angular interference patterns observed in LEED can be used to determine the structure of well-ordered crystals, in analogy to X-ray diffraction. Due to the small inelastic mean free path of electrons in this energy range, typically around 1 nm, LEED samples only the topmost atomic layers of a crystal and is, therefore, best suited for the analysis of surface geometries. X-ray photons, on the other hand, have a much larger mean free path, typically a few µm. Therefore, X-ray diffraction delivers crystallographic information about the bulk structure of a crystal. Another important difference is that multiple scattering plays an important role in the diffraction process of electrons at solid surfaces, which is not the case for photons. Therefore, the analysis of LEED data with respect to the exact positions of atoms at a surface is somewhat more complicated and requires fully dynamical quantum mechanical scattering calculations.

The use of LEED as a standard technique for surface analysis started in the early 1960s when large enough single crystals and commercial instruments became available for surface studies. At first, the technique was only used for qualitative characterization of surface ordering and the identification of two-dimensional superstructures. The quantitative information about the positions of the atoms within the surface is hidden in the energy-dependence of the diffraction spot intensities, the so-called LEED I-V, or I(E), curves. Computer programs and the computer power to analyze these data became available in the 1970s. With the ever growing speed of modern computers LEED-IV structure determination has been applied to increasingly complex surface structures. To date LEED is the most precise and versatile technique for surface crystallography.

For further information about the history, experimental set-up, and theoretical approaches of LEED refer to the books by Pendry [2], Van Hove and Tong [3], Van Hove et al. [4], and Clarke [5]. The present chapter makes extensive use of these works.

20.2
Basic Principles

The basic principle of a standard LEED experiment is very simple: a collimated mono-energetic beam of electrons is directed toward a single crystal surface and the diffraction pattern of the elastically back-scattered electrons is recorded using a position-sensitive detector. For electrons, as for all wave-like objects, the angular intensity distribution due to the interference of partial waves back-scattered from a periodic array is described by Bragg's law or, more conveniently, by a set of Laue equations, one for each dimension of periodicity, which predict a regular pattern of diffraction spots.

20.2.1
Surface Periodicity and Reciprocal Lattice

Because of the short penetration depth of low-energy electrons the diffraction process is determined by a small number of atomic layers at the crystal surface. The electrons do not probe the full crystal periodicity perpendicular to the surface. Therefore, the array of relevant scatterers is only periodic in two dimensions. The surface lattice can be described by a pair of lattice vectors a_1 and a_2, which are parallel to the surface plane, and the surface unit cell, that is, the contents of the parallelogramm spanned by a_1 and a_2. The surface consists of identical copies of the unit cell at every point

$$\mathbf{R} = m_1\, \mathbf{a_1} + m_2\, \mathbf{a_2} \tag{20.3}$$

with integer numbers m_1 and m_2. The left-hand side of Figure 20.1 illustrates common square, rectangular, and hexagonal surfaces and the lattice vectors defining their unit cells.

The two-dimensional Laue equations are based on reciprocal lattice vectors within the surface plane which are defined by the real space lattice vectors through a set of four simultaneous equations:

$$\mathbf{a_1} \cdot \mathbf{a^*}_1 = 2\pi \quad \mathbf{a_2} \cdot \mathbf{a^*}_2 = 2\pi \tag{20.4a}$$

$$\mathbf{a_1} \cdot \mathbf{a^*}_2 = 0 \quad \mathbf{a_2} \cdot \mathbf{a^*}_1 = 0 \tag{20.4b}$$

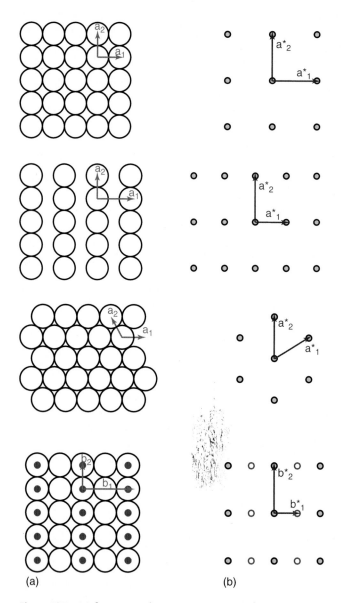

(a)   (b)

Figure 20.1 (a) from top to bottom, arrangement of atoms in the {100} (square), {110} (rectangular), and {111} (hexagonal) surfaces of a simple face centered cubic crystal lattice and a p(2 × 1) superstructure on a square surface; the diagrams include lattice vectors defining the surface unit cell and the corresponding reciprocal lattices (b).

In order for the scalar products in Equation 20.4a to be dimensionless, the reciprocal lattice vectors must have units of inverse length, nm^{-1}. As a consequence of Equation 20.4b a^*_2 and a^*_1 must be perpendicular to a_1 and a_2, respectively, which means that a rectangular real-space lattice will also have a rectangular reciprocal lattice. For non-rectangular lattices the angles are different in real space and reciprocal space. The right-hand column of Figure 20.1 shows the corresponding reciprocal lattices for each of the surfaces on the left. The reciprocal lattice vectors define the positions of the diffraction maxima through the *Laue equation* (Equation 20.5).

$$k_{||,out} (n_1, n_2) = k_{||,in} + n_1 a^*_1 + n_2 a^*_2 \tag{20.5}$$

$k_{||,out}$ is the component of the wave vector of the diffracted electrons, which is parallel to the surface plane (by convention, this is the *xy*-plane). $k_{||,in}$ is the parallel component of the wave vector of the incoming electron beam. Note that the Laue equation (Equation 20.5) defines a two-dimensional vector, hence it actually comprises two equations, one for each component. Each diffraction spot corresponds to the sum of integer multiples of a^*_1 and a^*_2. The integer numbers (n_1, n_2) are used as indices to label the spots.

Energy conservation demands that the length of the **k**-vector is the same, $(2m_e E_{kin}/h^2)^{1/2}$, for both the incoming and the elastically scattered electron wave. This defines the vertical or *z*-component, $k_{z,out}$, of the back-diffracted electrons in the (n_1, n_2) spot:

$$k_{z,out} (n_1, n_2) = \left[2m_e E_{kin}/h^2 - |k_{||,out} (n_1, n_2)|^2 \right]^{1/2} \tag{20.6}$$

Note that, unlike for X-ray diffraction, there is no Laue-condition for the *z*-component of k_{out}. The only condition for diffraction into a spot (n_1, n_2) is that $k_{z,out}$ has a real value, that is, the argument of the square root ($[]^{1/2}$) on the right-hand side of Equation 20.6 must not be less than zero. This condition is synonymous with the obvious fact that the length of the parallel component of **k** cannot be greater than the length of the entire vector, but it also limits the number of observable LEED spots. The number of observable spots increases with increasing electron energy while the polar emission angle with respect to the specular spot $(0,0)$ decreases for each spot. This is illustrated in Figure 20.2 for normal incidence $(k_{||,in} = 0)$; in this case $k_{||,out} (n_1, n_2) = n_1 a^*_1 + n_2 a^*_2$ is

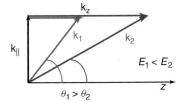

Figure 20.2 Relationship between k_z, $k_{||}$, and the emission angle for a diffracted electron wave at two different energies.

constant for a given pair of spot indices and only $k_{z,out}$ is affected by changes in the kinetic energy.

Only the specular spot does not change its position as a function of energy if the angle of incidence is kept constant.

20.2.2
Superstructures

Superstructures formed by adsorbates or rearrangements of the surface atoms can lead to a periodicity of the surface lattice greater than that of the bulk-truncated single crystal. In these cases, the lattice vectors for the superstructure, $\mathbf{b}_1$ and $\mathbf{b}_2$, can always be related to the lattice vectors of the bulk-truncated surface, $\mathbf{a}_1$ and $\mathbf{a}_2$, through

$$\mathbf{b}_1 = m_{11}\,\mathbf{a}_1 + m_{12}\,\mathbf{a}_2$$
$$\mathbf{b}_2 = m_{21}\,\mathbf{a}_1 + m_{22}\,\mathbf{a}_2 \tag{20.7}$$

the numbers m_{ij} are the coefficients of the *superstructure matrix* $M = [m_{11}\ m_{12};\ m_{21}\ m_{22}]$, which is a straightforward way of characterizing any superstructure. Depending on whether all m_{ij} are interger numbers or not the superstructure is either called *commensurate* or *incommensurate*. Superstructures lead to additional spots in the LEED pattern, for which fractional indices are used. The reciprocal lattice vectors for these spots can be calculated directly from the coefficients of the superstructure matrix according to the following set of equations [4]:

$$\mathbf{b}^*_1 = (m_{11} \cdot m_{22} - m_{12} \cdot m_{21})^{-1} \cdot (m_{22}\,\mathbf{a}^*_1 - m_{21}\,\mathbf{a}^*_2)$$
$$\mathbf{b}^*_1 = (m_{11} \cdot m_{22} - m_{12} \cdot m_{21})^{-1} \cdot (-m_{12}\,\mathbf{a}^*_1 - m_{11}\,\mathbf{a}^*_2) \tag{20.8}$$

The fractional indices of the superstructure spots are multiples of the prefactors of $\mathbf{a}^*_1$ and $\mathbf{a}^*_2$ in Equation 20.6

Another, less general notation according to Wood [6] specifies the lengths of the vectors $\mathbf{b}_1$ and $\mathbf{b}_2$ in units of $\mathbf{a}_1$ and $\mathbf{a}_2$, respectively, together with the rotation angle α between $\mathbf{b}_1$ and $\mathbf{a}_1$ (only specified if α is not zero):

$$p/c\ (|\mathbf{b}_1|/|\mathbf{a}_1| \times |\mathbf{b}_2|/|\mathbf{a}_1|)\ R\alpha \tag{20.9}$$

p indicates a "primitive" and c a "centred" surface unit cell. Examples are "$p(2 \times 1)$," "$p(\sqrt{3} \times \sqrt{3})\ R30°$," and "$c(2 \times 2)$." This notation is not applicable to all superstructures but it is more frequently used than the matrix notation because it is shorter. As an example, a $p(2 \times 1)$ superstructure on a square substrate surface is shown at the bottom of Figure 20.1. The corresponding superstructure matrix is $[2\ 0;\ 0\ 1]$ and the reciprocal lattive vectors are $\mathbf{b}^*_1 = 1/2 \cdot \mathbf{a}^*_1$ and $\mathbf{b}^*_2 = \mathbf{a}^*_2$.

20.2.3
Spot Intensity versus Energy

There is no Laue-condition for the z component of $\mathbf{k}_{out}$, that is, diffraction spots are allowed for a wide range of kinetic energies. This does not mean, however, that the intensities of spots are constant with the energy. Although the electrons do not experience the full periodicity of the crystal perpendicular to the surface, there is still interference of electrons scattered from different atomic layers parallel to the surface. For infinite penetration depth this would impose a third Laue condition for $k_{z,out}$ and therefore each (n_1,n_2) spot would have sharp intensity maxima ("*Bragg peaks*") for certain values of E_{kin} and zero intensity for all other energies. Since the penetration depth is very small, the back-scattered electrons only interact with a few layers of atoms, giving rise to broad maxima at the Bragg peak positions and non-zero intensities in the intermediate energy regimes of the intensity versus energy curve of each spot (also known as *intensity versus voltage* or *IV curve*). The combination of non-periodic layer distances near the surface, different atomic scattering potentials and multiple scattering events leads to shifts in the Bragg peaks and intensity maxima at other energies in the IV curves. All these effects are reproduced by fully dynamical quantum mechanical scattering calculations [2, 3]. An example is given in Figure 20.7.

20.2.4
Spot Profiles

While the spot positions and intensities carry information about the size and the local geometry within the surface unit cell, the spot profile, that is, the shape and width of a diffraction spot, is determined by the long-range relative arrangement of the unit cells at the surface. Vertical displacements of the surface unit cells (e.g., steps, facets) lead to split spots and changes in the spot profile as a function of electron energy. If all surface unit cells are in the same plane (over a length of at least 10 nm, which is a typical coherence width of LEED instruments), the spot profile does not change with energy.

A periodic arrangement of equal steps at the surface causes spot splitting at energies, which lead to destructive interference between electrons reflected from adjacent terraces ("out-of-phase condition", see Figure 20.3(b)). By measuring these energies the step height can be determined directly. For a more random arrangement of steps the analysis of energy-dependent changes in the spot profiles allows, in many cases, the determination of the mean step height and a characterization of the step distribution [7, 8]. Facets lead to extra spots which move in $k_{\parallel}$ upon changes in the kinetic energy.

Point defects, static disorder, and thermally induced displacements lead to an increase in the background intensity between the spots. Depending on the correlation between the scatterers, the background is either homogeneous (no correlation) or structured (correlation). If the coherently ordered surface areas (islands, domains) are small (<10 nm) and at the same vertical height, the width of these areas, Δw, is inversely proportional to the width of the LEED spots, $|\Delta k_{\parallel}|$:

$$|\Delta k_{\parallel}| = 2\pi/\Delta w \tag{20.10}$$

This relation holds independently for each direction parallel to the surface. It is particularly useful for determining the size of adsorbate islands which lead to extra

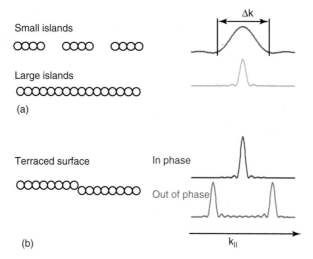

Figure 20.3 Effect of island size on the spot profile (a) and spot splitting induced by regular steps (b); in phase: constructive interference between electrons reflected from adjacent terraces; out of phase: destructive interference. (According to [9].)

superstructure spots (see Figure 20.3(a)). A good introduction (in German) into spot profile analysis is given in the book by Henzler and Göpel [9].

20.3
Experiment

The standard modern LEED system is of the "rear view" type, which is schematically depicted in Figure 20.4. The incident electron beam, accelerated by the potential V_0, is emitted from the electron gun behind a transparent hemispherical fluorescent screen and hits the sample through a hole in the screen. Typically, the electron beam has a current of around $1\,\mu A$ and a diameter of 0.5–1 mm. The surface is in the center of the hemisphere so that all back-diffracted electrons travel toward the LEED screen on radial trajectories.

Before the electrons hit the screen they have to pass a retarding field energy analyzer (RFA). This consists of four (sometimes three) hemispherical grids concentric with the screen, each containing a central hole, through which the electron gun is inserted. The first grid (nearest to the sample) is connected to earth ground as is the sample, in order to provide a field-free region between the sample and this grid. A negative potential $-(V_0 - \Delta V)$ is applied to the second and third grid, the so-called suppressor grids. These repel all electrons that have undergone non-elastic scattering processes and have lost more than $e\Delta V$ (typically around 5 eV) of their original kinetic energy. Thus, only elastically scattered electrons and those with small energy losses can pass through to the fluorescent screen. The

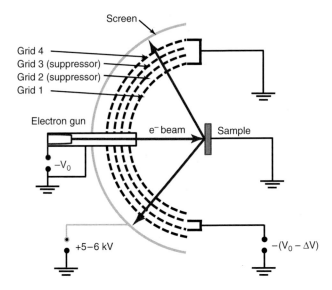

Figure 20.4 Schematic diagram of a typical LEED instrument.

fourth grid is usually on ground potential in order to reduce field penetration of the screen voltage to the suppressor grids. The screen is at a potential of the order of 5–6 kV; it provides the electrons with enough energy to make the diffraction pattern visible on the fluorescent screen. The pattern can be observed through a view-port from behind the transparent screen. Only the electron gun assembly (diameter <15 mm) limits the view slightly.

MCP-LEED systems with position sensitive "microchannel plate" (MCP) electron multipliers between the RFA grids and the fluorescent screen have become commercially available in recent years for applications that require low incident beam currents, either to avoid beam damage (e.g., organic molecules) or charging of insulating samples (e.g., oxides). These systems can be operated with electron currents as low as 1 nA. Typical LEED systems have diameters of around 140 mm.

The LEED pattern is recorded using a video camera with suitable image processing software. As with all methods that use electrons as probes, vacuum conditions are required because electrons cannot penetrate a gas atmosphere at normal pressures. In general, however, the vacuum conditions required to avoid contamination of clean surfaces are more rigorous (typically $<10^{-9}$ m bar) than those imposed by the use of electrons (typically $<10^{-6}$ m bar).

20.4
Applications

In this section we will discuss a small selection of typical applications of LEED in order to illustrate the different levels at which this technique yields information about surface geometries.

20.4.1
Leed Pattern: CO on Ni{111}

The adsorption of carbon monoxide on the {111} surface of nickel is a good example of how LEED diffraction patterns can be used for a simple characterization of adsorbate structures. With increasing coverage of CO adsorbed on Ni{111} four different LEED patterns are observed between about 0.30 and 0.62 ML (1 ML corresponds to 1 molecule per substrate surface atom):

- a diffuse [2 1; −1 1] or $p(\sqrt{3} \times \sqrt{3})$ $R30°$ pattern between 0.3 and 0.4 ML,
- a sharp [2 0; 1 2] or $c(2 \times 4)$ pattern for coverage around 0.5 ML,
- a sharp [3 1; −1 2] or $p(\sqrt{7} \times \sqrt{7})$ $R19°$ pattern between 0.56 and 0.60 ML,
- a more complicated [3 2; −1 2] pattern at the maximum coverage of 0.62 ML, which is described as "$c(2\sqrt{3} \times 4)rect$" in non-standard Wood notation.

Images of the first three patterns are depicted in Figure 20.5a together with the corresponding real-space unit cells (red arrows and dashed lines) Figure 20.5b. The middle row shows the complete (2×4) unit cell (in black). Note that the "c" in the Wood notation $c(2 \times 4)$ means that the center and the corners of the (2×4) unit cell are lattice points. Therefore the primitive unit cell is only half the size, as indicated by the red arrows. The matrix notation always refers to the primitive unit cell. The yellow arrows in the LEED patterns indicate the reciprocal lattice vectors corresponding to the unit cells marked in red.

For the $c(2 \times 4)$ and $p(\sqrt{7} \times \sqrt{7})$ $R19°$ structures it is not possible to reach all diffraction spots by adding integer multiples of these two vectors. This is because the observed pattern is a superposition of LEED patterns arising from different parts of the surface, where the ordered arrangements of molecules are the same, in principle, but may have different orientations. Such *rotation or mirror domains* are usually observed if the superstructure has lower symmetry than the underlying substrate alone. Any symmetry operation of the substrate surface (rotation or mirror) that is not shared with the superstructure will therefore convert the superstructure unit cell into a unit cell that is equivalent but has a different orientation. This new unit cell has a different reciprocal lattice with a new set of diffraction spots. All orientation domains are equivalent and will, therefore, cover equal areas of the surface. In the case of the $c(2 \times 4)$ superstructure, which has a rectangular unit cell, the missing symmetry is the threefold rotation of the hexagonal substrate surface; therefore, there are two additional rotational domains, indicated in green, each of which gives rise to a separate set of diffraction spots. The $p(\sqrt{7} \times \sqrt{7})$ $R19°$ superstructure has a threefold rotation symmetry but does not share the mirror symmetry plane with the substrate (dashed line) this leads to an extra mirror domain, again indicated in green, with a set of extra diffraction spots.

If the adsorbate coverage is known from other methods, as in the present example, it is straightforward to work out the number of molecules per unit cell: there is one molecule in the $p(\sqrt{3} \times \sqrt{3})$ $R30°$ unit cell (coverage 1/3), two in

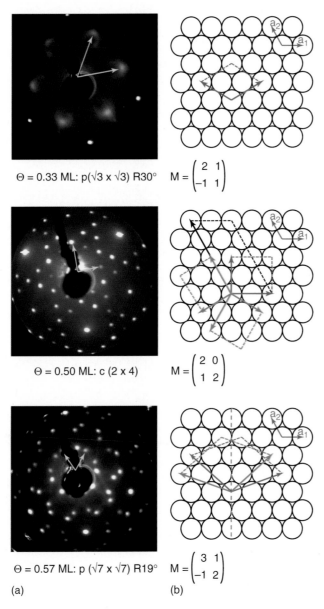

$\Theta = 0.33$ ML: $p(\sqrt{3} \times \sqrt{3})$ R30° $M = \begin{pmatrix} 2 & 1 \\ -1 & 1 \end{pmatrix}$

$\Theta = 0.50$ ML: $c(2 \times 4)$ $M = \begin{pmatrix} 2 & 0 \\ 1 & 2 \end{pmatrix}$

$\Theta = 0.57$ ML: $p(\sqrt{7} \times \sqrt{7})$ R19° $M = \begin{pmatrix} 3 & 1 \\ -1 & 2 \end{pmatrix}$

(a)   (b)

Figure 20.5 Experimental LEED patterns formed by CO adsorbed on Ni{111} (a) and corresponding real-space unit cells (b): $p(\sqrt{3} \times \sqrt{3})$ R30° (top, $E_{kin} = 98$ eV), $c(2 \times 4)$ (middle, $E_{kin} = 129$ eV), and $p(\sqrt{7} \times \sqrt{7})$ R19° (bottom, $E_{kin} = 117$ eV). Note that real space diagrams are rotated by about 30° with respect to the crystal orientation of the experiment; the dark structure extending from the top left to the middle of the LEED patterns is the shadow of the electron gun [10].

the $c(2 \times 4)$ (coverage 2/4) and four molecules in the $p(\sqrt{7} \times \sqrt{7})$ $R19°$ unit cell (coverage 4/7).

The diffraction spots of the $p(\sqrt{3} \times \sqrt{3})$ $R30°$ pattern are significantly broader than those of the other structures. This indicates that the ordered domains are considerably smaller than the coherence or transfer width of the LEED system. The radial spot width is about one-fifth of the length of the reciprocal lattice vectors, therefore the corresponding width of the domains is on average about five unit cells or 2 nm.

20.4.2
Spot Profiles

Figure 20.6 shows an example of energy-dependent changes in the spot profiles of terraced surfaces [8]. The data were collected from a vicinal Pd{100} surface, which is tilted by $1.1°$ with respect to the (100) plane. This leads to terraces with

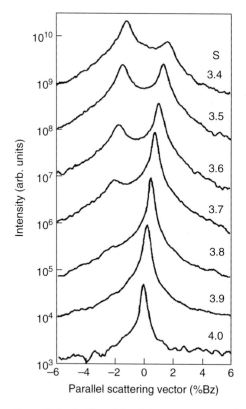

Figure 20.6 Profiles of the (0,0) spot from a terraced (vicinal) Pd{100} surface recorded perpendicular to the step edges. The abscissa units are percent fractions of $|a_2{}^*|$. Reprinted from Surface Science, 396, Wollschläger *et al.*, "Diffraction spot profile analysis for vicinal surfaces with long-range order", 94, Copyright (1998), with permission from Elsevier [8].

(100) orientation, as at the top of Figure 20.1, separated by steps parallel to the [011] direction (vector a_1 in the top diagram of Figure 20.1). The scan direction for the spot profiles is perpendicular to the step edges, that is, along a_2 in real space or a_2^* in reciprocal space, respectively. The abscissa units of Figure 20.6 are percent fractions of $|a_2^*|$. The parameter S is a dimensionless quantity, which is proportional to $k_{z,out}$ (n_1, n_2) and, hence, depends on the electron energy through Equation 20.6. S describes the phase difference between electron waves emerging from different terraces in a convenient way: an integer value of S indicates the in-phase condition or constructive interference for all terraces, whereas an integer value plus 0.5 corresponds to maximum destructive interference (out-of-phase condition) between terraces separated by mono-atomic steps.

The spot profile changes very dramatically from a single sharp peak at the expected spot position (0) for $S = 4.0$ to a double peak with a minimum at the actual spot position for $S = 3.5$. The separation between the two peaks is 2.8% of $|a_2^*|$, therefore the average terrace width in this direction is $(0.028)^{-1}$. $|a_2| = 36 |a_2|$ or 9.8 nm, which is the value expected for a tilt angle of $1.1°$. (Note that the factor 2π in Equation 20.10 is not needed when $|\Delta k_{||}|$ and Δw are expressed as multiples of real and reciprocal lattice vectors.) By fitting the peak shape, additional information about the width distribution and roughness of the surface can be obtained, which is described in detail by Wollschläger *et al.* [8].

Obviously, this kind of information can also be obtained by scanning probe microscopy (scanning tunneling microscopy, STM; atomic force microscopy, AFM) with less sophisticated data analysis. The advantage of LEED spot profile analysis is that the data acquisition is fast and can easily be performed while the surface undergoes structural changes (e.g., varying temperature, during adsorption). LEED also provides an average over much larger surface areas (typically 1 mm^2) than microscopic techniques can normally image simultaneously.

20.4.3
LEED-IV Structure Determination

As discussed in Section 20.2.3, the three-dimensional arrangement of atoms within the unit cell is responsible for the spot-intensity variations as a function of electron energy, the LEED-IV curves. Modern electron scattering programs reproduce all features observed in LEED-IV curves, however, the dominance of multiple scattering in electron diffraction does not normally permit determination of the surface geometry directly from a set of experimental IV curves. Instead, LEED-IV structure determination works on the principle of "trial and error." Theoretical IV curves are calculated for a large number of model geometries and compared with the corresponding experimental curves. The agreement is quantified by the means of a reliability factor or R-factor. There are several ways of defining such R-factors [4] with Pendry's R-factor, R_P, being the most common one [11]. By convention, R_P is 0 when the agreement is perfect and 1 for uncorrelated sets of IV curves. Usually, automated search procedures are used, which modify the model geometries until an *R*-factor minimum is found. The geometry with the lowest *R*-factor is the result of the structure determination.

The level of precision in the resulting crystallographic data depends on the lowest *R*-factor achieved and the total energy range of overlapping experimental and theoretical IV curves. The energy overlap is typically between 1000 and 3000 eV, depending on the number of observable spots. Typically, R_P values of around 0.1 can be expected for clean close-packed metal surfaces, for more complex metal and semiconductor surfaces and adsorption structures of simple molecules one can reach R_P-factors of around 0.15–0.25, and 0.25–0.35 for more complex molecular superstructures. The main reason for the gradually worse agreement between theoretical and experimental IV curves as the surface structures become more complex lies in the approximations in conventional LEED theory, which treat the atoms as perfect spheres with constant scattering potential in between ("muffin-tin potential"). This description is somewhat inaccurate for the scattering potential of more open surfaces and organic molecules. As a consequence, a precision of 1–2 pm can be achieved for atoms in close-packed metal surfaces, whereas the positions of atoms within organic molecules are typically determined within ±10–20 pm. The coordinates perpendicular to the surface are usually more precise than those parallel to the surface plane, because the main scattering direction is perpendicular to the surface.

Examples of experimental and best-fit theoretical IV curves for one of the previous examples, the $c(2 \times 4)$ structure of CO on Ni{111}, are shown at the top of Figure 20.7 [12]. The graph also lists the individual *R*-factors for each pair of theoretical and experimental IV-curves. The geometry with the lowest average *R*-factor, 0.172 (average weighted with the energy range of each individual IV curve), is shown at the bottom of Figure 20.7. The unit cell contains two CO molecules adsorbed on two different threefold hollow sites. The coordinates of the molecules and the first two layers of Ni atoms were determined within the structure analysis. The precision for the coordinates of the Ni atoms is between 3 pm (z) and 9 pm (x,y). Carbon and oxygen atoms are weak scatterers, therefore, their contribution to the intensity variations in the IV curves is smaller than that of the Ni atoms and, consequently, their coordinates are less precise, between 4 pm (z) and 20 pm (x,y).

Owing to the vast increase in available computer power, close to a thousand surface structures have been determined in the last three decades, the majority of which were clean metal and semiconductor surfaces and adsorbate structures of atoms and small molecules. Two review articles by Heinz *et al.* [13] and Over [14] provide good overviews and discussions of LEED structure determinations of clean and adsorbate-covered surfaces and further references. The "NIST Surface Structure Data Base" compiled by Watson *et al.* contains a complete list of all structures up to 2002 [15].

More recently, the capabilities of LEED-IV structure determination have been significantly extended to solve more complex surface structures, such as those of quasi-crystals [16], graphene overlayers [17], and adsorption structures of important organic molecules such as benzene [18] and C_{60} [19].

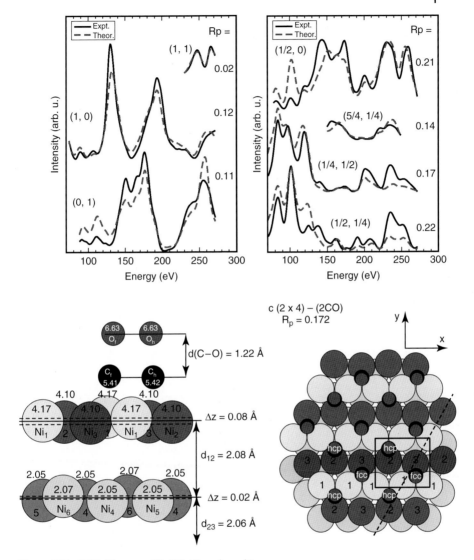

Figure 20.7 LEED-IV curves (70–270 eV) and resulting surface geometry of the $c(2 \times 4)$ superstructure of CO on Ni{111} [12].

20.4.4
LEED-IV on Disordered Layers

Usually, LEED-IV structure determination of layers of adsorbed atoms or molecules requires single crystal surfaces with long-range ordered adsorbate layers. Structural information for adsorbate-covered surfaces without long-range order can be obtained, however, in a similar way, when the energy dependence of the diffusely scattered intensity is analyzed (Diffuse LEED [20, 21]) or from the IV curves of integer-order spots, which are still observed even if the adsorbate layer is not ordered [22, 23]. In both cases, however, the data analysis must assume that the local adsorption geometry is the same for all adsorbates. The main problem in both approaches is the amount of data (energy overlap) available for the analysis. This problem can be solved by recording data for different angles of incidence. If the difference in incidence angles is sufficient, each angle will provide an independent set of IV curves, which can greatly improve the reliability and precision of the structure determination [24].

20.5
Conclusions and Perspectives

LEED is the most accurate and powerful experimental technique for surface crystallography at a level of precision that enables the chemical characterization of interatomic bonds. Often, scanning probe microscopy (e.g., STM, AFM) is seen as an alternative because it yields direct real-space images of surface structures at the atomic level, but the two techniques are really almost complementary. Scanning probe microscopy allows fast data acquisition and interpretation and the study of individual features, regardless of the degree of order, but it cannot deliver direct structural information about the three-dimensional arrangement of atoms at the pm level, in particular, not for atoms below the outermost surface layer. LEED can deliver precise crystallographic data but is restricted to relatively well-ordered surface structures. The results always reflect the properties of a large ensemble of surface unit cells. photoelectron diffraction (PED) [25] and surface X-ray diffraction (SXRD) [26] are related surface sensitive electron and X-ray diffraction methods, which deliver crystallographic information at a similar level of accuracy. PED is element specific and does not require long-range order; SXRD also works under high-pressure conditions where electrons cannot be used. These methods, however, require synchrotron radiation and are, therefore, not as readily available as LEED systems, which are part of the standard equipment of most surface science laboratories.

In the previous sections we have highlighted only a small fraction of the research that can be carried out by LEED with an emphasis on simplicity in order to explain the basics of the technique. A number of recent innovations have opened up the technique to a variety of technically important surface and interface systems with relevance to biology and nano-electronics.

Many of the recent developments in *LEED-IV structure determination* were directed toward improving the model calculations involved in the data analysis. This includes approximations that replace parts of the full quantum mechanical scattering calculations and thus speed up the optimization process (e.g., "Tensor LEED" [27], "molecular T matrix approach" [28]), "direct methods" aiming at a direct conversion of IV curves into a three-dimensional structure [29], and better mathematical descriptions of scattering potentials and thermal vibrations of semiconductors and organic molecules. To date, computer power is only a limiting factor for very large unit cells with many (>20) geometrical parameters to be optimized. The determination of a medium size structure can be performed on a modern personal computer within a matter of hours or a few days. Often the lack of sufficient experimental data for comparison with model calculations is a more severe limitation for the analysis of more complex surface structures with large unit cells. This limitation can be overcome by recording IV curves at different angles of incidence, each creating an additional set of data [24]. The sum of these improvements enables the accurate characterization of structures at the interfaces between inorganic substrates and large organic molecules, as found in biological interfaces or organic electronic devices, and thus open exciting new applications for surface and interface crystallography by LEED.

Another exciting perspective is offered by the low-energy electron microscope (LEEM), a combination of imaging electron microscope and LEED. This microscopic technique was developed by Bauer and Telieps already in the 1960s and 1970s [30, 31] but has become widely available only in the last decade or so. The combination of imaging and diffraction allows the characterization of surface areas of the size of micrometers to nanometers. One application, often referred to as "*micro-LEED*," is the collection of LEED-IV data from an area of a few micrometers in diameter or less. In this way, surface structures of single domains on single crystal surfaces [32], artificial nanostructures of semiconductor devices, or crystallites of polycrystalline material [33] can be determined, which enables surface structure determination for completely new classes of materials with a wide range of applications.

References

1. Davisson, C. and Germer, L.H. (1927) *Phys. Rev.*, **30**, 705.
2. Pendry, J.B. (1974) *Low Energy Electron Diffraction*, Academic Press, London.
3. Van Hove, M.A. and Tong, S.Y. (1979) *Surface Crystallography by LEED*, Springer, Berlin.
4. Van Hove, M.A., Weinberg, W.H., and Chan, C.-M. (1986) *Low-Energy Electron Diffraction*, Springer, Berlin.
5. Clarke, L.J. (1985) *Surface Crystallography – An Introduction to Low Energy Electron Diffraction*, John Wiley & Sons, Ltd, Chichester.
6. Wood, E.A. (1964) *J. Appl. Phys.*, **35**, 1306.
7. Henzler, M. (1977) in *Electron Spectroscopy for Surface Analysis* (ed. H. Ibach), Springer, Berlin, p. 117.
8. Wollschläger, J., Schäfer, F., and Schröder, K.M. (1998) *Surf. Sci.*, **396**, 94.
9. Henzler, M. and Göpel, W. (1991) *Oberflächenphysik des Festkörpers*, Teubner, Stuttgart.
10. Held, G., Schuler, J., Sklarek, W., and Steinrück, H.-P. (1998) *Surf. Sci.*, **398**, 154.
11. Pendry, J.B. (1980) *J. Phys. C*, **13**, 937.
12. Braun, W., Steinrück, H.-P., and Held, G. (2005) *Surf. Sci.*, **575**, 343.
13. Heinz, K. (1994) *Surf. Sci.*, **299–300**, 433.
14. Over, H. (1998) *Prog. Surf. Sci.*, **58**, 249.
15. Watson, P.R., Van Hove, M.A., and Hermann, K. (2002) NIST Surface Structure Database: Version 5.0.
16. Ferralis, N., Pussi, K., Cox, E.J., Gierer, M., Ledieu, J., Fisher, I.R., Jenks, C.J.,

Lindroos, M., McGrath, R., and Diehl, R.D. (2004) *Phys. Rev. B*, **69**, 153404.

17. Moritz, W., Wang, B., Bocquet, M.-L., Brugger, T., Greber, T., Wintterlin, J., and Günther, S. (2010) *Phys. Rev. Lett.*, **104**, 136102.

18. Held, G., Braun, W., Steinrück, H.-P., Yamagishi, S., Jenkins, S.J., and King, D.A. (2001) *Phys. Rev. Lett.*, **87**, 216102.

19. Li, H.I., Pussi, K., Hanna, K.J., Wang, L.-L., Johnson, D.D., Cheng, H.-P., Shin, H., Curtarolo, S., Moritz, W., Smerdon, J.A., McGrath, R., and Diehl, R.D. (2009) *Phys. Rev. Lett.*, **103**, 056101.

20. Heinz, K., Starke, U., and Bothe, F. (1991) *Surf. Sci. Lett.*, **243**, L70.

21. Heinz, K., Starke, U., Van Hove, M.A., and Somorjai, G.A. (1992) *Surf. Sci.*, **261**, 57.

22. Poon, H.C., Weinert, M., Saldin, D.K., Stacchiola, D., Zheng, T., and Tysoe, W.T. (2004) *Phys. Rev. B*, **69**, 035401.

23. Braun, W. and Held, G. (2005) *Surf. Sci.*, **594**, 203.

24. Held, G., Wander, A., and King, D.A. (1995) *Phys. Rev. B*, **51**, 17856.

25. Woodruff, D.P. (2007) *Surf. Sci. Rep.*, **62**, 1.

26. Feidenhans'l, R. (1989) *Surf. Sci. Rep.*, **10**, 105–188.

27. Rous, P.J. (1992) *Prog. Surf. Sci.*, **39**, 3.

28. Blanco-Rey, M., de Andres, P., Held, G., and King, D.A. (2005) *Surf. Sci.*, **579**, 89.

29. Seubert, A., Saldin, D.K., Bernhardt, J., Starke, U., and Heinz, K. (2000) *J. Phys.: Condens. Matter*, **12**, 5527.

30. Bauer, E. (1994) *Rep. Prog. Phys.*, **57**, 895.

31. Bauer, E. (1998) *Surf. Rev. Lett.*, **5**, 1275.

32. de la Figuera, J., Puerta, J.M., Cerda, J.I., El Gabaly, F., and McCarty, K.F. (2006) *Surf. Sci.*, **600**, L105.

33. Cornish, A., Eralp, T., Shavorskiy, A., Bennett, R.A., Held, G., Cavill, S.A., Potenza, A., Marchetto, H., and Dhesi, S.S. (2010) *Phys. Rev. B*, **81**, 085403.

Part IV
Biomolecules and Materials

Methods in Physical Chemistry, First Edition. Edited by Rolf Schäfer and Peter C. Schmidt.
© 2012 Wiley-VCH Verlag GmbH & Co. KGaA. Published 2012 by Wiley-VCH Verlag GmbH & Co. KGaA.

21
Femtosecond Vibrational Spectroscopies and Applications to Hydrogen-Bond Dynamics in Condensed Phases

Jörg Lindner and Peter Vöhringer

▪ Method Summary

Acronyms, Synonyms
- Infrared (IR)
- Coherent anti-Stokes Raman-scattering (CARS)
- Optical Kerr effect (OKE)
- Impulsive stimulated Raman scattering (ISRS)
- Optical heterodyning (OHD)
- Ultraviolet, visible (UV/VIS)
- Molecular dynamics (MD).

Benefits (Information Available)
- techniques expose molecular dynamics on femtosecond time scales
- vibrational phase and population relaxation (IR pump–probe)
- diagonal and off-diagonal vibrational anharmonicities (IR pump–probe)
- molecular reorientation dynamics (transient anisotropy)
- dynamics and mechanisms of vibrational line broadening (IR pump–probe)
- kinetics and mechanisms of equilibrium chemical exchange (IR pump–probe).

Limitations (Information Not Available)
- CARS, OHD-OKE: only linear spectroscopic information available
- CARS, OHD-OKE: only sensitive to Raman-active modes
- IR pump–probe: only sensitive to IR-active modes visible
- IR pump–probe: limited to pump-induced absorbances of ≥ 0.01
- IR pump–probe: observation of chemical exchange and molecular reorientation dynamics limited by the vibrational lifetimes.

Methods in Physical Chemistry, First Edition. Edited by Rolf Schäfer and Peter C. Schmidt.
© 2012 Wiley-VCH Verlag GmbH & Co. KGaA. Published 2012 by Wiley-VCH Verlag GmbH & Co. KGaA.

21.1
Introduction

Vibrational spectroscopies such as infrared (IR) absorption or Raman scattering are indispensable tools in all areas of modern chemistry and have, therefore, become standard analytical methods in almost every academic or industrial laboratory throughout the world [1, 2]. They can provide molecular structural information very rapidly (albeit not as detailed as NMR) on a wide variety of different samples (liquids, liquid films, fine particles and powders, surfaces, solid pellets, and so forth) and can easily be adapted to highly specialized instrumental requirements (remote sensing, microscopy, tomography, etc.).

What have not become standard tools in analytical chemistry yet are the time-resolved variants of vibrational spectroscopies [3, 4]. Commercial IR spectrometers are available that can be synchronized to external perturbations of the sample of interest brought about by rapid mixing of reagents, temperature jumps, or pulsed laser illumination. Using the so-called rapid-scan technique, time resolutions of, at best, 10 ms can be achieved and the step-scan approach enables researchers to resolve chemical events and short-lived intermediates on millisecond to nanosecond time scales [5].

Whereas such spectrometers rely on incoherent Globar and similar continuous light sources, time-resolved IR spectroscopy on even shorter time scales requires laser-based devices. Recent developments in the field of nonlinear optics and parametric processes have extended the reliable generation of ultrashort laser pulses into the mid-IR spectral region [6, 7], thereby allowing monitoring of chemical dynamics on femtosecond time scales by utilizing dipole-allowed IR transitions. Complementary, laser-based methods that exploit Raman-active rather than IR-active transitions have also been established and continue to be developed into ever more powerful techniques.

In general, one can distinguish two types of femtosecond time-resolved vibrational spectroscopies depending upon the nature of the excitation (or perturbation) process. In one class of experiments, the system of interest is excited *vibrationally* and the ensuing molecular dynamics (MD) is detected by means of vibrational spectroscopy. In such types of measurement, one is obviously interested in the vibrational dynamics of the system, such as vibrational energy and/or phase relaxation phenomena. In another class of experiments, the system under investigation is excited *electronically* rather than vibrationally but the resulting MD is again monitored through vibrational spectroscopic approaches. In such cases, the research interest lies in unraveling the photochemistry and photophysics that originate from the light-induced UV/VIS-transition of the system [8, 9].

In what follows, we will focus entirely on pure vibrational spectroscopies, that is, methods that involve an initial vibrational perturbation with a subsequent vibrationally-mediated detection of the resultant relaxation dynamics. We can generally distinguish between time-resolved IR and time-resolved Raman spectroscopies but mixed IR–Raman approaches have also been implemented and used with great success.

21.2
Vibrational Pump–Probe Spectroscopy

Conventional pump–probe spectroscopy is a stroboscopic technique for the time-resolved detection of atomic, molecular, or materials dynamics [10]. In the classical set-up, the system under investigation interacts with a sufficiently short light pulse (termed the *"pump" pulse*) thereby causing a perturbation and initiating a dynamical process of interest. The dynamical process can simply be the relaxation of the system back to its original equilibrium prior to the pump or it can represent the evolution of the system toward a new equilibrium when the pump pulse induces an irreversible net chemical transformation. Regardless of the nature of the system's response, the pump-induced processes give rise to a modified and dynamically evolving spectral response of the system. This spectral response at a well-defined time after the interaction with the pump field can, in turn, be detected through another interaction with a properly delayed and sufficiently short second pulse (termed the *"probe" pulse*). By repeating this pump–probe sequence for a great number of different time delays, one can reconstruct the full spectro-temporal evolution connected with the dynamics of the system initiated by the pump.

The nature of the dynamics of interest dictates the requirements for the light pulses to be used. First, the pump and probe pulses need to be much faster than the events to be monitored. Secondly, the pump and probe electric fields need to be in resonance with the vibrational transitions to be exploited. Obviously, when using for the pump process directly a dipole-allowed IR transition at the Bohr frequency, $\tilde{v}_{if} = (E_f - E_i)/hc$ (in wavenumber units, h equals Planck's constant, and c is the speed of light), the pulsed electric field must have spectral components at that frequency (Figure 21.1a,b).

However, when the same transition happens to be Raman-active, it can, in principle, also be driven by two coincident electric fields whose difference frequency

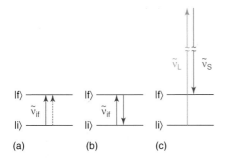

(a) (b) (c)

Figure 21.1 Time-ordered diagrams to second order in the pump-electric field representing the excitation schemes of time-resolved vibrational spectroscopy. Solid and dashed vertical arrows denote interactions of an electric field with the ket- and bra-side of the density matrix, respectively. In each diagram, time evolves from left to right. A direct infrared excitation generates an excited-state population (a) and a ground-state hole (b) to second-order in the pump field whereas a Raman-excitation induces a vibrational coherence (c) at the same level of nonlinear order.

matches the vibrational Bohr frequency, $\tilde{v}_{if}$. The two frequency components required are commonly termed the *"Laser"* $(\tilde{v}_L)$ and *"Stokes"* $(\tilde{v}_S)$ *frequency*, respectively, and are usually located in the visible region of the electromagnetic spectrum (see Figure 21.1c). These optical fields are, therefore, far from any vibrational or electronic resonance of the system. Because of the Fourier theorem, ultrashort light pulses exhibit a broad spectral width. This makes it possible to provide the Laser and the Stokes frequencies simultaneously by a single pulsed electric field provided its bandwidth, $\Delta\tilde{v}$, is larger than $\tilde{v}_{if}$. In this case, the vibration is said to be quasi-impulsively driven by the Raman scattering event.

There is, however, yet another fundamental difference between the two excitation scenarios. In Figure 21.1, we have already adopted the traditional diagrammatic representation introduced by Albrecht and coworkers to illustrate the time-evolution of the density matrix that is necessary to capture both population *and* coherence contributions to linear and nonlinear spectroscopies in the weak field limit [11]. In the direct IR scheme, a population is created in the upper vibrational state and, at the same time, the population of the lower vibrational state is depleted. The creation of the upper state population and the lower state "hole" occurs at second order of the density matrix perturbation expansion in the matter-field interaction [12]. At the same level of nonlinearity, the Raman-excitation scheme yields a vibrational coherence between the states $|i\rangle$ and $|f\rangle$ rather than a population or a depletion. As discussed below, this has profound implications for the nature of the signals to be detected with the subsequent probe pulse.

21.2.1
Infrared Pump–Probe Spectroscopy

We restrict ourselves to temporally well-separated pulses and disregard signal contributions originating from mixed/improper time-orderings when pump and probe electric fields overlap in time (coherent coupling) or when the probe pulse precedes the pump (perturbed free induction decay (FID)) [12]. Also, we limit our attention to the vibrational ground state of the system and its first two vibrationally excited states in a given mode.

The excited-state populated at second order in the pump electric field can be interrogated at a later time with the probe pulse having frequency components at $\tilde{v}_{12} = (E_2 - E_1)/hc$ (cf. Figure 21.2a). This third matter–field interaction sends the ket-vector of the density matrix to an energetically higher lying two-quantum vibrational state, thereby creating a coherence, $|2\rangle\langle 1|$, which oscillates at the frequency, $c\tilde{v}_{12}$. The resulting third-order polarization $P^{(3)}(t)$ radiates a signal field (wavy lines in Figure 21.2) that is out-of-phase with respect to the $\tilde{v}_{12}$-components of the probe electric field. If the signal field propagates collinearly with the probe field (as is the case in conventional pump–probe spectroscopy) a photo-detector will therefore monitor a reduced intensity of the probe pulse as compared to the probe intensity in the absence of the pump field. The signal is therefore called a *"transient"* absorption or a *(pump-)induced absorption*.

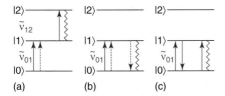

Figure 21.2 Classical pump–probe interactions in pure time-resolved vibrational spectroscopy. The excited-state population can be probed via transient absorption to the two-quantum excited state (a) or via stimulated emission back to the ground state (b). The ground-state "hole" appears as a nonlinear ground-state bleach (c).

Similarly, a probe electric field having frequency components at the fundamental transition, $\tilde{\nu}_{01} = (E_1 - E_0)/hc$, can project the bra back down to the ground vibrational state (see Figure 21.2b). The ensuing coherence, $|1\rangle\langle 0|$, will result in a signal field at the fundamental frequency that oscillates in-phase with the $\tilde{\nu}_{01}$-components of the probe electric field. As a result, the photodetector will register an increased intensity of the probe pulse, which is termed *"stimulated" emission* or *(pump-)induced emission*.

Finally, the ground-state hole (cf. Figure 21.2c) can also be examined by the same probe field having finite spectral amplitude at $\tilde{\nu}_{01} = (E_1 - E_0)/hc$. This interaction results in the very same coherence, $|1\rangle\langle 0|$, and, as a consequence, another in-phase signal field of equal amplitude and frequency as that connected with stimulated emission will be generated. This third contribution is commonly termed *ground-state bleach*.

Consider for a moment a purely harmonic system where the fundamental transition coincides spectrally with the transient absorption, that is, precisely $\tilde{\nu}_{01} = (E_1 - E_0)/hc = (E_2 - E_1)/hc = \tilde{\nu}_{12}$. In this instance, the square of the matrix elements of the transition dipole operator for absorptive transitions from $|v\rangle$ to $|v + 1\rangle$ are proportional to $v + 1$. As a result, the transient absorption is twice as large as the ground-state bleach, which in turn is of equal magnitude to the stimulated emission, because the matrix elements squared for emissive transitions from $|v\rangle$ to $|v - 1\rangle$ are proportional to v. Since, for the harmonic oscillator, the induced absorption, induced emission, and ground-state bleach are spectrally degenerate, all pump-induced signals will vanish. Pump-induced signals can be detected only because the vibrational modes are anharmonic! This is because anharmonicity lifts the spectral degeneracy between the fundamental $|0\rangle \rightarrow |1\rangle$ and the "hot" $|1\rangle \rightarrow |2\rangle$ transitions and modifies their associated dipole matrix elements. In other words IR-pump–IR-probe spectroscopy is a spectroscopy of molecular anharmonicities!

When using femtosecond IR-pulses, the spectral bandwidths of the pump and the probe pulses may be large enough to cover the fundamental and the "hot" transition simultaneously. In this case, it is advantageous to spectrally disperse the probe pulse after it has interacted with the pump pulse inside the sample. This can be accomplished with a monochromator whose exit slit is replaced by an array detector, thereby enabling the measurement of the frequency-dependent intensity of the probe pulse for a given pump–probe time-delay. An example of such a measurement is given in Figure 21.3 for the antisymmetric stretching mode of azide ions dissolved in liquid water.

The spectra display the pump-induced (or differential) optical density $\Delta OD(t, \tilde{\nu}_{pr})$ of the sample at the pump–probe time-delay, t, as a function of the probe frequency, $\tilde{\nu}_{pr}$. This quantity equals the difference between the optical densities of the sample experienced by the probe pulse in the presence and in the absence of the pump pulse, that is,

$$\Delta OD(t, \tilde{\nu}_{pr}) = OD(t, \tilde{\nu}_{pr})|_{\text{Pump on}} - OD(\tilde{\nu}_{pr})|_{\text{Pump off}}$$

The negative ΔOD feature around 2050 cm^{-1} corresponds to the combined signals of ground-state bleach and excited-state stimulated emission. The positive ΔOD feature originates from the excited-state absorption, which is shifted to lower probe frequencies due to the anharmonic character of the asymmetric stretching mode. In principle, the frequency spacing between these two features can help extract a value for the anharmonicity constant of this azide anion vibration.

Furthermore, Figure 21.3 reveals that both the anharmonically shifted transient absorption and the ground-state bleach/stimulated emission decay on a time scale of a few picoseconds. Referring again to Figure 21.2, we emphasize that following the two matter–field interactions involving the pump pulse, the system is left in an excited-state population or in a ground-state depletion. Between the second pump interaction and the probe interaction (the so-called population period, t_{23}) the system evolves freely so as to re-establish a Boltzmann distribution between the states $|0\rangle$ and $|1\rangle$. This relaxation depopulates the excited-state and refills the ground-state hole according to $\exp(-t_{23}/T_1)$, where T_1 corresponds to

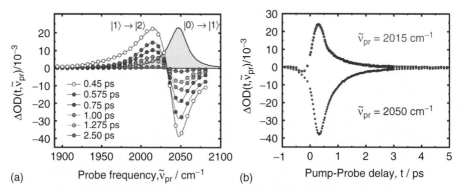

(a)

(b)

Figure 21.3 (a) Pump-induced femtosecond infrared spectra in the region of the antisymmetric stretching vibration of azide anions in liquid water at a pressure of 500 bar and a temperature of 60°C for various pump–probe time delays. The linear absorption band is shown as the gray-shaded spectrum. (b) Pump–probe time traces in the region of the transient absorption and of the ground state bleach emphasizing the kinetics of vibrational energy relaxation of the azide asymmetric stretching vibration.

the excited-state lifetime. From the decay of the pump-induced optical density along the pump–probe time-delay we can obviously retrieve information regarding the vibrational population dynamics. It should be stressed at this point, that such information cannot be obtained from stationary vibrational spectroscopy, for example, Fourier-transform IR spectroscopy. FTIR-spectra provide us with the vibrational structure (i.e., fundamental frequencies, overtones, combination bands, etc.) of the system and with the bandwidth associated with each vibrational transition. In principle, the dynamical information is hidden in the spectral shape of each vibrational resonance. However, it requires coherent non-linear IR spectroscopies such as IR-pump–IR-probe spectroscopy to uncover the underlying dynamical processes which give rise to a particular spectral line shape.

Another dynamical process that is readily available from the classical pump–probe set-up is related to the reorientational motion of the molecules. To this end, we exploit the vectorial properties of the pulsed electric fields and the spectroscopic transitions involved in the pumping and probing processes. Since the probability for absorption of a photon by a molecule depends on the square of the scalar product of the transition dipole moment and the electric field vector, the pump interaction creates a spatially anisotropic distribution of molecules in the excited state and in the ground state (cf. Figure 21.4) [13].

Neglecting inertial effects at early times, the temporal evolution of the angular distribution of molecules is reminiscent of diffusive motion. The solution of the rotational diffusion equation using the anisotropic distribution created by an infinitely short pump pulse as initial conditions is also shown in Figure 21.4 for representative delays. The excess orientation along the pump field vector at $t = 0$ constitutes a non-equilibrium situation and, consequently, the system relaxes toward equilibrium by reestablishing a fully isotropic angular distribution. When projecting the time-dependent distribution onto the plane perpendicular to the pump wave vector, a characteristic angle relative to the E-field can be identified under which all projections intersect. This angle is called the *"magic angle"* and is

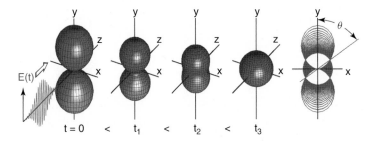

Figure 21.4 Temporal evolution of the orientational anisotropy of the molecular orientation. The plots show the probability for finding the molecules in the excited state as a function of the orientation in the laboratory frame after interaction of the sample at time zero with a pump pulse that was linearly polarized along the y-axis and propagated along the z-axis. Excited-state depopulation is neglected. The rightmost figure displays projections of the distributions onto the xy-plane for various times after excitation. All projections intersect at the magic angle $\theta = 54.7°$.

equal to $\theta = \arctan(\sqrt{2}) = 54.7°$. When probing the system with a light pulse that is polarized at the magic angle relative to the pump polarization, the measured differential optical density does not contain any contribution arising from the orientational dynamics of the molecules. Only under the magic angle can we record the pure population dynamics unperturbed by the reorientational motion.

On the other hand, it is possible to reveal the orientational dynamics by determining the so-called "transient anisotropy" [13]. For this purpose, the differential optical density needs to be measured twice, once with the probe pulse polarized along the y-axis (i.e., parallel to the pump polarization) and once again with the probe pulse polarized along the x-axis (i.e., perpendicular to the pump). These two quantities, $\Delta OD_{||}(t, \tilde{v}_{pr})$ and $\Delta OD_{\perp}(t, \tilde{v}_{pr})$, then determine the transient anisotropy according to

$$r\left(t, \tilde{v}_{pr}\right) = \frac{\Delta OD_{||}(t, \tilde{v}_{pr}) - \Delta OD_{\perp}(t, \tilde{v}_{pr})}{\Delta OD_{||}(t, \tilde{v}_{pr}) + 2\Delta OD_{\perp}(t, \tilde{v}_{pr})},$$

$$= \frac{2}{5}\left\langle P_2\left[\vec{\mu}_{Pump}(0) \cdot \vec{\mu}_{Probe}(t)\right]\right\rangle$$

where the expression in the denominator represents the isotropic signal, which in turn is equal to the differential optical density recorded under the magic angle divided by three. The enumerator constitutes the anisotropic signal and contains exclusively the reorientational dynamics. More accurately, the anisotropy measures the loss of orientational memory as quantified by the correlation function of the second-rank Legendre polynomial, P_2, of the scalar product between the pump transition dipole at time $t = 0$ and the probe transition dipole at a later time, t. Care has to be taken in the interpretation of the transient anisotropy when multiple transition dipoles contribute to the differential optical densities. In this case, $r(t, \tilde{v}_{pr})$ can contain additional coherent contributions that give valuable information pertaining to vibrational degeneracies and symmetries. Furthermore, artificial incoherent contributions can also arise that are not at all related to the orientational dynamics of interest.

It can be seen in Figure 21.4 that the pump-pulse creates an excess of excited state molecules that are oriented parallel to the pump electric field. Therefore, the ensuing angular randomization causes a majority of molecules to rotate away from the pump-pulse polarization. As a result, the differential optical density polarized parallel to the pump field will decay with time, even in the absence of population relaxation. Since the pump pulse has created an excess of excited-state molecules oriented along y, there is a net deficiency of excited state molecules with orientation along x. Therefore, the same randomization causes an effective replenishment of molecules aligned perpendicular to the pump electric field. Hence, a temporal rise will be observed in the absence of population relaxation if the differential optical density is recorded with a probe-pulse polarized perpendicular to the pump. In the presence of ground state recovery and excited state decay, the rising reorientational component to the perpendicular signal is often obscured, nevertheless, it still exists and a calculation of the anisotropy will fully reveal it.

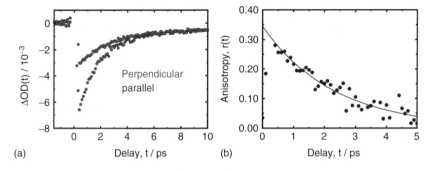

Figure 21.5 (a) Pump-induced optical density of HOD in H_2O at 500 bar and 298 K following OD-stretching excitation with 200 fs pulses at 2500 cm^{-1}. The detection frequency is 2544 cm^{-1}, which is located in the region of the ground-state bleach. (b) Temporal evolution of the transient anisotropy reconstructed from the data in (a). The solid curve corresponds to a fit of the data to a monoexponential decay with a time constant of 2.3 ps.

In Figure 21.5, we show femtosecond IR-pump–IR probe experiments on mono-deuterated water, HOD, dissolved in normal liquid water, H_2O. The pump and probe pulses were tuned into the OD-stretching resonance of HOD, which is centered around 2500 cm^{-1} and is very broad due to hydrogen-bonding (see Section 21.3).

In these data, the probe frequency was located in the ground state bleach region. It can be seen that around zero time delay the magnitude of the bleach experienced by a parallell polarized probe beam exceeds that of the perpendicularly polarized probe pulse by more than a factor of 2. Ideally, when the pump and probe transition dipoles are identical, as is the case for probing of the ground state bleach, the signals should differ by exactly a factor of 3 at zero delay and, as a result, the anisotropy $r(t = 0)$ is equal to 2/5. With increasing delay, the parallel and perpendicular optical densities asymptotically approach each other as the transition dipole orientation is gradually randomized and the angular anisotropy of the ground state hole is destroyed. Concurrently, the transient anisotropy decays to zero thereby revealing the dynamics of orientational randomization (see Figure 21.5b).

21.2.2
Coherent Anti-Stokes Raman Scattering

We now turn our attention to the time-ordered diagram displayed in Figure 21.1c and the interrogation of the vibrational coherence $|1\rangle\langle 0|$ created at second order upon exposure of the system to two off-resonant optical pulses with a relative time delay of zero. Their difference frequency is matched to the Bohr frequency of the fundamental $|0\rangle \rightarrow |1\rangle$ transition.

Another matter–field interaction with the temporally delayed probe field centered again at the Laser frequency creates a third-order polarization that oscillates at the optical frequency, $c\tilde{\nu}_{AS} = c\tilde{\nu}_L + (E_1 - E_0)/h$. The resulting signal field is termed the *"anti-Stokes" field* whose intensity can be detected with a conventional photodetector (homodyne detection [14]) as a function of time delay between the Laser–Stokes pair of pump pulses and the Laser probe pulse. This pump–probe scheme is denoted coherent anti-Stokes Raman scattering (CARS, see Figure 21.6).

Again, during the temporal pump–probe delay the system is allowed to evolve freely, resulting in loss of phase correlation among the molecular dipoles of the ensemble. The decay of the macroscopic vibrational dipole is composed of vibrational pure dephasing due to the fluctuating interactions of the individual oscillators with their microscopic environments, and vibrational population relaxation. The optical third-order polarization is related to the Raman-FID [12], which in turn is reflected in the decay of the measured anti-Stokes intensity according to $\exp(-2t_{23}/T_2)$ with increasing pump–probe delay, t_{23}. Here, T_2 represents the vibrational dephasing time and the factor of 2 takes into account the homodyne detection of the signal field. If one were to heterodyne the anti-Stokes field by measuring its interference with a local oscillator (LO) field (see also next section) one would indeed be able to reconstruct the electric field associated with the Raman-FID. Its Fourier transform would recover the linear vibrational line shape, as measured in a standard frequency-resolved Raman spectrum of the system.

This highlights the great disadvantage of coherent time-resolved third-order Raman spectroscopies versus time-resolved third-order IR spectroscopies: their inability to disentangle the elementary MD and their associated time scales that are responsible for the line shape of the linear resonance. If one wants to separate

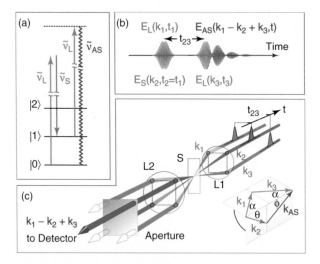

Figure 21.6 The vibrational coherence created by stimulated Raman-excitation can be probed by coherent scattering, here, for example, at the anti-Stokes frequency. (a) Time-ordered energy-level diagram for the CARS experiment. (b) Pulse sequence indicating the temporally coincident laser and Stokes pump fields, the temporally delayed laser probe field and the signal field emitted at the anti-Stokes frequency. (c) Experimental folded Boxcars geometry for quasi-phase-matching of the four-wave mixing process.

the pure dephasing dynamics from population decay using a Raman-technique, one has to resort to nonlinearities higher than third-order and to deal with the enormous instrumental complexities that come along with fifth and higher-order Raman spectroscopies.

In CARS spectroscopy, the input fields are characterized not only by their frequencies but also by their wave vectors (i.e., their propagation direction). Hence, for the nonlinear optical process to be efficient, phase matching needs to be fulfilled, that is, $\Delta k = k_1 - k_2 + k_3 - k_{AS} = 0$, where k_1 and k_2 are the wave vectors for the Laser and the Stokes pump whereas k_3 and k_{AS} are the wave vectors of the Laser probe and the anti-Stokes signal beams. Phase matching ascertains that a well collimated and directional signal beam is emitted from the sample with maximal intensity. Although the simplest way of accomplishing phase matching in CARS is when all three input beams and the CARS beam are fully collinear, it is often advantageous to use crossed beams, as shown in Figure 21.6 and to spatially filter off the coherent emission from undesired background [15]. It is important to note, however, that when working with broadband femtosecond pulses, there is always a finite wave vector mismatch in the Boxcars geometry that cannot easily be compensated for [15]. A detector facing the $k_1 - k_2 + k_3$ direction will record the time-integrated intensity (i.e., the modulus square) of the anti-Stokes field. We speak again of homodyne detection.

In Figure 21.7, time-resolved femtosecond CARS-signals of acetonitrile are displayed. The spontaneous Raman spectrum of neat liquid acetonitrile has an intense and narrow band at 2943 cm^{-1} that is caused by the CH$_3$-stretching vibration, ν_1, of symmetry A$_1$. Selecting the difference frequency of the Laser and Stokes pulses to be in resonance with this vibration one can detect a corresponding CARS signal at the proper anti-Stokes frequency. Such a signal decays single-exponentially (see

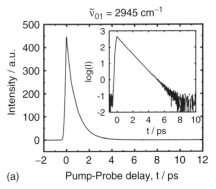
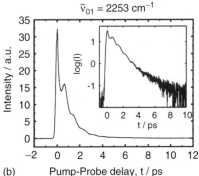
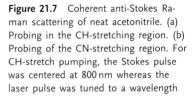

Figure 21.7 Coherent anti-Stokes Raman scattering of neat acetonitrile. (a) Probing in the CH-stretching region. (b) Probing of the CN-stretching region. For CH-stretch pumping, the Stokes pulse was centered at 800 nm whereas the laser pulse was tuned to a wavelength of 648 nm. Coherent anti-Stokes emission was then detected at 544 nm. Likewise for CN-stretch pumping the same Stokes pulse was used but the laser pulse was tuned to 678 nm. In this case, anti-Stokes emission was detected at a wavelength of 588 nm.

Figure 21.7) with a time constant of 840 fs corresponding to a dephasing time of $T_2 = 1.68$ ps. This value translates directly into a homogeneous Lorentzian line width $\Delta\tilde{\nu}_{\mathrm{FWHM}} = 1/(\pi c T_2)$ of 6.3 cm^{-1}, in excellent agreement with the isotropic spontaneous Raman line width [16].

In Figure 21.7, a CARS decay from the CN-stretching vibration, ν_2, of acetonitrile at 2253 cm^{-1} is also shown. We notice immediately that the decay is far from single exponential. Instead, a double-exponential decay with time constants of 2.45 ps and 760 fs is observed, on which a low-frequency oscillation with a period of 850 fs is superimposed. The bi-exponential decay gives evidence for at least two components that shape the spontaneous Raman band of the CN-stretching mode with line widths of approximately 4.3 and 14 cm^{-1}. The oscillatory component on the other hand is a beating phenomenon that is caused by the broad Raman excitation bandwidth. The experiments were carried out with 120-fs duration pulses having a Fourier-limited bandwidth well in excess of 100 cm^{-1}. This enables the creation of additional vibrational coherences with the ground state according to Figure 21.8.

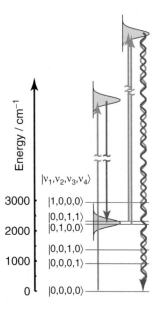

Figure 21.8 Energy level diagram explaining the polarization beats in coherent anti-Stokes Raman scattering of neat acetonitrile in the CN-stretching spectral region. At time $t = 0$, the laser (left upward pointing arrow) and Stokes pulses (downward pointing arrow) impinge on the sample preparing a vibrational coherence between the first excited CN-stretching state, ν_2, and the vibrational ground state. At the same time, the pump pulse pair also creates a vibrational coherence between the $(\nu_3 + \nu_4)$-combination state and the ground state. These two coherences oscillate at slightly different Bohr frequencies, thereby giving rise to a low-frequency modulation of the CARS intensity when intercepted with the variably delayed laser probe pulse. The beat frequency as a function of pump–probe time-delay is a measure of the energetic splitting of the vibrational states accessible within the Raman excitation bandwidth.

In the particular case of acetonitrile, a combination band of the C-C stretching mode (ν_4, symmetry A_1, fundamental frequency 917 cm^{-1}) with the CH$_3$-deformation mode (ν_3, symmetry A_1, fundamental 1374 cm^{-1}) gives rise to additional Raman activity at 2292 cm^{-1}. The corresponding vibrational state is shifted by only 39 cm^{-1} to higher frequencies relative to the CN-stretch. In other words, both the CN-fundamental and the $\nu_3+\nu_4$-combination are within the Raman excitation bandwidth and, as a result, both states are coherently coupled to the vibrational ground state by the pump pulse pair at time $t = 0$.

Each of the two coherences oscillates at its proper Bohr frequency and will be converted upon interrogation by the Laser probe pulse into a third-order polarization emitting at the optical frequency $c(\tilde{\nu}_L + \tilde{\nu}_2)$ and $c(\tilde{\nu}_L + \tilde{\nu}_3 + \tilde{\nu}_4)$, respectively. It is the homodyne detection, that is, the detection of the intensity of the corresponding anti-Stokes signal field, which is responsible for the periodic modulation along the pump–probe delay at the difference frequency, $c(\tilde{\nu}_3 + \tilde{\nu}_4 - \tilde{\nu}_2)$. Strictly speaking, the oscillatory feature observed in Figure 21.7 corresponds to an *inter*molecular vibrational polarization beat [17] and appears as a result of the simultaneous creation of the *intra*molecular coherences $|\nu_3 + \nu_4\rangle\langle 0|$ on one set of molecules in the liquid and $|\nu_2\rangle\langle 0|$ on another set of particles.

This phenomenon should formally be differentiated from a vibrational quantum beat, which is a true quantum interference that can only arise from the preparation of an *intra*molecular interstate coherence such as $|\nu_3 + \nu_4\rangle\langle\nu_2|$, that is, where the two states of the same molecule are coherently coupled by the pump fields. Since in femtosecond CARS spectroscopy both the Laser and the Stokes pump fields are vibrationally off-resonant, an *intra*molecular interstate coherence not involving the vibrational ground state can only be created at fourth order or higher with regard to matter–field interactions.

The above short discussion has demonstrated that, albeit being of tremendous analytical power (e.g., as a diagnostic tool of combustion processes, analysis and thermometry of gaseous samples [18], or more recently in the life sciences as a detection scheme for *in vivo* imaging methods such as endoscopy [19] or confocal microscopy [20, 21]) CARS spectroscopy fails to deliver new insights into molecular vibrational dynamics. This is because despite relying on the non-linear optical response of the medium it still provides linear spectroscopic information only that is entirely equivalent to that obtained from classical spontaneous Raman scattering in the frequency domain.

21.2.3
Raman-Induced Optical Kerr Effect and Impulsive Stimulated Raman Scattering

So, if coherent Raman spectroscopies such as CARS can only provide linear spectroscopic information and we desire to explore vibrational dynamics, why bother with these techniques at all? Frequency domain light scattering and linear IR absorption become increasingly difficult with decreasing vibrational frequencies to be detected. Indeed, linear absorption suffers from the fact that most materials have insufficient transparency in the far-IR spectral region from 10 to $\sim$500 cm^{-1}. Therefore, the design of suitable sample cells and of spectrometer accessories can become a critical issue. Also, conventional incoherent light sources are rather weak in this spectral range. This intensity problem has recently been alleviated somewhat by advances in laser-based or laser-driven terahertz (THz) technology [22–24], which in turn have sparked the field of condensed-phase THz spectroscopy [25–28]. However, the required higher-order (i.e., coherent nonlinear) THz-pump–THz-probe spectroscopy to elucidate condensed-phase vibrational *dynamics* has not been performed to date.

Similarly, Raman spectroscopy below 500 cm^{-1} (i.e., low-frequency Raman-scattering or Rayleigh-wing scattering) is hampered by the overwhelming elastically scattered Rayleigh background, which calls for highest resolution triple monochromators, narrow-band excitation lasers, and matched notch-filters with high damping. Although various reduced representations have been proposed for subtraction of the Rayleigh line, there is still considerable confusion regarding their appropriateness [29].

Although the frequency region below 500 cm^{-1} is difficult to access by Raman spectroscopy, it is, nonetheless, a highly informative window to the dynamics of liquid solutions. This is because, apart from low-frequency intramolecular modes such as torsions, soft bendings, or ring puckerings, this region also contains spectroscopic activity resulting from *inter*molecular degrees of freedom. Knowledge of this intermolecular region should improve our understanding of the mechanisms and time scales of interactions between the particles in dynamically disordered condensed phases (see Section 21.3) [30].

We have seen in the previous section that it is possible to measure the absolute magnitude squared of the Raman-FID of a vibrational transition when the difference frequency of the driving fields is in resonance. If we wish the CARS method to be sensitive to vibrational modes of the intermolecular spectral region, the frequencies of the Laser and the Stokes fields have to approach one another. Indeed, since femtosecond pulses are intrinsically broadband, there is eventually no need for separate lasers to deliver the two driving fields. Instead, it is sufficient to derive the Laser and Stokes pulses from the very same femtosecond source, provided its spectral bandwidth is larger than the vibrational frequency of the mode to be observed. If the same source is then also used for the final probe interaction, all three pulses become spectrally degenerate with respect to each other and with respect to the signal field emitted from the third-order polarization. This general scheme is, therefore, called degenerate four-wave mixing (DFWM, cf. Figure 21.9) [12].

If the pulses are Fourier-limited and their bandwidths exceed the vibrational level spacing, the pulse durations are shorter than the vibrational period and the matter–field interactions are said to be "impulsive". Since we are interested in observing Raman-active vibrational modes, the degenerate CARS scheme is also termed impulsive stimulated Raman scattering (ISRS) [31].

ISRS can be phase matched in a perfectly symmetrical folded Boxcars geometry, where it can be viewed as a diffraction experiment. The first two degenerate and coincident pump pulses with wave vectors k_1 and k_2 interfere in the sample to

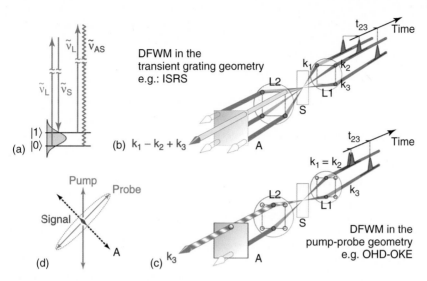

Figure 21.9 (a) Time-ordered diagram for non-resonant degenerate four-wave mixing. (b) Folded boxcars geometry for transient grating scattering. Note that all three input fields and the signal field have the same center frequency. (c) The pump–probe geometry emerges from the folded boxcar geometry in the limiting case $k_1 = k_2$. In this case, the probe pulse scatters in the forward direction and the probe pulse itself heterodynes the third-order polarization. (d) In the optical Kerr effect setup, the probe pulse is polarized 45° with respect to the pump. The aperture, A, is then replaced by an analyzing polarizer whose transmission axis is set to −45° with respect to the pump. In the absence of the pump, there will be no leakage through the analyzer. The interaction of the pump-pulse with the sample induces an ellipticity of the probe polarization, which will be able to leak through the crossed analyzer.

create a spatially periodic modulation of its refractive index. When the third pulse propagating along k_3 impinges on this phase grating it will be diffracted in the Bragg direction, $k_S = k_1 - k_2 + k_3$. Again, monitoring the intensity of the signal field in this particular direction is equivalent to a homodyne detection of the Raman-FID, which oscillates around the optical driving frequency and decays due to the simultaneous dephasing of all the driven Raman-active vibrations whose frequencies fall within the bandwidth of the driving fields. As the pump–probe time delay is scanned, one can observe redistribution of the spectral amplitude of the probe pulses during each cycle of the vibrational coherence corresponding to periodic out-of-phase Stokes and anti-Stokes shifting within the probe spectral bandwidth [31].

But what happens if we gradually decrease the relative angle of incidence between the two pump fields irradiating the sample until they propagate collinearly? Obviously, the scattering angle continuously decreases until we are finally back in the pump–probe geometry. When k_1 equals k_2 the grating constant (in units of m^{-1}) goes to zero and the probe pulse is diffracted in the forward direction. In other words, the third-order polarization emits the signal field in the direction of the probe beam itself. Positioning the photodetector in the probe beam path yields an electrical signal that is proportional to the modulus square of the sum of the probe and signal fields, that is,

$$|E_3(t) + E_S(t)|^2 = |E_3(t)|^2 + |E_S(t)|^2 + 2|E_3(t) \cdot E_S(t)|$$

The first two terms correspond to the intensity of the probe field and the non-linear optical signal field. The last term however represents an interference between the signal and the probe field. Obviously, the probe pulse itself serves as a local oscillator (LO) for heterodyning the third-order polarization [14]. Using lock-in detection, we can extract exclusively this interference term, and we can thereby measure the third-order polarization in a linear fashion rather than its squared modulus as in CARS.

For signal-to-noise reasons it is highly advantageous to be able to adjust the LO intensity independently. A particular set-up that enables exactly this in the pump–probe geometry is the optical Kerr-effect (OKE) arrangement (see Figure 21.9d) [14, 32]. Here, the probe beam before entering the sample is sent through a linear polarizer (and often a quarter-wave retardation plate for compensation of residual static birefringence of the lenses and of the sample cell) thereby setting its polarization at an angle of $45°$ relative to the pump pulse. An analyzer transmitting at an angle of $-45°$ with respect to the pump will totally reject the probe beam. A detector monitoring the k_3 direction will, therefore, not be able to detect any photons; its electrical signal is zero.

However, when the pump hits the sample, it exerts a torque onto the anisotropically polarizable molecules. These in turn respond by slightly rotating in an effort to orient the major principal axis of their polarizability tensor in a direction parallel to the pump electric field. As a result, the sample's refractive index becomes spatially anisotropic, almost like in uniaxial nonlinear optical crystals. Therefore, when the probe beam travels through this birefringent sample its polarization becomes slightly elliptical. It is the component of the probe electric field, which is now polarized orthogonally with respect to the input polarization, that represents the signal field emitted by the third-order polarization (see Figure 21.9d). This component is able to pass through the crossed analyzer and is finally registered by the photodetector. A heterodyne detection of $P^{(3)}(t)$ is made possible by slightly rotating the input polarizer in the probe beam path in front of the sample, that is, by slightly rotating the input polarization such that there is static leakage of the probe beam through the analyzer, even in the absence of the pump. The photodetector now registers the intensity of the superposition of the signal field with the leaking part of the probe pulse, the latter of which serves as the LO (heterodyne detection). In the optically heterodyne detected optical Kerr effect (OHD-OKE) set-up, the relative phase between the LO and the signal field is such that, by varying the pump–probe time delay, one is able to measure the real part of the full third-order nonlinear response function. Fourier transformation of this quantity yields the imaginary part of the frequency-domain susceptibility, that is, the low-frequency Raman spectrum of the material, an example of which is shown in Figure 21.10.

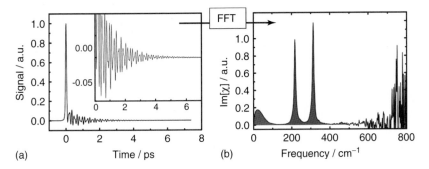

Figure 21.10 (a) Time-resolved OHD-OKE trace for liquid CCl$_4$ measured with 30 fs light pulses having a center wavelength of 800 nm. (b) Low-frequency depolarized Raman spectrum obtained by Fourier transformation of the OHD-OKE trace. The Raman peaks at frequencies above 150 cm^{-1} correspond to intramolecular vibrations of the molecules while the band below 100 cm^{-1} is due to intermolecular vibrational motion in the liquid.

The OHD-OKE signal of liquid carbon tetrachloride is composed of an instrument-limited peak at zero time delay, which corresponds to the electronic hyper-polarizability response and which does not contain information regarding the vibrational modes of the system. At later times, strong oscillatory features are observed that result from the impulsively created coherences of intramolecular vibrational modes whose frequencies are smaller than the bandwidth of the optical pulses. Finally, an extremely small overdamped response with a very low frequency is also discernible. This feature is entirely due to the intermolecular interactions in the liquid that we are interested in.

The relationship of this time-resolved OKE experiment to the linear Raman spectrum becomes apparent upon Fourier transformation of the time-domain data (see Figure 21.10). The resultant spectrum is reliable for frequencies up to ~600 cm^{-1}, beyond which the driving fields have insufficient bandwidth and the noise becomes overwhelming. Also note that it is devoid of the Rayleigh peak at 0 cm^{-1}, which renders this method superior to light scattering for a characterization of low-frequency vibrational modes in condensed-phase systems. The OKE-spectrum can be divided into two distinct regions. Below 150 cm^{-1}, the intermolecular degrees of freedom of the liquid give rise to a broad and structureless feature. It is due to rotational and translational motions of the molecules in their local solvation structure, almost like a molecular "rattling" motion in a cage. These motions can be collective in nature without being correlated over large distances and many molecules. There are also diffusive degrees of freedom, which may be described by hydrodynamic theories of the liquid [33].

Above 150 cm^{-1}, two sharp lines are seen at 218 and 314 cm^{-1}. These features are due to the intramolecular ν_2 and ν_4-modes of CCl$_4$, corresponding to the doubly degenerate scissoring (symmetry E) and triply degenerate asymmetric deformation (F$_2$). The question arises as to why the totally symmetric stretching

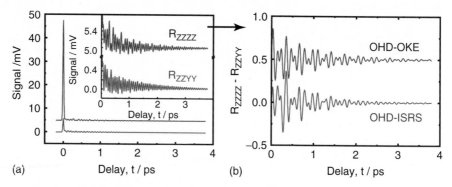

Figure 21.11 (a) Time-resolved OHD-ISRS traces for liquid CCl_4 addressing different tensor elements of the non-resonant third-order response function. (b) Comparison of the OHD-OKE response with the depolarized non-resonant Raman response reconstructed from OHD-ISRS data.

vibration (ν_1, symmetry A_1) around $459\,\text{cm}^{-1}$ is not observed. The answer lies in the tensorial nature of the time-dependent nonlinear response function (or correspondingly, the frequency-dependent nonlinear susceptibility, χ) measured by third-order spectroscopies. A careful theoretical analysis demonstrates that the OHD-OKE measurement is sensitive to the difference of the tensor components, $\chi_{ZZZZ} - \chi_{ZZYY}$. This quantity is entirely equivalent to the depolarized Raman spectrum of the material, thus rendering the OKE technique incapable of revealing highly polarized Raman-active vibrations.

To observe these modes, we need to go back to transient grating scattering, set the polarizations of all three electric fields as well as the signal field parallel to each other, and identify a method that allows for heterodyning in the Bragg direction (OHD-ISRS [34]). In this case, one would measure exclusively the tensor element χ_{ZZZZ}, which must contain all totally symmetric vibrations. The consistency of the data can then be tested for by rotating the probe pulse polarization by $90°$ and repeating the transient grating scattering experiment to independently measure the anisotropic tensor component, χ_{ZZYY}. The difference of these two measurements, $\chi_{ZZZZ} - \chi_{ZZYY}$, should recover the independently obtained OHD-OKE result. A comparison between OHD-OKE and OHD-ISRS confirming these concepts for liquid carbon tetrachloride is shown in Figure 21.11 where optical heterodyning in the transient grating scattering geometry was achieved by deriving a LO from an additional thermal grating that was also created by the two pump pulses (for further details see Ref. [34]).

Heterodyne-detected DFWM experiments, such as those shown in Figure 21.11, form the basis of all modern multi-dimensional electronically resonant optical and vibrationally resonant IR spectroscopies.

21.3
Applications

21.3.1
Hydrogen Bonding

Hydrogen bonding (H-bonding) is probably the most important non-covalent interaction in nature and arises at both the *intra*molecular and the *inter*molecular level. For example, H-bonds between particles are pivotal for the physico-chemical properties of associated liquids. The famous anomalies of water can be traced back to the formation of a unique network of H-bonds that is highly random in both space and time. Furthermore, intermolecular hydrogen-bonding is very often at the core of biochemical processes such as molecular recognition and the formation of supramolecular architectures, the most famous examples of which are the helical duplex assemblies of nucleic acids carrying the genetic code of every individual biological organism.

Hydrogen-bonding has also profound implications for the vibrational structure and the vibrational spectroscopy of molecular systems [35]. Consider the general H-bonded system D–H···A, where D–H denotes an electron deficient hydrogen-atom covalently bonded to an electronegative atom, the so-called "hydrogen donor", and where A represents an electronegative atom or an electron-enriched molecular moiety, the "hydrogen acceptor". The coupling between these two groups is primarily governed by dipole–dipole interactions but, depending upon the strength of the coupling, the participation of van-der-Waals interactions as well as covalent charge transfer effects need to be invoked.

In the intuitive terms of a chemist, the transfer of electrons from non-bonding orbitals of the acceptor (lone pairs) to anti-bonding orbitals of the donor leads to a weakening of the covalent D–H bond and, consequently, to a decrease in its stretching frequency. In fact, for intermolecularly H-bonded donor–acceptor pairs this behavior can easily be verified by comparison of the IR absorption spectrum in the intramolecular stretching region of D–H in the absence of the acceptor (e.g., in the dilute gas phase of pure DH or in highly diluted solutions of DH in weakly interacting liquid solvents) and in the presence of A allowing for hetero-aggregation with formation of D–H··· A-complexes.

The influence of hydrogen bonding on the stretching vibrational frequency is exemplified in Figure 21.12 by comparing the IR absorption spectra of ethanol monomers and oligomers dissolved in a nonpolar solvent [36]. The spectra highlight the OH-stretching spectral region around $3400 \, \text{cm}^{-1}$. For monomeric ethanol in liquid carbon tetrachloride, the IR-absorption consists primarily of a single and spectrally very narrow band centered around $3615 \, \text{cm}^{-1}$, which is prototypical for isolated hydroxyl groups weakly interacting with their nonpolar solvent (so-called α-OH with frequency ν_α). When the solute concentration is increased to about $0.2 \, \text{M}$ aggregates are formed, which are believed to consist of extended chains and rings of ethanol molecules. These give rise to a second, spectrally very broad absorption that increases with growing solute concentration at the expense of the

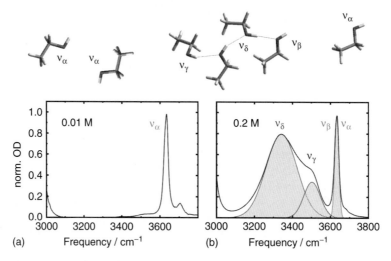

Figure 21.12 Infrared absorption spectra of the hydroxyl-stretching region of ethanol monomers (a) and oligomers (b) in liquid carbon tetrachloride at room temperature. The oligomer spectrum can be decomposed into multiple contributions from different types of hydroxyl oscillators.

α-OH resonance. It appears strongly red-shifted relative to ν_α and can formally be decomposed into two individual resonances. The first is centered at 3340 cm^{-1} and is attributed to hydroxyl groups in the interior of the chain and ring-like aggregates that can act as both donors and acceptors at the same time (so-called δ-hydroxyls). The second is centered at 3500 cm^{-1}. It is attributed to hydroxyl groups at the end of the oligomer chain that donate their H-atoms into a hydrogen bond to a neighboring ethanol (so-called γ-OHs). They have to be distinguished from the dangling hydroxyl groups, that is, those that also terminate the oligomer chain but that do not donate their H-atom to a hydrogen bond. The resonance of such terminal β-OHs acting as acceptors only merges with the α-band of the isolated molecules.

From Figure 21.12 it is obvious that the hydroxyl stretching absorption is a highly sensitive spectroscopic marker band that can be used to probe MD in hydrogen-bonded systems. Quantitatively, this H-bond induced frequency shift of the DH-stretching resonance can be described by an empirical potential originally developed by Lippincott and Schroeder [37]. Herein, the potential energy of the D–H$\cdots$A-complex, $V_{tot}(r_{DH}, r_{DA})$, as a function of the DH-bond length, r_{DH}, and the distance between the donor and the acceptor atom, r_{DA}, is composed of four terms: (i) The covalent D–H bond is given by a Morse potential, $V_{DH}(r_{DH})$. (ii) At any given donor–acceptor distance, the non-covalent H$\cdots$A contact is also described by a Morse potential, $V_{HA}(r_{DA} - r_{DH})$ but here, along the H$\cdots$A distance, $r_{HA} = r_{DA} - r_{DH}$. (iii) An electrostatic attraction, $V_{el}(r_{DA} - r_{DH})$, is assumed to exist between the donor atom D and the acceptor group A, and finally (iv), a van-der-Waals repulsion, $V_{rep}(r_{DA} - r_{DH})$, between D and A completes the total potential energy

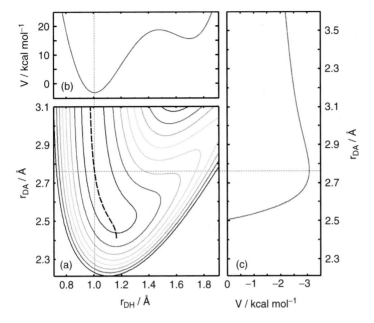

Figure 21.13 (a) Contour representation of the Lippincott–Schroeder potential describing the proton translocation as a function of the donor-hydrogen bond length, r_{DH}, and the donor acceptor distance, r_{DA}. (b) Cut through the potential along, r_{DH}. (c) Cross-section along r_{DH}.

surface relevant to proton translocation. An example of $V_{tot}(r_{DH}, r_{DA})$ is shown in Figure 21.13 for the collinear encounter geometry D–H···A where the donor and acceptor atoms are assumed to be equal, as in the case of a water dimer.

Note that for large D–A separations there are actually two potential minima along the proton coordinate (see Figure 21.13b). These can be thought of as being consistent with the two resonance structures

$$O-H \cdots\cdots\cdots O \quad\longleftrightarrow\quad \overset{(-)}{O}\cdots\cdots H\overset{(+)}{-}O$$

From the potential surface the D–H-stretching frequency, ν_{DH}, and the hydrogen-bond energy, E_{HB}, as a function of donor–acceptor distance can be obtained by applying the condition $(\partial E_{tot}/\partial r_{HA}) = 0$. In addition, one can extract the force constant, k_{DA}, and hence, the frequency for the relative translational donor–acceptor motion. We will come back to this motion in the following sections.

The donor–hydrogen equilibrium distance and the resultant DH-stretching frequency are shown in Figure 21.14 as a function of donor–acceptor separation. The potential predicts that upon approach of the donor by the acceptor, the D–H bond length gradually extends until at some critical D–A distance the two minima merge and the proton is evenly shared between the two

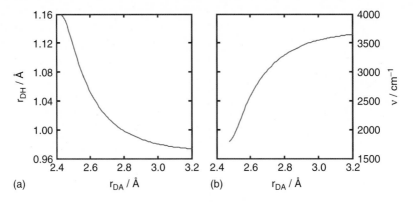

Figure 21.14 Equilibrium DH-bond distance (a) and frequency of the DH-stretching vibration as a function of the donor–acceptor separation (b).

moieties. At the same time, the frequency of the DH-stretching mode continuously decreases, thereby accurately tracking the donor–acceptor geometry. With this intimate relationship between *intramolecular* stretching frequency and *intermolecular* geometry, time-resolved vibrational spectroscopies become very powerful diagnostic tools for MD in hydrogen-bonded systems. Structural fluctuations of the local hydrogen-bond geometries will immediately be reflected in the dynamics of vibrational line broadening and vibrational relaxation. We can explore these phenomena straightforwardly by performing nonlinear femtosecond mid-IR and Raman spectroscopy on the intramolecular DH-stretching resonance, as discussed in the previous sections.

For associated liquids like water, the interrelation between the D–H frequency and the donor–acceptor separation can be revealed in a very intuitive fashion by measuring the linear absorption spectrum in the OH-stretching spectral region as a function of the thermodynamic state variables, temperature, and density [38]. Representative spectra are reproduced in Figure 21.15.

Starting with supercritical water at a temperature in excess of 650 K and a density reminiscent of a compressed gas, the OH-stretching spectrum is found at 3600 cm^{-1}. Although the bulk density is four times smaller than in the liquid under ambient conditions, there is still a considerable shift of the absorption band of more than 100 cm^{-1} relative to the true gas phase. This is indicative of considerable hydrogen-bonding, even under these extreme conditions. As predicted, a compression of this dilute fluid results in a decrease of the average inter-particle distance and, as a consequence, the absorption spectrum shifts toward lower frequencies. This clearly demonstrates that, on average, the hydrogen bridges become gradually stronger and shorter.

To make a quantitative connection between an experimentally observable spectrum and the detailed hydrogen-bond geometry detailed MD simulations may prove very helpful [38]. Such calculations can also unravel the coupling between

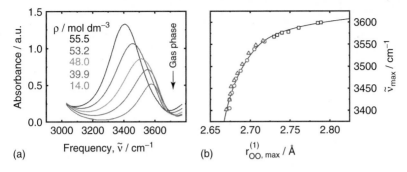

Figure 21.15 OH-stretching absorption band of mono-deuterated water in liquid and supercritical heavy water (a) and frequency of maximal absorbance as a function of the nearest-neighbor oxygen–oxygen distance (b) obtained from molecular dynamics simulations.

the donor–acceptor stretching motion and the hydrogen-bond bending coordinate – an intermolecular vibrational coordinate that can also be included in the Lippincott–Schroeder potential.

21.3.2
Optical Kerr Effect of Water

Let us now turn our attention to observing exactly these intermolecular modes, for example, the relative translational displacement of the donor and the acceptor and their relative angular orientational motion. An optimized Lippincott–Schroeder potential for water predicts a frequency for motion along the O–O coordinate of about 204 cm^{-1} (or equivalently, 6 THz) corresponding to a vibrational period of 167 fs [37]. This mode basically represents a restricted translational motion of the centers-of-mass of neighboring water particles in a direction parallel to the H-bond. It is, therefore, also called hydrogen-bond stretching vibration. It can be related to the longitudinal acoustic (LA) phonon of hexagonal and cubic ice [39]. There is also an intermolecular vibration in liquid water that corresponds to the motion of the oxygen atoms in a direction perpendicular to the H-bonds. This hindered translation is called the hydrogen-bond bending mode and can be associated with the transverse acoustic (TA) phonon of the above cited polymorphs of ice [39].

These intermolecular degrees of freedom of liquid water have been characterized in detail by femtosecond OHD-OKE measurements by a number of groups [40–44]. The time-resolved Kerr-response of ice and liquid H$_2$O close to the melting point under ambient pressure is reproduced in Figure 21.16. Apart from the electronic hyper-polarizability contribution, the nuclear response of the solid is dramatically different to that of the liquid. For ice, the nuclear response consists of a number of weakly damped oscillations the most prominent of which has a frequency of 209 cm^{-1}, in excellent agreement with the O-O-stretching vibration predicted by the model potential. The restricted translational density of states of ice Ih and Ic as

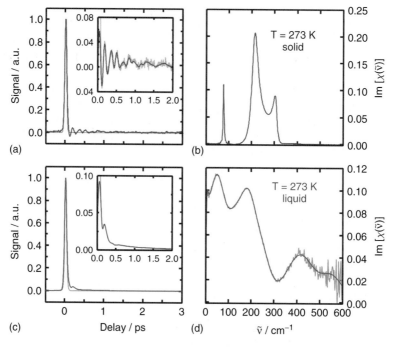

Figure 21.16 Time-resolved OHD-OKE transients of solid (a) and liquid (c) water near the normal melting point and corresponding low-frequency depolarized Raman spectral density (b,d) obtained through Fourier transformation of the time-domain data. The spectrum for the solid was obtained by performing a linear prediction singular value decomposition prior to Fourier transformation to improve the signal-to-noise ratio. The experimental spectrum for the liquid (noisy data) was also fitted to a stochastic line broadening model (for further details see Ref. [43]).

measured by OHD-OKE and depolarized low-frequency Raman scattering can, in principle, be understood in terms of lattice dynamical calculations in the presence of orientational disorder [45, 46].

When going to the liquid, the nuclear OHD-OKE response becomes heavily damped in the time-domain. As a result, the frequency domain features broaden and shift to lower frequencies as compared to the solid. Walrafen has discussed the spectral density of the liquid in terms of acoustic and optic phonons just like for the solid without implying any long-range periodic order [39]. The band around 50 cm^{-1} represents hydrogen-bond bending while that at 180 cm^{-1} originates from H-bond stretching. The resonances are enormously broadened because of the overwhelming disorder as compared to the solid, and the concomitant dephasing among the intermolecular vibrators of the ensemble. An analysis of the liquid spectral density has been performed using a modified version of Kubo's stochastic theory of the line shape, that takes into account a discrete size distribution of water aggregates which can dynamically interconvert by breaking and making of hydrogen bonds [42, 43]. Such a simulation then reveals a characteristic correlation time for hydrogen bond dynamics which is of the order of 1 ps at a temperature of 273 K.

The OHD-OKE spectrum reveals yet another band for frequencies in excess of 300 cm^{-1}. This resonance can be assigned to the librational modes of the liquid, which correspond to the restricted rotational motion of the water molecules about their three principal axes of inertia. Since these modes involve the motion of the light hydrogen atoms, the librational band exhibits a pronounced isotope shift. Finally, the OHD-OKE response contains additional zero-frequency (i.e., non-oscillatory) components, which can be fitted phenomenologically by multi-exponential or stretched exponential decays. They are related to collective orientational and translational dynamics that do not obey a classical Arrhenius temperature dependence. Attempts have been made to discuss and analyze these long time tails in terms of mode coupling theory for glass-forming liquids. Furthermore, for liquid water between the normal melting and boiling points a detailed comparison has been made between the complementary techniques of OHD-OKE and THz time-domain spectroscopy [42, 43].

21.3.3
Vibrational Relaxation in Water

Vibrational energy relaxation (VER) plays a central role in chemical reaction dynamics [47, 48]. Femtosecond mid-IR pump–probe spectroscopy has been used extensively in the past to understand the mechanisms and the time scales of VER, in particular in hydrogen-bonded liquids such as water and alcohols [49, 50]. In pure liquid H_2O, the fundamental OH-stretching transition was excited and the subsequent relaxation of the excess vibrational energy, ultimately leading to a canonically heated laser focal volume, has been studied in great detail by Bakker and coworkers [51] and others [52–55]. It was found that, immediately after photo-excitation, the vibrational energy transfers and delocalizes resonantly over a large number of neighboring water particles via dipole–dipole coupling similar to Förster's resonant energy transfer. This initial process occurs within the first 200 fs and is evidenced by an ultra-rapid loss of memory of the transition dipole orientation as measured by the transient anisotropy (see Section 21.2.1). Further complications arise from the strong Fermi coupling between the OH stretching fundamental and the first overtone of the bending vibration.

To circumvent this problem, researchers have investigated the dynamics of VER of highly diluted mono-deuterated water (HOD) in light (H_2O) [56–59] and heavy (D_2O) water [59–62]. These systems provide a highly dilute solute with spectroscopically distinct OH (or OD) oscillators that are immersed in a bath of OD (or OH) oscillators. For example, one can excite the OH stretching vibration of HOD around 3400 cm^{-1} and one can probe the ensuing relaxation processes in the same OH-stretching spectral window without any perturbing signal contributions that may arise from the surrounding D_2O solvent. The OD-stretch resonance is located around 2500 cm^{-1} and, hence, resonant dipole–dipole coupling is suppressed. In addition, the fundamental of the HOD bending mode is located at 1450 cm^{-1} and

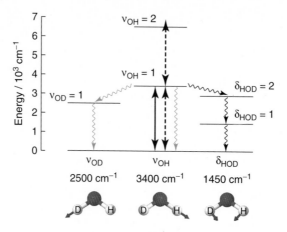

Figure 21.17 Vibrational manifold of mono-deuterated water (HOD) with the OD-stretch (left), the OH-stretch (middle), and the bending vibration (right). Spectroscopic transitions exploited in femtosecond mid-IR pump and probe experiments are highlighted as vertical arrows. Possible VER pathways are indicated by the wavy arrows.

we can expect its second excited state to be Fermi resonant with neither the OH-nor the OD-stretching fundamental. The solute vibrational manifold relevant to femtosecond-mid-IR experiments is summarized in Figure 21.17.

MD simulations by Hynes and coworkers suggest that a fundamental OH-stretching excitation of HOD in D_2O will initially relax into the second excited state of the bending vibration, and that this step is rate-limiting [63]. The first pump–probe experiment under ambient conditions provided a lifetime of the first excited state of the OH stretch in the range between 500 fs and 1 ps, depending upon the choice of the pump and the probe wavelengths.

Note the importance of hydrogen bonding for the rate and the mechanism of vibrational relaxation as the time constant for the stretch-to-bend overtone transition is ~50 ps for monomeric water in deuterated chloroform solution [64]. Of course, the intermolecular interactions of the H_2O molecules with their nonpolar surroundings are much weaker. But, furthermore, for the solute particles to relax the OH-stretching quantum into the second excited state of the bending vibration they are required to deposit the energy mismatch into solvent degrees of freedom. The energy gap between $\nu_{OH} = 1$ and $\delta_{HOD} = 2$ is approximately 500 cm^{-1}, which matches nicely the intermolecular spectral density of restricted translational and restricted rotational modes of an H_2O and D_2O solvent. Those energy accepting modes are, however, entirely absent in a nonpolar non-associating liquid. Compare, for example, the depolarized intermolecular Raman spectral density of liquid CCl$_4$ (see Figure 21.10) with that of liquid water (see Figure 21.16). This fundamental difference easily explains the orders of magnitude slower OH-stretching lifetime in halogenated hydrocarbons.

Coming back to HOD in D_2O, the broad range of lifetimes was attributed to a distribution of hydrogen-bonded configurations of the solute molecule, whose OH-stretching resonance appears to be partially inhomogeneous on the time scale of VER. Accordingly, the relaxation is faster for a sub-ensemble of strongly hydrogen-bonded OH oscillators absorbing at the red edge of the steady state OH stretch band as compared to a weakly hydrogen-bonded sub-ensemble excited at the blue edge. As a consequence, the spectro-temporal OH-stretching response of HOD in D_2O is very complex and highly reminiscent of spectral diffusion dynamics [65].

Indeed, three-pulse IR-echo peak shift measurements by Tokmakoff and coworkers [66] indicated that the decay of the correlation function for OH-frequency fluctuations is biphasic with a fast ~70-fs component bringing about roughly 80% of the overall decay and a slower 1.2-ps component accounting for the remainder. In addition, a heavily damped oscillation with a period of ~160 fs was seen to contribute to the correlation function, which was assigned to coupling of the OH-stretching transition dipole to the intermolecular H-bond stretching mode (see previous section). The long-time tail of the correlation function was ascribed to collective structural reorganizations involving hydrogen-bond breakage and formation in the solvent surrounding the OH-oscillator. Since, under ambient conditions, vibrational relaxation occurs on similar time scales as the structural reorganization, the pump–probe data are highly non-exponential in time.

The question arises as to whether these two physically distinct processes can be disentangled cleanly by conventional pump–probe spectroscopy. With this challenge in mind, Schwarzer *et al.* have studied the detailed temperature and density dependence of the mid-IR response of HOD dissolved in liquid-to-supercritical D_2O [67, 68]. Figure 21.18 displays representative differential absorbance spectra for this system as a function of the delay at an elevated temperature and pressure. The expected general features of ground state bleach and excited state absorption can be seen, both of which decay in concert on a time scale of 2 ps. Most strikingly, however, these two signal contributions are separated by a well defined isosbestic point. Probing at 3350 cm^{-1} yields a time-resolved transient that is indeed delay-independent within the signal-to-noise ratio. This finding is clearly in contradiction to the room temperature, normal pressure data, which are heavily perturbed by spectral diffusion. Furthermore, over a wide range of temperatures and densities, the excited state absorption decays strictly in a single-exponential fashion – again, in remarkable disagreement with the data obtained under ambient conditions. These findings immediately indicate that, regardless of the actual pathways, depopulation of the first excited state of the OH stretch is indeed rate-limiting and any possible intermediate vibrational state of the solute (such as, e.g., the second excited state of the bending mode) must be short-lived so as to become negligibly occupied at all times. In fact, the time constants for ground state recovery are within the error margins identical to the excited state absorption decay. Furthermore, the isosbestic point and the single-exponential decays rule out any signatures of spectral diffusion dynamics, quite in contrast to ambient conditions [67, 68].

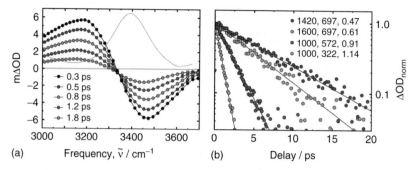

Figure 21.18 Femtosecond mid-IR pump–probe spectroscopy in the OH-stretching spectral region of 2% mono-deuterated water (HOD) in light water (H$_2$O) with excitation at 3530 cm^{-1}. (a) Pump-induced spectra for representative pump–probe time delays. The data were recorded at 398 K and 500 bar corresponding to a density 53.4 mol l^{-1}. The gray curve represents the linear absorption spectrum of the sample under these conditions. (b) Decay of the excited state absorption for various thermodynamic conditions. The numbers indicate pressure in bar, temperature in K, and density in mol l^{-1} of the sample.

The temperature-dependent inverse lifetime of the OH-stretching excited state can be plotted as a function of the bulk density, as shown in Figure 21.19. For a given temperature the data can, in principle, be fitted surprisingly well by an isolated binary collision (IBC) model using an attractive hard sphere solute embedded in a Lennard-Jones solvent, as exemplified for the 697 K isotherm (Figure 21.19(a), curve covering densities from 0 to 0.8 g cm^{-3}). However, as discussed in detail in Refs [67, 68], IBC theory fails miserably in reproducing even qualitatively the temperature dependence of the isochoric VER rate coefficient. A closer inspection of Figure 21.19 reveals that, at constant density, the rate coefficient increases with decreasing temperature. In other words, VER accelerates upon cooling of the liquid! This intriguing discovery is yet another manifestation of the famous anomalous behavior of water and finds its explanation in the temperature and density dependence of the local hydrogen-bonded solvation structure around each H$_2$O (here: HOD) molecule in the liquid.

According to Lawrence and Skinner [69], the dominant effect responsible for the apparent density dependence is the density dependence of the stretch-to-bend overtone energy gap, that is, a solvent shift of the vibrational eigenstates of the solute. Such an interpretation is supported by the observation that the OH-stretching absorption spectrum exhibits a pronounced hypsochromic shift with decreasing density, as demonstrated nicely by Figure 21.15, while at the same time the fundamental bending resonance remains more or less at the same spectral position. Interestingly, when plotting the VER rate coefficient against the density and temperature dependent dielectric constant, ε, all data collapse onto a single straight line. Furthermore, the frequency of the maximal OH-stretching absorption also follows a nearly perfect linear correlation with ε. To understand the physical rationale behind this correlation it is instructive to refer to MD simulations on water that were originally intended to calculate ε from ambient to supercritical

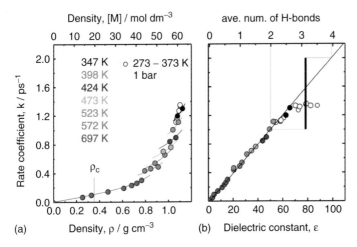

Figure 21.19 Temperature-dependent rate coefficient for vibrational energy relaxation of HOD in H_2O as a function of the solvent density (a) and the solvent dielectric constant (b). The vertical bar indicates the range of rate constants found at room temperature, ambient pressure. The top axis converts the temperature and density dependent dielectric constant of water into an average number of hydrogen bonds per water molecule.

conditions [70]. These simulations relied on a fluctuating charge potential for water (TIP4P-FQ) and employed very stringent structural and energetic criteria for identifying hydrogen bonds. The MD calculations suggest that the dielectric constant is an accurate measure for the average number of hydrogen bonds a water molecule is engaged in. Obviously, this number is a much better probe for the local density in the solvation structure than the bulk density. If indeed the dominant density effect is a modification of the stretch-to-bend overtone energy gap (as the linear spectra indeed suggest) then it comes as no surprise that the VER rate coefficient correlates so nicely with the dielectric constant.

We finally point out that the data seem to depart from the straight $k(\varepsilon)$-line at a dielectric constant of about 70. This deviation is not an artifact of the experiment but is fully reproducible. It appears that, for sufficiently high temperatures, spectral diffusion is much faster than energy transfer. The spectra show an isosbestic frequency, the experimentally observed decays are single-exponential and their inverse time constants represent ensemble-averaged population relaxation rates. As already mentioned, under ambient conditions where $\varepsilon = 78$, spectral diffusion due structural reorganization within the hydrogen-bond network is no longer faster than VER but instead occurs on the same time scale. It can be shown easily that under such circumstances, deviations from the linear $k(\varepsilon)$-relation must occur because the long-time decay of the $\nu_{OH} = 1$ population is biased toward slower relaxing HOD molecules that are only weakly connected to the hydrogen-bond network. Hence, the rate constants obtained by numerically fitting the tail of the non-exponential signals fall below the $k(\varepsilon)$-line, which in turn represents a rate that is averaged over the entire ensemble rather than over a sub-ensemble of weakly-bonded species [67, 68].

Conceptually similar studies have been carried out on the OD-stretching vibration of HOD in H_2O, including the full temperature and density dependence of the energy relaxation dynamics [71]. Since, under ambient conditions, the lifetime of the first excited OD-stretching state of HOD in light water is already a factor of 2 longer as compared to its OH counterpart in heavy water, the measured rate should represent an ensemble-averaged quantity, even at the lowest temperatures. As a result, the determined rate coefficients indeed follow a strictly linear $k(\varepsilon)$-correlation giving undisputable evidence for a clear separation of the time scales of VER and spectral diffusion that persists in the HOD/H_2O system from gas-like densities all the way up to compressed liquid-like densities. In passing, we finally mention that on varying the thermodynamic conditions, the mechanism for OD-stretching relaxation of HOD in H_2O can be fully controlled. Referring to Figure 21.17, the excited OD-stretching state decays primarily through the second excited bending state ($\delta_{HOD} = 2$) when the density is low and the OD-stretch-to-bend overtone energy gap is as small as the thermal energy. In contrast, increasing the density widens this energy gap and the direct decay of $\nu_{OD} = 1$ into the vibrational ground state prevails [71].

21.3.4
Vibrational Relaxation in Ammonia

An ammonia (NH_3) molecule has three hydrogen atoms and a lone electron pair. Just like in the case of water (H_2O), these features enable the NH_3 molecules to act as a hydrogen-bond donor and a hydrogen-bond acceptor simultaneously. Therefore, liquid ammonia is often cited in chemistry textbooks as another example for associated liquids forming extended networks of hydrogen-bonds that are highly random both in space and time, just like in water. Indeed, X-ray diffraction reveals a crystal structure of solid NH_3 in which each molecule is H-bonded to six nearest neighbors, thereby acting even as a triple H-donor and a triple H-acceptor at the same time [72]. Yet, experimental verification for the existence of H-bonds in the liquid is much more difficult to obtain.

To test the significance of H-bonding for VER in this fluid, and to provide a direct comparison to the complementary studies on water, Schäfer *et al.* decided to carry out femtosecond-MIR-spectroscopy on the stretching vibrations in liquid-to-supercritical NH_3. Since the NH-stretching region of NH_3 is heavily perturbed by a Fermi resonance between the stretching fundamental, ν_1, and the first overtone of the anti-symmetrical bending mode, $2\nu_4$, it was focused on the ND-stretch of NH_2D (the solute) in NH_3 (the solvent) [73]. The mono-deuteration lifts the twofold degeneracy of the anti-symmetric stretching and deformation modes of NH_3 and results in pairs of stretching or deformation fundamentals whose components have either a_1 or b_2 vibrational symmetry, that is, they are either symmetric or anti-symmetric with respect to the C_S plane of symmetry of the mono-deuterated ammonia (see Figure 21.20a) [74].

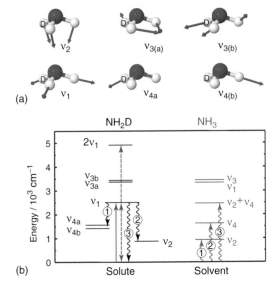

Figure 21.20 (a) Vibrational modes of ammonia. (b) Comparison of the vibrational manifolds of mono-deuterated ammonia (NDH$_2$, solute) and light ammonia (NH$_3$, solvent). Spectroscopic transitions exploited in femtosecond mid-IR pump and probe experiments are highlighted as vertical arrows. Possible VER pathways are indicated by the wavy arrows.

There are, in principle, three possible relaxation pathways for an initial ND-stretching excitation. It turns out that all of them are almost resonant with a vibrational excitation in the solvent. The solute ND-stretching fundamental can decay into either fundamental of the asymmetric deformations, with the energy release being absorbed by the solvent umbrella mode (pathway 1 in Figure 21.20). Alternatively the solute may depopulate the excited ND-stretching state by relaxing into its own umbrella fundamental. This requires the solvent deformation mode to take up the vibrational excess energy (pathway 2). Finally, the direct recovery of the solute ground state is also possible because the solvent exhibits an umbrella-deformation combination state that can be excited to compensate for the energy mismatch. Since all three pathways are resonant within thermal excitations, and the vibrationally excited solute is strongly coupled to the solvent via hydrogen-bonding, we would anticipate the relaxation dynamics to be ultrafast – that is, similar to water under comparable thermodynamic conditions.

Inspecting first the linear mid-IR absorption spectrum of the system NDH$_2$ in NH$_3$ (see Figure 21.21) [75, 76], we notice a red-shift of the ND-stretching band with increasing density and, hence, with decreasing inter-particle distance, r_{bulk}. As mentioned above, such behavior is indicative of hydrogen bonds but it does not prove their existence unequivocally. The overall r_{bulk}-dependence of the frequency of maximal absorbance is considerably weaker as compared to the complementary HOD/H$_2$O system, which can simply be understood in terms of their different hydrogen-bond energies. After all, ammonia is still a gas under ambient conditions.

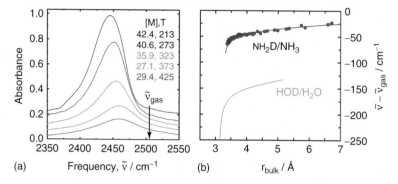

Figure 21.21 (a) ND-stretching absorption spectrum of NH_2D in NH_3. The numbers indicate the density in $mol\,l^{-1}$ and the temperature in K. (b) Dependence of the spectral position of the ND-stretching resonance of NH_2D in NH_3 on the mean inter-particle distance obtained from the bulk density. For comparison, the corresponding dependence of the system HOD in D_2O is also shown.

A representative pump–probe spectrum is shown in Figure 21.22 where excited state absorption and ground-state bleach decay on time scales of ~10 ps [76]. Again an isosbestic probe frequency separating absorption from bleach/emission can be observed, indicating that the depopulation of the initially prepared $v = 1$ ND-stretching state is obviously rate-limiting for recovery of the vibrational ground state. Furthermore, spectral diffusion appears to be unimportant. Figure 21.22 also shows the density dependence of the rate coefficient for various temperatures. We note that in the thermodynamic range investigated, ND-stretch VER in ammonia is

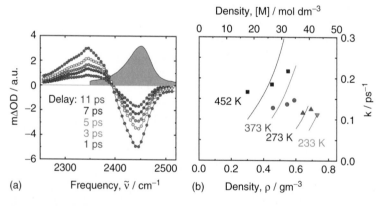

Figure 21.22 (a) Mid-infrared pump–probe spectra of NH_2D in NH_3 following an initial ND-stretching excitation with femtosecond pulses centered at 2450 cm^{-1}. The solute concentration was 6 mol% and the sample was kept at 233 K and 1050 bar, corresponding to a solvent density of 43 $mol\,l^{-1}$. The linear absorption spectrum of the sample is reproduced in gray. (b) Density-dependence of the relaxation rate coefficient for various temperatures. The solid curves represent model calculations based on Landau–Teller theory. For details, see Ref. [76].

indeed quite fast – the lifetime varies between 4 and 10 ps, which is about an order of magnitude faster than the OH-stretch relaxation of water in a nonpolar solution. The ND-stretching lifetimes of NDH_2 in NH_3 are comparable in magnitude to those of the OD-stretching lifetime of HOD in H_2O although at considerably lower temperature [76].

A comparison of the isochoric temperature dependences of the VER rates in the two systems is highly informative. Whereas for water the rates were increasing with decreasing temperature, quite the opposite behavior is observed in ammonia. A thermal acceleration of the relaxation dynamics is what would be expected on the basis of IBC theory. In its simplest form, IBC predicts the relaxation rate to be proportional to the collision frequency. Since the latter quantity scales linearly with the mean relative velocity of the colliding particles, we would expect the relaxation rates to be proportional to the square root of the temperature (neglecting any T-dependence of the transition probability per collision). Quite surprisingly, the data seem to be in qualitative agreement with such a simple interpretation [76].

Following up on such a notion, Schäfer *et al.* analyzed the experimental relaxation rates using the classical Landau–Teller theory for vibrational relaxation [76]. The theory requires the spectral density of the fluctuating solvent forces projected onto the vibrational coordinate as additional input. These were taken from MD simulations on a simple spherically symmetric oscillator (a "breathing sphere" mimicking the vibrationally excited solute) embedded in a Lennard-Jones solvent. The results of these simulations are confronted with the experimental data in Figure 21.22 (b) (solid curves).

It can be seen that the temperature dependence is reproduced qualitatively rather well, considering the simplicity of the model. Notice that the model relied on totally unspecific intermolecular interactions between the solute and the solvent. Hydrogen-bonding was not even taken into account at all. From this qualitative agreement, we come to the surprising conclusion that hydrogen bonding is rather unimportant for the dynamics of VER in ammonia [76].

The theory fails to reproduce the isothermal density-dependence of the relaxation rate. This is, however, not a consequence of the theory being inappropriate for hydrogen-bonded systems but rather due to highly specific solvent-induced shifts of the vibrational eigenstates. It appears as if upon decreasing the density at constant temperature the experimental data approach a finite intercept. A nonzero rate at vanishing ρ is however entirely unphysical, from which we are led to infer that $k(\rho)$ must exhibit an inflection point at low densities. Obviously, by decreasing the density, the quantum mechanical resonance between solute and solvent energy gaps, which actually facilitates an efficient transfer of vibrational energy between the two particles, becomes gradually better, thereby effectively slowing down the approach of $k(\rho) \rightarrow 0$ upon $\rho \rightarrow 0$ [76].

Which of the three pathways for energy relaxation sketched in Figure 21.20 dominates VER in the NH_2D/NH_3 system? This final question can, in principle, again be addressed by Landau–Teller theory combined with classical MD simulations. However, they need to go beyond a simple breathing sphere model and have to include realistic force fields for flexible solute and solvent particles.

21.3.5
Vibrational Dynamics of H-Bond Wires

We have seen in the previous section that it is possible to obtain highly detailed information regarding the rates and mechanisms of VER in hydrogen-bonded liquids by means of femtosecond mid-IR spectroscopy. In neat liquid water, a very peculiar complication arose from the fact that an excited OH-stretching oscillator is surrounding by other hydroxyl groups that can resonantly accept the vibrational excitation. In fact, the excitation may very quickly hop from one molecule to another, and so on. The question may then duly be raised as to whether the OH-stretching vibration is actually localized on an individual molecule or fully delocalized over many particles.

To address such an issue, it is necessary to control the spatial extension of the H-bond network and to study the vibrational dynamics of a water dimer, trimer, tetramer, and so on, until one gradually approaches the bulk. Femtosecond mid-IR pump–probe spectroscopy of alcohol oligomers surrounded by nonpolar liquids (see Figure 21.12 for ethanol aggregates in CCl_4) has indeed been carried out recently [77, 78]. These systems are, however, stoichiometrically ill-defined and it is impossible to differentiate between the various oligomers of different sizes that may coexist in the sample. In addition, chain oligomers may coexist with cyclic aggregates. Therefore, a system that allows for full control over the number of hydroxyl groups in the H-bond network and the network dimensionality would be ideal.

Seehusen *et al.* have proposed diastereomeric poly-alcohols as quasi-one-dimensional model systems for studying the effect of network length on the dynamics of OH-stretching VER and spectral diffusion. As described by Paterson and coworkers, saturated hydrocarbons bearing a 1,3-*anti*-methyl substitution pattern were used as a scaffold to support the H-bond wire [79]. The repulsive interactions between the bulky methyls guide the hydrocarbon backbone into an extended rigid conformation similar to syndiotactic polypropylene (see Figure 21.23a,b). The remaining C-atoms along the backbone can now be used to stereoselectively attach hydroxyl groups, either in an all-*syn* or an all-*anti* orientation (Figure 21.23c and d). In the former case, an extended H-bond wire can easily be established by the molecule, whereas in the latter case the methyls seem to inhibit such a network formation.

The linear mid-IR absorption spectra of such polyalcohols in a weakly interacting nonpolar solvent are shown in Figure 21.24 for the case of the all-*syn* orientation and for various numbers of hydroxyl groups attached to the hydrocarbon backbone [80]. It can be seen that the OH-resonance shifts markedly to lower frequencies with increasing chain-length. Considering our general discussion of the influence of hydrogen bonding on the stretching frequencies of intramolecular oscillators (see Section 21.3.1), one would interpret this finding with an average H-bond strength that is increasing with increasing network length. But in going from a diol and a tetrol to a hexol we are not really increasing the average H-bond energies. Instead, only alcoholic repeat units were added to the molecule, which clearly demonstrates

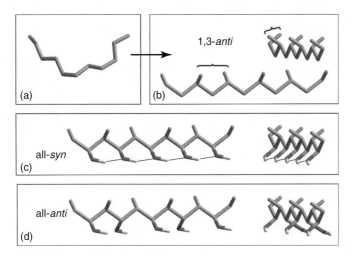

Figure 21.23 Synthetic hydrogen-bond wires based on conformationally controlled hydrocarbons (a and b). The all-*syn* stereochemical orientation of the OH-groups in combination with the all-*anti* methyl substitution pattern facilitates network formation (c) whereas an all-*anti* orientation inhibits it (d).

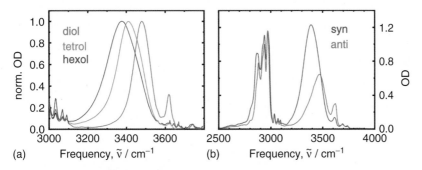

Figure 21.24 (a) Linear absorption spectra in the OH-stretching spectral region of synthetic hydrogen-bond wires derived from all-*syn* polyalcohols (diol to hexol from right to left) as a function of the network length and (b) comparison of the linear spectra of the all-*syn* and the all-*anti* tetrol. Deuterated chloroform was used as the solvent in all cases.

that the intramolecular OH-stretching oscillators are much better described as normal modes that are fully delocalized over the H-bond network. The mechanism by which the OH-oscillators are coupled (i.e., through-bond or through-space via transition dipole coupling) remains to be investigated. A clarification of this issue will certainly convey some very important implications regarding the assignments of the OH-stretching resonances of chain-like oligomers of alcohols in nonpolar solvents (see Figure 21.12) as well as regarding the nature of the OH-stretching spectra of liquid water and aqueous solutions.

The spectra of the polyalcohols are enormously broad and the hexol OH-stretching band already has a bandwidth that is almost identical to the corresponding resonance of HOD in heavy water [80]. Surprisingly, the all-*anti* counterparts are characterized by two resonances rather than one. A sharp feature above 3600 cm^{-1} is clearly reminiscent of "free" OH-oscillators that are not donating into a hydrogen-bond to its adjacent hydroxyl group. Instead, they are primarily interacting with the nonpolar surroundings. Whether they are actually accepting a hydrogen-bond from its other neighbor remains unclear at this stage. The free-OH resonance appears at the expense of the broad red-shifted absorption band prototypical for hydrogen-bonded OH, which in turn is diminished by about a factor of 2 as compared to the all-*syn* tetrol. The fact that a "bound-OH" band is observed is surprising in itself as the formation of a hydrogen-bond network was initially not at all anticipated [80].

Density functional theory (DFT) reveals an optimized structure for the all-*syn* tetrol that is fully consistent with the predictions from Figure 21.23 [80]. On the other hand, the same DFT methodology identifies indeed a minimum energy structure for the all-*anti* diastereomer that supports the unexpected experimental result and reveals an H-bond network that is highly strained. Room temperature Langevin MD simulations (cf. Figure 21.25) show that the hydrogen-bond networks of the two diastereomeric systems are entirely different in their dynamics [80]. Whereas the all-*syn* system supports an extended H-bond wire that is rigid and highly stable over tens of picoseconds, ultrafast H-bond breakage and formation is observed for the all-*anti* counterpart on time scales as short as 100 fs. These marked differences are also reflected in the spectro-temporal responses of the polyalcohols following an initial ultrafast OH-stretching excitation [80].

Inspecting first the transient spectra of the all-*anti* polyalcohol (see Figure 21.26, here: tetrol), we notice the expected signal contributions: a ground state bleach/stimulated emission at the pump frequency and the anharmonically

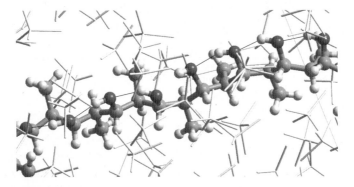

Figure 21.25 Snapshot of a classical molecular dynamics simulation (AMBER force field) of an all-*syn* octol (ball-and-stick rendering) in chloroform (gray sticks) at room temperature highlighting an instantaneous defect in the hydrogen-bond wire.

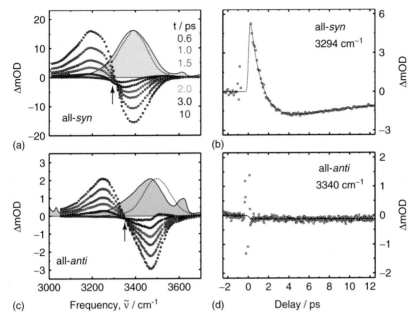

Figure 21.26 Pump-induced transient spectra (a,c) and time-resolved pump–probe transients (b,d) for the all-*syn* (a,b) and the all-*anti* tetrol (c,d). Deuterated chloroform was used as the solvent in all cases. The gray shaded spectra are the linear absorption spectra of the poly-alcohols and the dashed curves represent the pump spectral profiles. Notice the dynamic red-shift of the bleach/emission in the all-*syn* species causing the isosbestic point to be absent and indicating the mid-infrared induced disruption of the wire with breakage of an H-bond.

shifted absorption of the OH-stretching excited state. Both components decay in concert and form an isosbestic point at a frequency of 3340 cm. A time-resolved transient at that detection frequency is stationary within the signal-to-noise ratio, as required. The decay kinetics occur with a time constant of 1.2 ps and reflect VER by intramolecular vibrational redistribution, presumably into the nearby CH-stretching vibration and low-frequency backbone modes of the tetrol. We can exclude intermolecular energy transfer into both degrees of freedom because in a nonpolar solvent like deuterated chloroform these dynamics are at least an order of magnitude slower (see also Section 21.3.3) [80].

Turning our attention to the all-*syn* tetrol, we again see the expected general features but we also notice marked differences. First, bleach/emission and transient absorption are no longer separated by a clear-cut isosbestic point. In fact, detecting at a frequency close to the zero-crossing region yields a time-resolved pump–probe that is clearly not stationary as it should be for an isosbestic point. Rather, a pump-induced absorption is seen at early times that turns over into a bleach/stimulated emission. The latter in turn recovers on a time scale in excess of 10 ps.

Secondly, an unexpected absorption is observed at later delays, which exhibits a pronounced shift to higher probing frequencies relative to the stationary OH-stretching resonance. In other words, following an ultrafast OH-stretching excitation of the rigid hydrogen-bond wire produces a species whose OH-stretching resonance is blue-shifted compared to the resonance of the network at equilibrium. Obviously, the IR pulse must have caused a substantial disruption of the network by breaking one of the hydrogen-bonded contacts thereby localizing the OH-stretching excitations on shorter fragments and causing a blue-shifted absorption. The data analysis retrieves a time constant for mid-IR induced H-bond fission of 800 fs, that is, the vibrational lifetime is significantly shorter than for the highly flexible network of the all-*anti* tetrol [80].

A detailed comparison with thermal difference spectra of the all-*syn* polyol demonstrates that once H-bond fission has occurred, the system is spectroscopically indistinguishable from the canonically heated molecule, that is, the excess vibrational energy initially deposited into the OH-stretching mode is statistically distributed over all degrees of freedom of the molecule.

Taking all findings together, we can conclude that VER is faster when hydrogen-bond breakage is involved. Whether this means that VER is mediated by H-bond fission, such as in a vibrational predissociation mechanism, remains to be clarified [80]. Theoretical calculations of the non-adiabatic coupling for the transition $v_{OH} = 1 \rightarrow v_{OH} = 0$ using a Fermi Golden Rule treatment and an optimized Lippincott–Schroeder potential have been performed by Staib and Hynes [81]. The theoretically predicted OH-stretching lifetimes are, however, about an order of magnitude too slow as compared to the vibrational lifetimes observed experimentally. Based on this discrepancy one might be tempted to dismiss entirely a vibrational predissociation mechanism for VER in this system. However, one of the most significant shortcomings of the theory is the neglect of the delocalized character of the OH-stretching vibrations in linear arrays of hydroxyl groups such as those discussed here.

At any rate, we close this section by re-emphasizing that in neat liquids the intermolecular modes of the molecule might also very well be delocalized over a large number of particles. Contemporary theory and simulations on water have recently begun to pursue such a notion of delocalized normal modes of the liquids from the local vibrational hydroxyl chromophores whose transition dipoles are considerably coupled (vibrational excitons) in the presence of dynamic disorder [82–84].

21.4
Summary

We have highlighted only a small fraction of the research that can be carried out with femtosecond vibrational spectroscopies and focused only on vibrational dynamics in hydrogen-bonded systems.

Highly impressive experiments using multidimensional variants of IR and Raman spectroscopies from several laboratories deal, for example, with dynamical structure elucidation of small globular peptides containing only a few amino acids to larger proteins, protein fibers, and bundles [85], as well as membrane-bound proteins [86]. These efforts also encompass the study of the folding dynamics of proteins containing electronic chromophores thereby rendering them photo-switchable systems [87]. There, the underlying principle is always the utilization of the couplings between the numerous modes of a polyatomic system for projecting their resonances onto two frequency axis rather than only one. Just like in two-dimensional NMR, the observation of cross peaks lets us establish correlations among the vibrational degrees of freedom from which the structure can then be deduced.

Dynamic exchange phenomena have been explored on a variety of chemical transformations ranging from solute–solvent complexation [88] and dissociation to fluxional rearrangements (i.e., pseudo-rotations) in transition metal compounds [89]. The dynamics may be spontaneous, such as equilibrium carbon-carbon single bond rotations [90], or they may be photo-induced, such as ligand migration in proteins [91]. Here, the interconversion between spectroscopically distinct species gives rise to cross peaks from which the system of coupled elementary reactions and the system of coupled rate equations can be established. Together, they provide the microscopic mechanism of the chemical transformation being studied.

Femtosecond vibrational spectroscopies, and in particular their multidimensional variants, have matured tremendously over the last few years but nevertheless technological and conceptual advancements continue to be made at an impressive rate. For example, with acousto-optic devices researchers are able to synthesize arbitrary mid-IR fields, like pairs or even trains of pulses, each with a well-defined spectro-temporal shape and width [92]. Full control over the phase of the driving fields brings multi-dimensional IR spectroscopy conceptually much closer to modern nuclear magnetic resonance. It is, in particular, this intriguing analogy with NMR, combined with an ultimate time-resolution well into the sub-100 fs regime, that makes these vibrational spectroscopies most promising techniques for a great number of exciting applications in fields as diverse as structural biology, template chemistry, or materials science.

References

1. Chalmers, J.M. and Griffiths, P.R. (eds) (2002) *Handbook of Vibrational Spectroscopy*, John Wiley & Sons & Inc., New York.

2. Schrader, B. (ed.) (1995) *Infrared and Raman Spectroscopy: Methods and Applications*, Wiley-VCH Verlag GmbH, Weinheim.

3. Siebert, F. and Hildebrandt, P. (2008) *Vibrational Spectroscopy in the Life Science*, Wiley-VCH Verlag GmbH, Weinheim.

4. Johnson, T.J. and Zachmann, G. (2000) *Introduction to Step-scan FTIR*, Bruker Optik GmbH.

5. Johnson, T.J., Simon, A., Weil, J.M., and Harris, G.W. (1993) *Appl. Spectrosc.*, **47**, 1376.

6. Petrov, V., Rotermund, F., and Noack, F. (2001) *J. Opt. A: Pure Appl. Opt.*, **3**, R1.

7. Cerullo, G. and De Silvestri, S. (2003) *Rev. Sci. Instrum.*, **74**, 1.

8. Nibbering, E.T.J., Fidder, H., and Pines, E. (2004) *Annu. Rev. Phys. Chem.*, **56**, 337.

9. Kukura, P., McCamant, D.W., and Mathies, R.A. (2007) *Annu. Rev. Phys. Chem.*, **58**, 461.

10. Weiner, A.M. (2009) *Ultrafast Optics*, Wiley-VCH Verlag GmbH, Hoboken, NJ.

11. Lee, D. and Albrecht, A.C. (1985) in *Advances in Infrared and Raman Spectroscopy*, vol. 12 (eds R.J.H. Clark and R.E. Hester), John Wiley & Sons, Inc., New York, p. 179.

12. Mukamel, S. (1995) *Principles of Nonlinear Optical Spectroscopy*, Oxford University Press, New York.

13. Fleming, G.R. (1986) *Chemical Applications of Ultrafast Spectroscopy*, Oxford University Press, New York.

14. Levenson, M.D. and Kano, S.S. (1988) *Introduction to Nonlinear Laser Spectroscopy*, Academic Press, London.

15. Romanov, D., Filin, A., Compton, R., and Levis, R. (2007) *Opt. Lett.*, **32**, 3161.

16. Choe, J.C. and Kim, M.S. (1986) *Bull. Korean Chem. Soc.*, **7**, 63.

17. Leonhardt, R., Holzapfel, W., Zinth, W., and Kaiser, W. (1987) *Chem. Phys. Lett.*, **133**, 373.

18. Beaud, P., Frey, H.M., Lang, T., and Motzkus, M. (2001) *Chem. Phys. Lett.*, **344**, 407.

19. Legare, F., Evans, C.L., Ganikhanov, F., and Xie, X.S. (2006) *Opt. Express*, **14**, 4427.

20. Evans, C.L., and Xie, X.S. (2008) *Annu. Rev. Anal. Chem.*, **1**, 883.

21. Volkmer, A. (2010) *Emerging Raman Applications and Techniques in Biomedical and Pharmaceutical Fields: Biological and Medical Physics, Biomedical Engineering*, (eds P. Matousek and M. Morris), Springer-Verlag, Berlin, p. 111.

22. Reimann, K. (2007) *Rep. Prog. Phys.*, **70**, 1597.

23. Davies, A.G., Linfield, E.H., and Johnston, M.B. (2002) *Phys. Med. Biol.*, **47**, 3679.

24. Feurer, T., Stoyanov, N.S., Ward, D.W., Vaughan, J.C., Statz, E.R., and Nelson, K.A. (2007) *Annu. Rev. Mater. Res.*, **37**, 317.

25. Schmuttenmaer, C.A. (2004) *Chem. Rev.*, **104**, 1759.

26. Kindt, J.T. and Schmuttenmaer, C.A. (1997) *J. Chem. Phys.*, **106**, 4389.

27. Kindt, J.T. and Schmuttenmaer, C.A. (1999) *J. Chem. Phys.*, **110**, 8589.

28. Rønne, C., Åstrand, P.-O., and Keiding, S.R. (1999) *Phys. Rev. Lett.*, **82**, 2888.

29. Nielsen, O.F. (2001) in *Handbook of Raman Spectroscopy* (eds I.R. Lewis and H.G.M. Edwards), Marcel Dekker Inc., New York.

30. Egelstaff, P.A. (2002) *An Introduction to the Liquid State*, Oxford University Press, London.

31. Nelson, K.A. and Ippen, E.P. (1989) *Adv. Chem. Phys.*, **75**, 1.

32. Smith, N.A. and Meech, S.R. (2002) *Int. Rev. Phys. Chem.*, **21**, 75.

33. Kivelson, D. and Madden, P.A. (1980) *Annu. Rev. Phys. Chem.*, **31**, 523.

34. Vöhringer, P. and Scherer, N.F. (1995) *J. Phys. Chem.*, **99**, 2684.

35. Grabowski, S.J. (ed.) (2006) *Hydrogen Bonding – New Insights*, vol. 3, Springer, Amsterdam, p. 519.

36. Graener, H., Ye, T.Q., and Laubereau, A. (1989) *J. Chem. Phys.*, **91**, 1043.

37. Lippincott, E.R. and Schroeder, R. (1955) *J. Chem. Phys.*, **23**, 1099.

38. Kandratsenka, A., Schwarzer, D., and Vöhringer, P. (2008) *J. Chem. Phys.*, **128**, 244510.

39. Walrafen, G.E. (1990) *J. Phys. Chem.*, **94**, 2237.

40. Castner, E.W. Jr., Chang, Y.J., Chu, Y.C., and Walrafen, G.E. (1995) *J. Chem. Phys.*, **102**, 653.

41. Palese, S., Schilling, L., Miller, R.J.D., Staver, P.R., and Lotshaw, W.T. (1994) *J. Phys. Chem.*, **98**, 6308.

42. Winkler, K., Lindner, J., Bürsing, H., and Vöhringer, P. (2000) *J. Chem. Phys.*, **113**, 4674.

43. Winkler, K., Lindner, J., and Vöhringer, P. (2002) *Phys. Chem. Chem. Phys.*, **4**, 2144.

44. Torre, R., Bartolini, P., and Righini, R. (2004) *Nature*, **428**, 296.

45. Mazzacurati, V., Pona, C., Signorelli, G., Briganti, G., and Ricci, M.A. (1981) *Mol. Phys.*, **44**, 1163.

46. Amoruso, A., Benassi, P., Crescentini, L., and Mazzacurati, V. (1998) *Phys. Rev. B*, **57**, 7415.

47. Chesnoy, J. and Gale, G.M. (1984) *Ann. Phys. Fr.*, **9**, 893.

48. Owrutsky, J.C., Raftery, D., and Hochstrasser, R.M. (1994) *Annu. Rev. Phys. Chem.*, **45**, 519.

49. Elsaesser, T. and Bakker, H.J. (2003) *Ultrafast Hydrogen Bonding Dynamics and Proton Transfer Processes in the Condensed Phase*, Kluwer Academic Publisher, Dordrecht.

50. Nibbering, E.T.J. and Elsaesser, T. (2004) *Chem. Rev.*, **104**, 1887.

51. Woutersen, S. and Bakker, H.J. (1999) *Nature*, **402**, 507.

52. Lindner, J., Vöhringer, P., Pshenichnikov, M.S., Cringus, D., Wiersma, D.A., and Mostovoy, M. (2006) *Chem. Phys. Lett.*, **421**, 329.

53. Ashihara, S., Huse, N., Espagne, A., Nibbering, E.T.J., and Elsaesser, T. (2007) *J. Phys. Chem. A*, **111**, 743.

54. Lindner, J., Cringus, D., Pshenichnikov, M.S., and Vöhringer, P. (2007) *Chem. Phys.*, **341**, 326.

55. Chieffo, L., Shattuck, J., Amsden, J.J., Erramilli, S., and Ziegler, L.D. (2007) *Chem. Phys.*, **341**, 71.

56. Gale, G.M., Gallot, G., Hache, F., Lascoux, N., Bratos, S., and Leicknam, J.C. (1999) *Phys. Rev. Lett.*, **82**, 1068.

57. Woutersen, S., Emmerichs, U., and Bakker, H.J. (1997) *Science*, **278**, 658.

58. Nienhuys, H.K., Woutersen, S., van Santen, R.A., and Bakker, H.J. (1999) *J. Chem. Phys.*, **111**, 1494.

59. Deàk, J.C., Rhea, S.T., Iwaki, L.K., and Dlott, D.D. (2000) *J. Phys. Chem. A*, **104**, 4866.

60. Steinel, T., Asbury, J.B., Corcelli, S.A., Lawrence, C.P., Skinner, J.L., and Fayer, M.D. (2004) *Chem. Phys. Lett.*, **386**, 295.

61. Asbury, J.B., Steinel, T., Stromberg, C., Corcelli, S.A., Lawrence, C.P., Skinner, J.L., and Fayer, M.D. (2003) *J. Phys. Chem. A*, **108**, 1107.

62. Bakker, H.J., Rezus, Y.L.A., and Timmer, R.L.A. (2008) *J. Phys. Chem. A*, **112**, 11523.

63. Rey, R., Møller, K.B., and Hynes, J.T. (2002) *J. Phys. Chem. A*, **106**, 11993.

64. Seifert, G., Patzlaff, T., and Graener, H. (2006) *J. Chem. Phys.*, **125**, 154506.

65. Bakker, H.J., Nienhuys, H.K., Gallot, G., Lascoux, N., Gale, G.M., Leicknam, J.C., and Bratos, S. (2002) *J. Chem. Phys.*, **116**, 2592.

66. Fecko, C.J., Eaves, J.D., Loparo, J.J., Tokmakoff, A., and Geissler, P.L. (2003) *Science*, **301**, 1698.

67. Schwarzer, D., Lindner, J., and Vöhringer, P. (2006) *J. Phys. Chem. A*, **110**, 2858.

68. Schwarzer, D., Lindner, J., and Vöhringer, P. (2005) *J. Chem. Phys.*, **123**, 16105.

69. Lawrence, C.P. and Skinner, J.L. (2003) *J. Chem. Phys.*, **119**, 3840.

70. Yoshii, N., Miura, S., and Okazaki, S. (2001) *Chem. Phys. Lett.*, **345**, 195.

71. Schäfer, T., Lindner, J., Vöhringer, P., and Schwarzer, D. (2009) *J. Chem. Phys.*, **130**, 224502.

72. Olovsson, I. and Templeton, D.H. (1959) *Acta Crystallogr.*, **12**, 832.

73. Schwartz, M. and Wang, C.H. (1973) *J. Chem. Phys.*, **59**, 5258.

74. Snels, M., Hollenstein, H., and Quack, M. (2006) *J. Chem. Phys.*, **125**, 194319.

75. Buback, M. (1974) *Ber. Bunsen-Ges. Phys. Chem.*, **78**, 1230.

76. Schäfer, T., Schwarzer, D., Lindner, J., and Vöhringer, P. (2008) *J. Chem. Phys.*, **128**, 064502.

77. Laenen, R., Rauscher, C., and Laubereau, A. (1998) *Chem. Phys. Lett.*, **283**, 7.

78. Gaffney, K.J., Davis, P.H., Piletic, I.R., Levinger, N.E., and Fayer, M.A. (2002) *J. Phys. Chem. A*, **106**, 12012.

79. Paterson, I. and Scott, J.P. (1997) *Tetrahedron Lett.*, **38**, 7445.

80. Seehusen, J., Lindner, J., Schwarzer, D., and Vöhringer, P. (2009) *Phys. Chem. Chem. Phys.*, **11**, 8484.

81. Staib, A. and Hynes, J.T. (1993) *Chem. Phys. Lett.*, **204**, 197.

82. Auer, B.M. and Skinner, J.L. (2008) *J. Chem. Phys.*, **128**, 224511.

83. Paarmann, A., Hayashi, T., Mukamel, S., and Miller, R.J.D. (2009) *J. Chem. Phys.*, **130**, 204110.

84. Falvo, C., Palmieri, B., and Mukamel, S. (2009) *J. Chem. Phys.*, **130**, 184501.

85. Strasfeld, D.B., Ling, Y.L., Shim, S.H., and Zanni, M.T. (2008) *J. Am. Chem. Soc.*, **130**, 6698.

86. Mukherjee, P., Kass, I., Arkin, I., and Zanni, M.T. (2006) *Proc. Natl. Acad. Sci. U.S.A.*, **103**, 3528.

87. Bredenbeck, J., Helbing, J., Behrendt, R., Renner, C., Moroder, L., Wachtveitl, J., and Hamm, P. (2003) *J. Phys. Chem. B*, **107**, 8654.

88. Fayer, M.D. (2009) *Annu. Rev. Phys. Chem.*, **60**, 21.

89. Cahoon, J.F., Sawyer, K.R., Schlegel, J.P., and Harris, C.B. (2008) *Science*, **319**, 1820.

90. Zheng, J., Kwak, K., Xie, J., and Fayer, M.D. (2006) *Science*, **313**, 1951.

91. Bredenbeck, J., Helbing, J., Kolano, C., and Hamm, P. (2007) *ChemPhysChem*, **8**, 1747.

92. Shim, S.H. and Zanni, M.T. (2009) *Phys. Chem. Chem. Phys.*, **11**, 748.

22
THz Technology and THz Spectroscopy: Modeling and Experiments to Study Solvation Dynamics of Biomolecules

David M. Leitner, Martin Gruebele, and Martina Havenith-Newen

Method Summary

Acronyms, Synonyms

- Terahertz (THz)
- Kinetic terahertz absorption spectroscopy (KITA)

Benefits (Information Available)

- THz absorption spectroscopy probes collective low frequency modes of the sample.
- THz absorption technique is a new technique to probe sensitively (label free) fast (sub-ps, ps) solvation dynamics.
- It is very sensitive to probe any changes in the solvation dynamics as caused by the solute.
- It is possible to determine the size of the dynamical hydration shell including all water molecules that are affected in their low frequency (THz motions) – this might include many more water molecules than the static hydration shell which includes only bound water molecules.
- Kinetic THz absorption spectroscopy monitors – in real time – the changes in solvation dynamics during a biological process.

Limitations (Information Not Available)

- THz absorption spectroscopy yields the overall change in solvation dynamics and is not site specific.

22.1
THz Technology

Recently a new window onto water dynamics around biomolecules has opened: terahertz (THz) light, at frequencies between microwaves (MWs) and the infrared (10^{12} Hz = 1 THz). Although far-infrared spectroscopy has been practiced for

Methods in Physical Chemistry, First Edition. Edited by Rolf Schäfer and Peter C. Schmidt.
© 2012 Wiley-VCH Verlag GmbH & Co. KGaA. Published 2012 by Wiley-VCH Verlag GmbH & Co. KGaA.

decades, THz technology is only now becoming an emerging field with a rapidly increasing number of groups participating. The recent development of a new generation of strong THz sources and applications has led to a rapid growth of the field [1, 2].

THz technology has been a great challenge for a long time. The spectral region between MW and IR technology was known as the *"THz gap"*, indicating a lack of powerful THz radiation sources. The previous inherent lack of bright THz sources can be attributed to the fact that classical electronic devices end and photonics devices such as diode lasers do not work in the relevant frequency range. The state of the art of THz sources has been reviewed recently by Tonouchi [3]. An overview of all sources is given in Figure 22.1.

Why has this region been so difficult to access? One reason is that the corresponding frequency is smaller than the band gap of any semiconductor. The last half century has witnessed enormous success in the development and application of diode lasers. The lasing mechanism in the diode laser can be described in the following way. In the same way that atoms and electrons have quantized specific states that they can populate, each diode has a valence band, and a conduction band. At a certain temperature all bands will be filled with electrons up to a certain energy level, which is called the *Fermi energy*. Only the valence band will be populated in the p-doped zone of a diode, but both the conduction and the valence band in

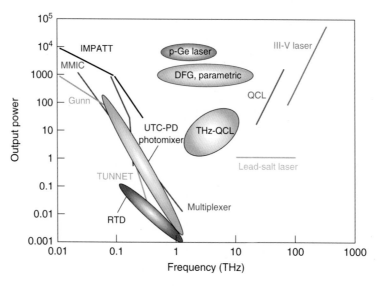

Figure 22.1 The region between 100 GHz and 30 THz is called the *THz region*. Emission power as a function of frequency is plotted on a logarithmic scale. Solid lines are for the conventional THz sources; IMPATT diode stands for impact ionization avalanche transit-time diode, MMIC stands for microwave monolithic integrated circuit, TUNNET stands for tunnel injection transit time and the multiplexer is an SBD frequency multiplier. Ovals denote recent THz sources. The values of the last two are indicated by peak power; others are by continuous power output. Taken from: M. Tonouchi, *Nature Photonics*, **1** [3].

the n-doped zone. Stimulated emission as needed for lasing requires an inverse population, which means that an upper energy level must be more populated than a lower energy level. No lasing can occur as long as only the lower level is populated, or when both the upper level and the lower level are populated equally. However, this situation changes for the diode laser when a voltage is applied. Due to diffusion within the transition zone between the p and n-doped regions, we find a small zone in which the conduction band (corresponding to the upper level) is populated, while the valence band is not fully populated. When a photon with the right energy hits this region it will induce stimulated emission of another photon with the same energy (frequency) and phase. The lasing frequency ν_{Laser} is determined by the energy gap between the conduction and the valence band, $\Delta E = h\nu_{Laser}$. Within the last decades progress in laser engineering has brought about diode lasers with an energy gap designed such that they are emitting radiation from the blue to the red. In the mid-infrared a lead salt diode laser can be tuned between 700 and 2700 cm^{-1}. However, the energy gap in any semiconductor is too large for emission below 700 cm^{-1} in the far-infrared or THz range.

The only lasers now being explored that emit in this frequency range are THz quantum cascade lasers or the p-Ge laser. Quantum cascade lasers use specifically designed periodic structures. They are designed such that adjunct materials have distinct properties. In this way the emitting frequency can be specifically tuned and can be smaller than the natural energy gaps for a diode laser.

The only high power laser source in the THz frequency range is the p-Ge laser with an intrinsic laser process within a subband (not between the conduction and valence bands). In this way the frequency is not restricted by the energy gap but by the energy differences within a band, which is smaller. The lasing mechanism is described in detail in [4]. In short it can be described as follows: Light and heavy holes (orbitals missing an electron) within a subband will be accelerated in crossed electric and magnetic fields. Their maximum energy is given by

$$E_{L,H} = \frac{1}{2} m^*_{L,H}{}^{-1} (p_{L,H})^2 \text{ and } p_{L,H} = m^*_{L,H} E/B \tag{22.1}$$

where $m^*_{L,H}$ is the effective mass of light and heavy holes (a hole is really electrons moving in the opposite direction, and their mobility determines the hole's effective mass), and E and B are the applied electric and the magnetic fields, which are perpendicular to each other. In summary, we obtain $E_{L,H} = 1/2 \, m^*_{L,H}(E/B)^2$, which implies that heavy holes reach a higher energy than light holes. The laser is designed such that the maximum energy of the heavy holes reaches the energy of optical phonons, which leads to a depopulation of heavy hole states by relaxation. In contrast, under the specific lasing conditions (cooling by a helium cryostat below 4 K), the energy of the light holes is too low to undergo any relaxation. In the crossed electric and magnetic fields the light holes gain energy, which can be reemitted by a transition to an empty state corresponding to a heavy hole. This transition frequency can be tuned by tuning the crossed electric and magnetic fields (see equation above), and allows laser operation between 30 and 140 cm^{-1}.

The p-Ge laser is a table top radiation source which exceeds the power of other table top THz radiation sources by several orders of magnitude [3]. Rapid heating upon laser emission stops the lasing process. As a consequence laser emission is achieved only in a pulsed mode. The p-Ge laser is a light source with the following specifications: duty cycle of up to 5%, peak laser power of up to 10 W. Using a special feed back design, a line width of less than 1 MHz can be achieved [4].

A second type of THz source is based on optical rectification: In the same way a diode can rectify an alternating current, a laser pulse can be rectified. If the envelope of the laser pulse is in the picosecond range, the rectified radiation will be THz light. Such an optical electronic rectifier is based upon the generation of radiation by a rapid acceleration and deceleration of charges. The frequency of the emitted radiation will be inversely proportional to the time scale of the generation, acceleration, retardation, and recombination of the accelerated charges. Microwave technology is a well established technology based upon acceleration of electrons, which leads to the emission of tunable MW radiation. However, for the generation of pulses in the THz regime, conventional electronic devices are not suitable because this requires the acceleration and deceleration of charges within less than a picosecond. Grischowski introduced a new concept, which started the THz age of time domain THz pulses obtained by rectification [5]. A review of applications of these THz time domain systems has been given by Schmuttenmaer [1] and Markelz [6].

The generation and detection of THz pulses by THz time domain spectroscopy is illustrated in Figure 22.2.

Short THz pulses of about picosecond total duration (several hundred femtoseconds full width at half maximum) spanning the THz frequency range can be generated by photoconductive switching of near-infrared pulses from a Ti:sapphire

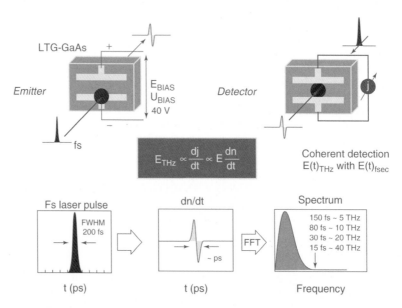

Figure 22.2 Generation and detection of THz pulses by THz time domain spectroscopy.

laser on a low temperature-grown (LTG) gallium arsenide photoconductive emitter. The required short time scales for acceleration are achieved by the short time scales for the optically induced generation of charges. Fast recombination times are realized, as determined by the specific material properties which guarantee fast deceleration.

The specifications of the input pulses are 800 nm wavelength and typically 10–100 fs duration at 500 mW average power. For THz emission, the carriers are accelerated by an external electric field of 30 V. A short photocurrent $J(t)$ flows across the photoconductive gap and an electric field E_{THz} in the THz frequency region is irradiated. The amplitude of the THz field is proportional to the derivative of the photocurrent $J(t)$:

$$E_{THz} \sim \partial J(t)/\partial t \tag{22.2}$$

The following expression can be derived for the peak amplitude of the irradiated electric THz field:

$$E^{peak}_{THz} \sim eJ(1 - R)P_{laser}\, V_b\, T_{int}/(h\nu D^2) \tag{22.3}$$

with e being the elementary charge unit, $h\nu$ the photon energy, P_{laser} the average laser power of the femtosecond laser, V_b the applied bias voltage, J the time averaged conductivity, R the optical reflectivity of the LTG-GaAs semiconductor surface, T_{int} the interval of the pump laser pulses, and D the width of the photoconductive gap.

In order to maximize the THz peak signal, the mobility of the carriers within the semiconductor substrate must be high and the photoconductive gap width must be small. Optimized geometries with respect to high THz peak power and bandwidth yield the following parameters: $W = 10\ \mu m$ for the beam waist, and $D = 5\ \mu m$ for the photoconductive gap. LTG-GaAs is used as the substrate with a very high carrier mobility μ of $100–300\ cm^2\ V^{-1}\ s^{-1}$.

The generated average THz output power lies typically in the µW to nW range (with logarithmically decreasing power for increasing frequencies), in picosecond duration pulses at 92 MHz repetition rate. The generated THz pulses are transmitted through a probe, thereby attenuated and phase shifted, and then probed by a coherent detection scheme.

The detector can be the same kind of LTG GaAs serving as an antenna. The electric field of the THz pulse is probed by the electric charges which are created by the simultaneous irradiation with a delayable 800 nm optical pulse derived from the same Ti:sapphire laser that was rectified to make THz light. The interaction of the THz and near-IR pulses generates a gated output signal. By scanning the time delay of the 800 nm reference pulse relative to the THz pulse on a translation stage ($\approx 0.6\ mm\ ps^{-1}$ for a single pass geometry up and down the stage), the THz electric field is mapped out precisely in time.

THz Time Domain Spectroscopy

In THz time domain spectroscopy a very short (typically 100 fs) pulse is sent into a probe and the change in the electric field of this THz pulse is then recorded on the detector as a function of time. The THz pulse can be described as an electromagnetic wave $E = E_o \sin(\omega t + \varphi)$ with E_o, ω, and φ describing the amplitude, frequency, and phase respectively. Due to the long wavelength, the short pulse contains few to single waves.

Each probe can be described by two specific properties which will affect the propagating THz pulse: the index of refraction (n) and the absorption coefficient (α) which are both frequency dependent and material specific. The dielectric properties of the sample can be equivalently described in the representation of the complex refractive index:

$$\hat{n} = n + ik$$

where the real part n describes the index of refraction which causes a deceleration (dispersion) of the traveling electromagnetic wave; the imaginary part of the complex refractive index is also referred to as the *extinction coefficient,* κ and exhibits damping (absorption) of the wave. In dielectric spectroscopy the properties of a sample are described by the complex electric permittivity ε, which can be directly related to the complex refractive index by:

$$\varepsilon'_{probe} = n^2 - k^2$$
$$\varepsilon''_{probe} = 2n\,k$$

If the electric field of the THz pulse which propagates through a probe is recorded as a function of time, the index of refraction can be deduced from the time delay of the detected pulse with respect to a reference pulse. The absorption coefficient can be obtained by recording the decrease in the amplitude of the THz pulse with respect to the free propagating probe pulse.

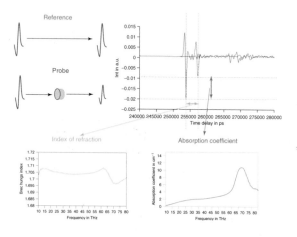

Only when both pulses, the 800 nm femtosecond laser and the THz pulse, arrive at the same time are the free carriers, as generated by the probe pulse, accelerated by the incoming THz field, inducing a measurable current across the photoconductive gap. For a delay time τ between the THz field and the femtosecond pulse, the current $I(\tau)$ measures the overlap between the incoming electric field E_{THz} and the time-dependent conductivity $g(t)$, as induced by the free carriers.

$$I(\tau) = 1/T \int E_{THz}(t)\, g(t+\tau)\, dt \qquad (22.4)$$

where I, E_{THz}, and T are the current at the detector, the electrical field of the THz pulse, and the repetition time, respectively. The amplitude of the detected signal is proportional to the amplitude of the THz field during the free carrier lifetime. The carriers are chosen such that recombination will occur shortly ($<500\,fs$) after excitation, providing a small sensitive window (gate) for THz detection. Due to the short detector gating, the black body radiation as well as other electrical and optical noise at these frequencies is effectively suppressed. The electric field $E_{THz}(t)$ can be transformed to:

$$E(\omega) = 1/2\pi \int \exp(-i\,\omega\, t)\, E(t) dt \qquad (22.5)$$

The absorption signal $I(\omega) = I_0 \exp(\alpha(\omega)l)$, with $\alpha(\omega)$ and l being the absorption coefficient and the intensity, is given as : $I(\omega) = |E(\omega)|^2$.

The first applications of THz time domain spectroscopy included imaging of thin films [7, 8], medical (see TeraView Ltd.), and spectroscopy of crystalline biomolecules [9].

The development of new sources within this so-called "THz gap" was crucial for scientific progress in the field of chemistry and material science. Other novel THz sources which have been developed recently are quantum cascade lasers, THz generation by photo mixing, and free electron laser [1, 4, 10–16].

22.2
THz Spectroscopy

In general, laser spectroscopy is a widely used and extremely successful technique in chemistry and biophysics for structure determination and imaging. A summary of the typical excitations of molecules and proteins ordered according to their excitation energy is shown in Figure 22.3. In general, the frequency of a harmonic oscillator is given as: $\omega = (k/\mu)^{1/2}$, with k being the force constant and μ describing the reduced mass. This means that with increasing reduced mass and decreased force constant the frequency of the mode will shift from the IR to lower and lower

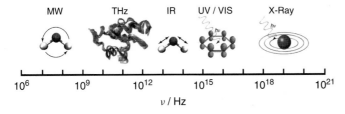

Figure 22.3 Typical molecular excitations versus the excitation energy.

frequencies. Therefore, all intermolecular modes (small k) or large amplitude modes involving many atoms (large μ) can be found in the THz range.

The THz range covers the important region of the collective motions of complex molecular systems. These include skeletal motions of proteins, which leave the intramolecular bonds intact, and phonons. They are also responsible for the dynamics of molecules within complex molecular systems. These characteristic modes cause large amplitude deformations of parts of the system. For this motion the reduced masses are large and the binding constant is usually smaller than for chemical bonds.

Whereas IR spectroscopy gives access to localized, intermolecular vibrations on the length scale of 0.001–0.5 nm, typical excitations in the THz regime involve motions over a length scale of 0.3–1 nm. This implies that the typical excitations go beyond a single chemical bond and even beyond a molecule. For solid state probes these are phonon bands or lattice vibrations, and for molecules in a solvent they are solvent modes or large amplitude modes of the molecule. In Figure 22.4 we show an example for anthracene which has an intramolecular mode at 1.9 THz. THz

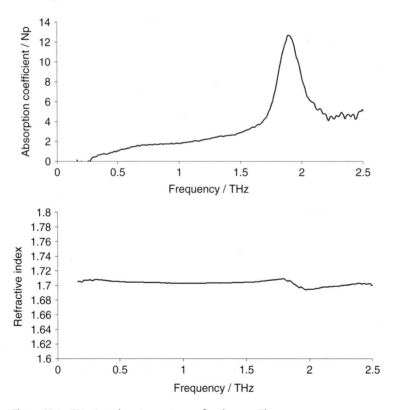

Figure 22.4 THz time domain spectrum of anthracene. The measured absorption coefficient and the refractive index are shown. Both show a resonance around 1.9 THz.

spectroscopy allows direct experimental access to all low frequency bending and twisting modes and, thereby, to the anharmonic part of the force constants. Both are properties which are of fundamental interest for predictions using a molecular dynamics (MD) simulation.

22.3
THz Spectroscopy of Solvated Probes

Previously, water absorption, low light intensities, thermal background noise, severe interference effects from the cell windows, and insensitive detectors made direct absorption spectroscopy measurements of solvated biomolecules in the THz virtually impossible. Any one of these obstacles is not deleterious by itself, but taken together, they posed a formidable problem. The applications of THz spectroscopy for the investigation of biomolecules in their natural surroundings – water – were, therefore, a technical challenge, due to a lack of powerful radiation sources in the so-called THz gap.

Modern THz instruments, building on decades of progress in far-infrared spectroscopy of solid samples and films, are now powerful enough to penetrate even water layers and thus look at fully solvated proteins, carbohydrates, lipids, or nucleic acids [17–20].

In general, a precise measurement of THz absorbance of (strongly absorbing) liquids is very difficult because it requires data acquisition over a series of very short and precisely determined path lengths. Moreover, when measuring a liquid fixed between two windows, these windows will also absorb and reflect part of the radiation. While the reflection factor at the window surfaces remains unchanged when varying the layer thickness, the absorption will increase exponentially with the layer thickness of the liquids.

There are two ways to overcome this difficulty: One is to measure in a reflection mode, and another is to measure a sufficiently thick layer of the liquid such that absorption dominates. However, this requires sufficient laser power in order to be able to still penetrate through the sample to the detector.

In Bochum we have realized a new table-top THz spectrometer using a p-Ge laser as a radiation source [21], with an output power that is increased by an order of magnitude over other available table-top THz sources.

The absorption upon transmission of a sample can then be simply described by Beer's law,

$$I(\omega) = I_o \, \exp(-\alpha(\omega) \, d) + C \tag{22.6}$$

with I_0, $\alpha(\omega)$, d, and C corresponding to the intensity before the probe, the absorption coefficient of the probe, the layer thickness of the probe, and the detector offset, respectively. Even with absorption coefficients of the order of $\alpha = 100\text{–}400 \text{ cm}^{-1}$, as is typical for solvated solutes in the THz range, our new set-up yields good signal-to-noise spectra in a matter of minutes. We are able to penetrate water to a depth of over 200 µm, thereby allowing reliable measurement

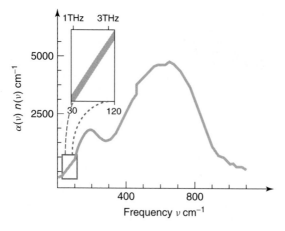

Figure 22.5 Spectrum of water in the THz and IR regions. The absorbance of water increases linearly between 1 and 3 THz (inset), the adequate frequency region for the observation of fast picosecond water dynamics.

of the absorption coefficients of water or solute–solvent mixtures in the THz with an accuracy of less than 1%, not previously possible. In Figure 22.5, we show the THz spectrum of bulk water.

In effect, THz light extends down to the dielectric regime, including two bands one of which has its maximum around $200\,\mathrm{cm}^{-1}$. They reflect collective water network motions covering nanosecond motions down to picosecond motions, such as librational and diffusional motions of hydration water.

Water molecules in the bulk hydrogen bond with three to four other water molecules at any given time, as deduced from, for example, neutron diffraction studies, but these hydrogen bonds are in a constant state of flux [22]. Diffusion of water molecules occurs on a picosecond time scale. Water molecules rotate in solution on a subpicosecond time scale, so that within a picosecond a hydrogen bond between any two molecules may break and reform. Over the longer time scale a given hydrogen bond between two water molecules may no longer reform, as reorientation and translation of a given water molecule favor new bond formation. These two main contributions which lead to a fluctuation in the water network are visualized in Figure 22.6.

How is this fast motion of the water correlated to the THz spectrum? For this we have to calculate the average autocorrelation function, C_M, of the system's dipole moment, $\mathbf{M}(t)$,

$$C_M(t) = \langle \mathbf{M}(0).\mathbf{M}(t) \rangle \tag{22.7}$$

This describes the correlation between the total dipole moment at a time 0 with the total dipole moment at a later time t. The ensemble average dipole autocorrelation function (C_M) can then be used to calculate the absorption cross section $\alpha(\omega)$

Rotational HB break
+ Rotational HB reforming

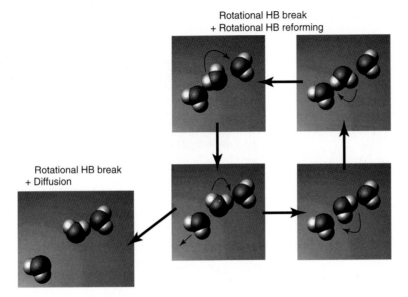

Rotational HB break
+ Diffusion

Figure 22.6 In bulk water we find a constant breaking and formation of hydrogen bonds due to the rotation and diffusion of the water molecules in bulk water.

according to

$$\alpha(\omega) = \frac{4\pi^2 \omega \left[1 - e^{-\hbar\omega/k_B T}\right]}{6\pi\, hcn(\omega)} \int\limits_{-\infty}^{\infty} dt\, e^{-i\omega t} C_M(t) \tag{22.8}$$

where n is the index of refraction and k_B is Boltzmann's constant. Thus THz absorption spectra are directly connected with the breaking and reformation of hydrogen bonds and are a measure of these fluctuations.

As a result, any change in the fast water network oscillations is connected with a change in THz absorbance, as seen when changing temperature [23]. Whereas for a 100 μm layer of water at 97 °C, 40% of the radiation at 1.5 THz is absorbed, at the freezing point only 0.4% of the radiation will be absorbed, which exceeds any pure density changes by orders of magnitude. The relative change of the absorption with temperature exceeds the relative changes in any other spectral region by far, because this frequency range probes directly the water network motions. Due to this sensitivity, precise measurements of the THz absorption coefficient provide a sensitive tool to detect induced changes in the fast water dipole fluctuations around solutes.

Vibrational Density of Biomolecules

Complex macromolecules have a great number of vibrations at different frequencies ω, so it makes sense to invoke as a central concept the spectral density $\rho(\omega)$, rather than accounting for every one of thousands of vibrational peaks separately. A plot of $\rho(\omega)$ for the five-helix bundle protein,

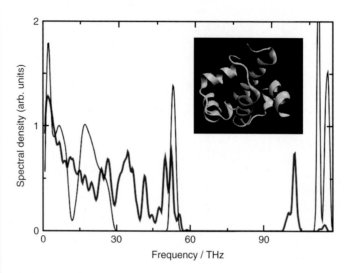

Figure 22.7 The spectral density (thick line) of the five-helix bundle protein λ^*_{6-85}, shown in the inset, describes how its vibrations are distributed as a function of frequency or energy. Only modes below a few THz correspond to collective motions of many atoms or residues within the molecule. Such motions are prime candidates for folding and function. Also shown is a spectral density for bulk water (narrow line), for comparison.

λ^*_{6-85}, in the region from 0 to 110 THz is shown in Figure 22.7. At the highest frequency, one finds vibrations from the lightest nuclei (hydrogen) and most localized vibrational modes. At lower frequencies (∼30 THz) are various small amplitude stretching and bending modes of the backbone and sidechains. Only below 10 THz do we begin to get modes delocalized to large numbers of atoms. In the region between 1 and 5 THz, water absorption (shown as a thin black line in Figure 22.1) plays a significant role. At the very lowest frequencies, the absorption declines toward zero as some power of the frequency, or $A \sim \omega^a$ [24–26]. Such a power-law scaling of the spectral density has been observed in low-temperature electron-spin relaxation measurements [25].

At first glance, a power-law variation of the absorbance at low frequency is not particularly surprising; the Debye theory of elastic materials predicts just such a variation, where the power, a, is just 1 less than the dimension of the material; for a three-dimensional object a is 2. What is surprising is that experiment and theory reveal an altogether different value of a, between about 0.3 and 1.0, for vibrational frequencies up to about 3 THz. This very simple functional form and the power, a, associated with it contain very interesting information. The fractional powers observed in low-temperature electron-spin relaxation measurements and computational studies of the

low-frequency motion of protein molecules indicate that vibrational couplings and energy transport in the biomolecule are lower-dimensional than
one might think – certainly not three-dimensional [24–26].

22.4
Biomolecule Solvation and Terahertz Dynamics: Important Concepts

THz absorption spectra of solvated biomolecules are nearly featureless in the THz
regime [27–29] (see Figure 22.8).

In order to explain this overall featureless change in absorption coefficient we
have to discuss two main concepts (see Figure 22.9): The first key concept in
THz spectroscopy of biomolecules in water is the "THz defect." The absorption
coefficient α, of a protein molecule varies with frequency, ω, as a power-law,
$\alpha \sim \omega^a$, where $a \approx 0.3–1$ [26]. If we naively separate protein and water absorbance,
then such a modest rise in the absorbance with increasing frequency is not enough
to compete with the much stronger absorbance of water at 1–3 THz. The missing
absorption arises because biomolecules, although large and extended, do not extend
in the same way as the full water network. Being bound in size, they lack some of
the lowest frequency motions that can be found in the bulk aqueous solvent. The
difference can range from a small percentage to nearly 100%. In the latter case, a
simple model assuming that biomolecules behave like empty cavities in the water
gives good agreement with experimental data!

A second key concept is the "THz excess." Despite the fact that pure biomolecule
solids or films generally absorb less than bulk water between 1 and 3 THz, there

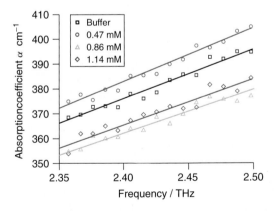

Figure 22.8 Part of the THz spectrum of a solvated protein. The THz absorption is shown for buffer and the solvated five helix bundle protein λ^*_{6-85} for different protein concentrations. The frequency dependence of the THz absorption coefficient is linear between 2.25 and 2.55 THz (15 °C, buffer, and two protein concentrations). We observed a THz excess for concentrations of 0.47 mM and a THz defect for concentrations of 0.86 and 1.14 mM.

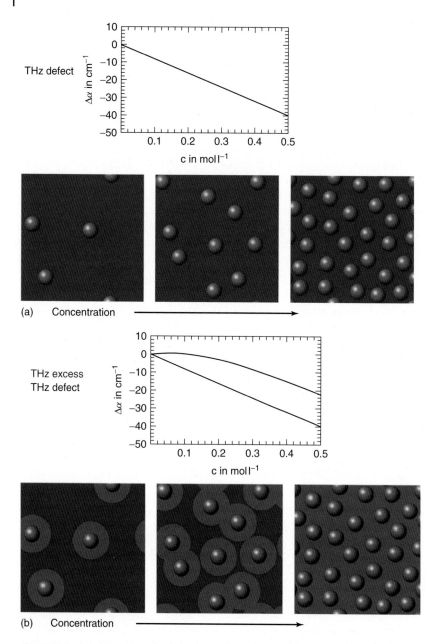

Figure 22.9 Important concepts for a description of THz absorption: (a) THz defect and (b) THz excess. The THz defect describes the linear change of THz absorbance with increasing solute concentration dissolved in water. If each solute is surrounded by a dynamical hydration shell with increased THz absorption compared to bulk water, we find an onset of non-linearity for the concentration dependent absorption coefficient as soon as the dynamical hydration shells overlap.

are still many situations where the biomolecule/water mixture at low concentration does absorb more than either the biomolecule or a bulk water sample. This can be explained only by invoking a third substance – biological or hydration water. If the presence of biomolecules perturbs nearby water molecules, this could have an effect on many of the measurable properties of water: density, relaxation rates, and reorientation rates.

Water has a built-in probe of its orientation: its dipole moment, with a negative charge at the oxygen end, and a positive charge at the hydrogen end of the molecule. The dynamical reorientation of the water dipole moment turns out to be affected over particularly long distances, up to several nanometers from the surface of a biomolecule. This reorientation arises as water molecules within the hydrogen bonding network tumble around and diffuse, constantly making and breaking hydrogen bonds. Couple that with the radius-squared increase in the number of water molecules as one moves outward to more remote solvation shells, and huge numbers of water molecules can be affected by a single biomolecule. THz measurements and simulations, and theories studying the THz spectrum, are exquisitely sensitive to the dynamical reorientation of dipole moments on the picosecond time scale – this is precisely what a THz spectrum measures (1 THz $= 10^{12}$ Hz $= 1$ ps^{-1}). A simple picture of a biomolecule in water thus has to include the protein, nucleic acid, or carbohydrate (causing a THz defect), bulk water (if far enough away), and hydration water with new physical properties, including a propensity for enhanced THz absorption – the THz excess. The thus-defined dynamical hydration shell water is not identical to the sterically bound water molecules which are seen by X-ray crystallography, NMR, or neutron crystallography. The dynamical hydration shell includes all water molecules that show water network dynamics distinct from the bulk, and thus a distinct THz absorbance. We have to note that the influence of the protein on the fluctuations of the water network motions can reach much further than the static hydration radius, since it involves only a change in the motions and not a fixed H-bond to the protein.

THz fluctuations of the collective dipole moment of the solvent water, and thus the THz spectrum, are exquisitely sensitive to the presence of nearby protein molecules. However, any influence on the time-dependent fluctuations of the dipole moment will change the THz absorbance, as is obvious when we consider the previous paragraph. One way to quantify hydrogen bond rearrangement is to give time-dependent survival probabilities for bonds between water molecules [30]. The THz fluctuations of the hydrogen bond network of water, arising in part from the breaking and reformation of hydrogen bonds between water molecules, are found to be considerably influenced by the presence of solute. MD simulations reveal that hydrogen bonds between exposed O atoms on a protein or saccharide and surrounding water molecules survive, on average, significantly longer than the hydrogen bonds between water molecules in the bulk [31, 32], mimicking the effect of cooling of the water near the protein surface. The relatively sluggish rearrangement of hydrogen bonds is not merely limited to bonds between the water and the solute. Computer simulations reveal a retardation of the rearrangement of hydrogen bonds between water molecules out to three or four solvation layers from a protein molecule, even to about two layers for a small saccharide [33].

THz absorption spectra cannot be decomposed into a spectrum of water and a spectrum of the protein, since the THz network motions are extending over protein into the water. They reflect the coupling between the large amplitude motions of the protein and the water network motions. Molecular simulations reveal that the THz oscillations of the water dipole moments are tuned by the presence of a protein as far as 20 Å away, or perhaps even further [27]. If the protein concentration is sufficiently highm such that the distance between protein surfaces in solution is about 20 Å or less, the collective THz oscillations in the water network no longer resemble the fluctuations of water in the bulk. The collective protein–water dynamics give rise to a new medium in which bulk water is absent.

22.5
Precise Measurements of the THz Absorption Coefficient

The absorption can be measured by varying the absorption path length and fitting the intensity, as recorded by the detector, to Beer's law [34]. However, in order to decrease systematic errors due to temperature fluctuations and detector performance a different set-up has proven to be superior. This measures the variation in the THz absorption directly with respect to a reference [33] and is shown in Figure 22.10. Whereas the absorption coefficients, α, of bulk water and the solvated biomolecule are both large (of the order of $400\,\text{cm}^{-1}$), the solute induced change is relatively small ($\Delta\alpha$ is of the order of $10-20\,\text{cm}^{-1}$), but carries the essential information.

As discussed above, the THz spectrum of the solvated biomolecules, to a first approximation, increases smoothly with frequency, similar to what was found for

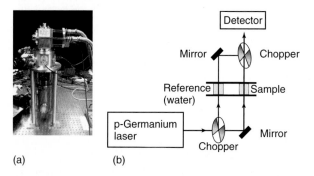

(a)   (b)

Figure 22.10 Set-up for THz measurements. (a) The 3 K cryostat in which the p-germanium laser is hosted. (b) The radiation path of the THz radiation. The mirror-chopper splits the radiation into two distinct beams. One part passes through the bulk water, the other through the biomolecule solution. Both are then focused onto the detector.

bulk water. Over a narrow range (e.g., 2.1–2.8 THz) the spectra can be fitted linearly, as observed in Figure 22.8. However, the important information is not contained in the variation of THz absorption with frequency, but instead in the variation of the THz absorption with increasing biomolecule concentration.

If the solute did not affect the water in its surroundings, we would expect a linear change from the large THz absorption coefficient of bulk water to a smaller absorption coefficient for the solute itself when raising the solute concentration from zero to 100%. In the case of $\alpha_{solute} \ll \alpha_{buffer}$, the following model would fit the absorption coefficient as a function of concentration:

$$\alpha = \alpha_{solute} \frac{V_{solute}}{V} + \alpha_{buffer} \frac{V - V_{solute}}{V} \approx \alpha_{solute}(1 - \rho_{solute} c_{solute}) \tag{22.9}$$

However, when each solute is surrounded by a dynamical hydration shell which has a THz absorption different from bulk water – due to changes in the water dynamics – then we expect a deviation from this linear behavior.

Note that the total volume of the dynamical hydration shells increases linearly with solute concentration at low concentrations. If the THz absorbance of the water within this dynamical hydration shell exceeds the absorbance of the bulk water displaced by the shell and protein, the overall absorption will at first increase linearly with protein concentration. Eventually, the hydration shells overlap, and the absorption coefficient α will no longer scale linearly with the solute. If the solute absorbs less than water at high concentration we expect a turnover in the absorption coefficient with increasing solute concentration. This deviation from linearity for the case of hydration shells is displayed in Figure 22.11, in which we have chosen trehalose–water mixtures as example.

The observed THz absorption can thus be described by:

$$\alpha = \alpha_{solute} \frac{V_{solute}}{V} + \alpha_{shell} \frac{V_{shell}}{V} + \alpha_{buffer} \frac{V - V_{solute} - V_{shell}}{V} \tag{22.10}$$

In a first proof of principle experiment we measured the solute-induced change in the THz absorption for a disaccharide, lactose [34]. For disaccharides, the three-component model provides an excellent description of the measured data, which are shown in Figure 22.11 by the dashed curve for trehalose.

The model allowed us to fit the size of the solvation shell because the solvation around sugars is rather homogeneous. The solid curve shows the results of a simulation using Equation 22.10 when varying the two free parameters, V_{shell} and α_{shell}. In Figure 22.11c the final parameters α_{shell} and V_{shell} are used, which were fitted to the measured data (dots). The simulated curve and the fitted curve are superimposed on each other.

The dotted line in Figure 22.11 corresponds to the pure THz defect (i.e., α_{shell} and V_{shell} are both set to zero). If the absorption coefficient of the water in the hydration shell is smaller than that of the buffer ($\alpha_{shell} < \alpha_{buffer}$), the predicted curve, represented by the solid line, will lie below the dotted line (Figure 22.11e). If the absorbance of the dynamical hydration shell exceeds the absorbance of the bulk water displaced by the shell and protein (THz excess: $\alpha_{shell} > \alpha_{buffer}$), the overall absorption will at first increase linearly with protein concentration and will

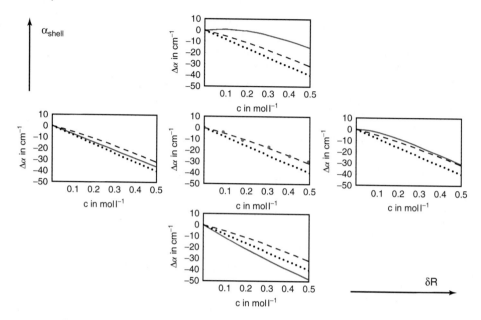

Figure 22.11 Predicted dependence of the THz absorption for given values of α_{shell} and V_{shell}, both for trehalose. The dotted curve shows the pure THz defect. The dotted line reflects the predicted THz absorption assuming an empty cavity of the size of the protein (α_{solute} is set to zero). The dashed curve in (c) corresponds to a simulation which is based upon the fitted concentration dependence of a lactose–water solution and remains fixed for comparison. The dots present the experimentally determined values. For comparison: the other curves were obtained when varying α_{shell} ((a) and (e)) and V_{shell} ((b) and (d)) by increasing or decreasing δR.

lie above the dotted curve. Eventually, the dynamical hydration shells overlap, and their volume actually decreases relative to the increasing volume of protein. In the extreme limit the hydration water will displace all bulk water in the interstitial spaces. A further increase in trehalose will then only result in an increase in $V_{trehalose}$. However, the absorption coefficient of trehalose is smaller than that of water by a factor of 10. This implies that the THz absorption coefficient has to reach this value when extrapolated to 100% trehalose. As a result, there is a turnover in the absorption coefficient. The measured absorption plotted against concentration will then show a nonlinear trend.

In Figure 22.11b to d the extent of the dynamical hydration shell (and thereby the volume V_{shell}) is increased. As we can see, the onset of non-linearity (the turning point) is shifting toward smaller and smaller concentrations, when increasing the size of the dynamical hydration shell. This is a consequence of the fact that for larger dynamical hydration shells an overlap of adjacent hydration shells occurs already at lower concentrations than for very small dynamical hydration shells. The concentration at which the onset of non-linearity occurs is directly correlated with the smallest concentration at which the dynamical hydration shells start to overlap. We can also clearly distinguish between the case where the water in

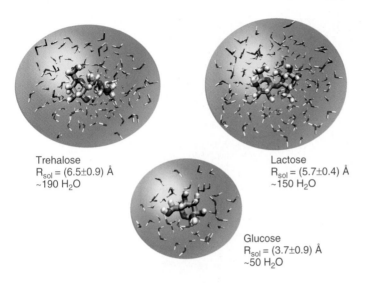

Trehalose
$R_{sol} = (6.5 \pm 0.9)$ Å
~190 H_2O

Lactose
$R_{sol} = (5.7 \pm 0.4)$ Å
~150 H_2O

Glucose
$R_{sol} = (3.7 \pm 0.9)$ Å
~50 H_2O

Figure 22.12 Dynamical hydration shell around a saccharide. Comparison of the size of the dynamical hydration shell of sugars as probed by THz spectroscopy according to Ref. [33].

the dynamical hydration shell has an increased (THz excess) or decreased THz absorption compared to the bulk.

If we compare the induced THz absorption changes for disaccharides we find that they can be systematically correlated with the number of sugar–water contacts [33].

By a fit of the measured THz absorption coefficients we were able to determine the size of the dynamical hydration shell for various sugars. An overview of our results is given in Figure 22.12. The results show that trehalose, which is the most efficient bioprotector, has the largest dynamical hydration shell.

22.6
THz Spectroscopy of Solvated Proteins

We have studied the THz spectrum of the five helix bundle protein λ^*_{6-85} which also yielded an unexpected non-monotonic trend in the measured THz absorbance as a function of the protein:water molar ratio [27]. The measurements could be explained by overlapping dynamical hydration shells of an unexpectedly long range around the proteins, more than 20 Å, which is greater than the pure structural correlation length usually observed. The data showed that about 1000 water molecules are directly influenced in its flow, a dramatic effect. The protein with its dynamical hydration shell is shown in Figure 22.13.

The change in the water network dynamics was found to be most pronounced for the wild-type and much less for the partially unfolded protein. The THz

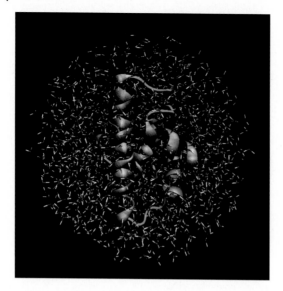

Figure 22.13 Dynamical hydration shell around a protein. The five helix bundle protein λ^*_{6-85} is shown surrounded by 1000 water molecules in the dynamical hydration shell.

absorbance for the partially unfolded protein is also different from that of the native protein, which can be explained by changes in the surrounding water network [19]. This raises new speculations about the biological significance of the long range influence of the water dynamics on protein dynamics and protein function [27], and the question whether proteins have evolved to dynamically restructure their solvent.

22.7
KITA: Kinetic THz Absorption Spectroscopy

Very recently, we have used changes in THz absorbance to monitor the hydration layer around a protein in real time during the water network rearrangement [35]. This has been called kinetic terahertz absorption (KITA) spectroscopy. The KITA experiment follows two different time scales. On a picosecond time scale, we monitor the shape of the THz electric field passing through the protein sample. This shape is attenuated and phase-shifted by the presence of the biomolecules, and changes as the proteins fold. The second time scale follows the kinetics of reaction, as we look how the electric field shape evolves over longer times while the protein folds (see Figure 22.14). So far, we have extended the observation over the millisecond to second time scale.

As with any technique which probes kinetics we need to initiate the reaction at a certain time, which will be the time $t = 0$. In KITA, protein folding is initiated by a stopped flow mixer. The THz pulse is time-delayed with respect to mixing

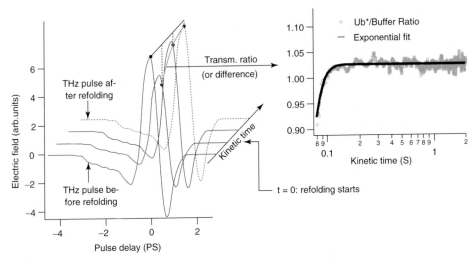

Figure 22.14 Kinetic terahertz absorption spectroscopy (KITA) monitoring the time scales for the rearrangement of the coupled protein–water network motions, as monitored by a change in the THz absorption.

to monitor the progress of the refolding kinetics. The stopped flow mixer takes a denatured protein solution (for example, at low pH, or after addition of a denaturant such as urea) and mixes it with a buffer that brings the protein back to folding conditions (e.g., by restoring pH to near 7, or reducing the denaturant concentration). The THz pulses report on how fast the protein refolds after the mixing has occurred.

To calibrate the new KITA technique, we also compared it with other techniques which are more sensitive to protein backbone dynamics: Time-resolved fluorescence spectroscopy, time-resolved circular dichroism (TRCD) and time-resolved small angle X-ray scattering (SAXS). KITA reports very different results from fluorescence spectroscopy for the same folding process. As the protein folds, native-like fluorescence is not acquired until about 1 s into the folding process. This is how long it takes for the tryptophan amino acid residue being monitored by fluorescence to be packed into the lower dielectric constant interior of the fully folded protein. THz absorption in contrast equilibrates in about 10 ms, and shows no significant variation of signal between 100 ms and 1 s. Thus the hydrogen bonding network around the protein rearranges on a much faster time scale than that required for complete folding. Later stages of the folding process are dependent on the early events when water molecules rearrange around the protein.

TRCD is sensitive to secondary structure formation, which involves breaking of protein backbone–water hydrogen bonds, and remaking them as protein–protein hydrogen bonds. SAXS follows the collapse of the protein to a smaller radius of gyration, and such a collapse may already begin while the solvation network and protein rearrange. Indeed, we find that TRCD in all cases, and SAXS for one mutant of the protein, also show a millisecond phase, in addition to the slower phase seen by fluorescence. Thus KITA shows the fastest overall response, fluorescence the slowest, and TRCD/SAXS a combination of both.

In the future, it will be very interesting to extend KITA from the millisecond to the microsecond time scale because the solvation network clearly reacts very rapidly during folding. A number of proteins with very fast folding time scales have been discovered, and may be amenable to investigation by KITA [36].

22.8
Further Application of THz Absorption Spectroscopy

The observation of specific intermolecular modes by THz spectroscopy was used by Jepsen and coworkers to examine pellet samples of lactose, aspirin, sucrose, and tartaric acid at room temperature. This could be used to develop a recognition coefficient based on unique features that successfully discriminated against these biomaterials [37, 38]. This has led to further investigations as to whether THz-imaging methods can also be successfully applied for detection of illicit drugs.

A nice example of the application of THz spectroscopy demonstrates the potential for secondary structure characterization. Siegrist *et al.* have investigated three different forms of the alanine tripeptide, $NH_3^+-Ala_3-O^-$ [9]. Depending on the conditions during recrystallization, trialanine is known to exist in one of two β-sheet forms, a dehydrated parallel β-sheet (p-Ala$_3$), and a hydrated anti-parallel β-sheet (ap-Ala$_3$-H$_2$O). The distinct forms gave rise to distinct, structure-specific THz spectra ($0.6-100\,cm^{-1}$) obtained at 4.2 K which were probing the collective intermolecular motions. Using THz spectroscopy even a third form was discovered and attributed to the dehydrated form, *ap*-Ala$_3$. This specificity in the THz spectrum is in sharp contrast to the mid-IR region where the FTIR spectrum of *ap*-Ala$_3$-H$_2$O is nearly identical to that of the *p*-Ala$_3$ and very similar to the dehydrated form. This shows that solvation and the rearrangement of hydrogen bonds have a significant influence on the THz spectrum. It shows the potential of THz spectroscopy to serve as a new analytical tool to observe the formation or destruction of long-range ordering in crystals, at surfaces and in thin films, which can be monitored as a change in the THz (intermolecular fingerprint) spectra. By probing phonon-like bands THz spectroscopy allows a quantitative measure of co-crystallization or formation of H-bonds [39]. The possibility to distinguish different structures makes it interesting for the pharmaceutical industry, especially since THz radiation will penetrate plastic, which allows the monitoring of specific compositions of pills even when sealed in plastic.

The sensitivity of THz spectroscopy as an analytical probe of biomolecular binding has been successfully demonstrated. Pioneering work showed a significant change in the absorption coefficient as well as in the refraction index for a single stranded versus a hybridized double-stranded DNA [40]. The detection is based upon changes in the intermolecular modes and is further explored to serve as a label-free biosensing method.

Recently, the light-induced THz emission of bacteriorhodopsin was detected with femtosecond resolution [41]. The detected THz emission could be attributed to an excited state intramolecular electron transfer in the retinal chromophore which leads to the emission of THz radiation. The generation process is similar to the acceleration of charges in the LTG GaAs crystal and has led to the generation of THz pulses in the THz time domain system.

The recent results have opened new research directions: THz absorption can detect subtle changes in the dynamical orientation of water molecules that are washed out in radial distribution functions from scattering experiments, it can report on distance scales beyond current NMR experiments [19, 27, 33, 34]. Mutagenesis can be used to ask questions about site specificity, but is not required to obtain a signal [19]. Using THz radiation sources we can excite collective motions of solvent molecules and of biomolecules whose time scales are of the order of a picosecond [18, 42, 43] and have, thereby, opened a new window onto the coupled protein–water dynamics.

Laser techniques will now allow the observation of changes in the water network fluctuations with sub-picosecond time scales, which are orders of magnitudes faster than neutron scattering or NMR experiments. This will enable us to follow the dynamics in real time during fast structural changes such as protein folding or signaling processes.

The development of quantitative models for the THz spectra will make it possible to understand the solvation dynamics of proteins at the molecular level. This will shed new light on the link between structural rearrangement and solvent dynamics, whether this influence has functional advantages that could have evolved, and pinpoint which protein properties cause dynamical reordering of the solvation shell.

Acknowledgment

We want to acknowledge the help and valuable contributions of all people who contributed significantly to this review: S. Ebbinghaus, M. Heyden, M. Krüger, B. Born, St. Funkner, and E. Bründermann. This work is supported by the Wissenschaftsministerium NRW, the Ruhr-Universität Bochum, the Human Frontier Science Programme, and the VW Stiftung.

References

1. Schmuttenmaer, C.A. (2004) *Chem. Rev.*, **104**, 1759.
2. Pickwell, E. and Wallace, V.P. (2006) *J. Phys. D: Appl. Phys.*, **39**, R301.
3. Tonouchi, M. (2007) *Nat. Photonics*, **1**, 97.
4. Bründermann, E. (2004) in *Long-Wavelength Infrared Semiconductor Lasers* (ed. H.K. Choi), John Wiley & Sons, Inc., Hoboken, pp. 279–350.
5. Fattinger, C. and Grischowsky, D. (1989) *Appl. Phys. Lett.*, **54**, 490–492.
6. Markelz, A.G. (2008) *IEEE J. Sel. Top. Quantum Electron.*, **14**, 180–190.
7. Turner, G.M., Beard, M.C., and Schmuttenmaer, C.A. (2002) *J. Phys. Chem. B.*, **106**, 11716.

8. Whitmire, S.E., Wolpert, D., Markelz, A.G., Hillebrecht, J.R., Galan, J.et al. (2003) Biophys. J., 85, 1269–1277.

9. Siegrist, K., Bucher, C.R., Mandelbaum, I., Hight, A.R., Walker, R., Balu, R., Gregurick, S.K., and Plusquellic, D.F. (2006) J. Am. Chem. Soc., 128, 5764.

10. Carr, G.L., Martin, M.C., McKinney, W.R., Jordan, K., Neil, G.R. et al. (2002) Nature, 420, 153–156.

11. Carr, G.L., Martin, M.C., McKinney, W.R. et al. (2002) Nature, 429, 153.

12. Colson, W.B., Johnson, E.D., Kelley, M.J., and Schwettman, H.A. (2002) Phys. Today, 55, 35–41.

13. Colson, W.B. (2001) Nuclear Instruments & Methods in Physics Research, A475, 397–400.

14. Hoffmann, S., Hofmann, M., Bründermann, E., Havenith, M., Matus, M. et al. (2004) Appl. Phys. Lett., 84, 3585–3587.

15. Kleine-Ostmann, T., Knobloch, P., Koch, M. et al. (2001) Electron. Lett., 37, 1461.

16. Stöhr, A. and Jäger, D. (2009) Photonic Oscillators for THz Signal Generation, in Microwave Photonics – Devices and Applications, (ed. Iezekiel, S.), John Wiley & Sons Canada Ltd., pp 85–110, ISBN 9780470848548.

17. Chen, J.-Y., Knab, J.R., Cerne, J., and Markelz, A.G. (2005) Phys. Rev. E, 72, 040901– 040904.

18. Dexheimer, S. (2007) Terahertz Spectroscopy: Principles and Applications, Taylor & Francis, London.

19. Ebbinghaus, S., Kim, S.J., Heyden, M., Yu, X., Gruebele, M. et al. (2008) J. Am. Chem. Soc., 130, 2374–2375.

20. Xu, J., Plaxco, K.W., and Allen, S.J. (2006b) Prot. Sci., 15, 1175–1181.

21. Bergner, A., Heugen, U. Bründermann, E. Schwaab, G. Havenith, M. et al. (2005) Rev. Sci. Instrum., 76, 063110.

22. Kumar, R., Schmidt, J.R., and Skinner, J.L. (2007) J. Chem. Phys., 126, 204101–204112.

23. Ronne, C., Thrane, L., Astrand, P.-O., Wallqvist, A., Mikkelsen, K.V. et al. (1997) J. Chem. Phys., 107, 5319–5330.

24. Elber, R. and Karplus, M. (1986) Phys. Rev. Lett., 56, 394–397.

25. Herrick, R.C. and Stapleton, H.J. (1976) J. Chem. Phys., 65, 4778–4785.

26. Leitner, D.M. (2008) Annu. Rev. Phys. Chem., 59, 233–259.

27. Ebbinghaus, S., Kim, S.J., Heyden, M., Yu, X., Heugen, U. et al. (2007) Proc. Natl. Acad. Sci. U.S.A., 104, 20749–20752.

28. Plusquellic, D.F., Siegrist, K., Heilweil, E.J., and Esenturk, O. (2007) Chem. Phys. Chem., 8, 2412–2431.

29. Zhang, C.F. and Durbin, S.M. (2006) J. Phys. Chem. B, 110, 23607–23613.

30. Bagchi, B. (2005) Chem. Rev., 105, 3197–3219.

31. Lee, S.L., Debenedetti, P.G., and Errington, J.R. (2005) J. Chem. Phys., 122, 204511–204510.

32. Tarek, M. and Tobias, D.J. (2002) Phys. Rev. Lett., 88, 138101–138104.

33. Heyden, M., Bründermann, E., Heugen, U., Niehues, G., Leitner, D.M. et al. (2008) J. Am. Chem. Soc., 130, 5773–5779.

34. Heugen, U., Schwaab, G., Bründermann, E., Heyden, M., Yu, X. et al. (2006) Proc. Natl. Acad. Sci. U.S.A., 103, 12301–12306.

35. Kim, S.J., Born, B., Havenith, M., and Gruebele, M. (2008) Angew. Chem. Int. Ed., 120, 6586–6589.

36. Liu, F. and Gruebele, M. (2008) Chem. Phys. Lett., 461, 1–7.

37. Fischer, B., Hoffmann, M., Helm, H., Modjesch, G., and Jepsen, P.U. (2005) Semicond. Sci. Technol., 20, S246.

38. Walther, B.M.F. and Jepsen, U. (2003) Chem. Phys., 288, 261.

39. Nguyen, K.L., Friščić, T., Day, G.M., Gladden, L.F., and Jones, W. (2007) Nat. Mater., 6, 206.

40. Bruchseifer, M., Nagel, M., Bolivar, P.H., Kurz, H., Bosserhoff, A., and Buttner, R. (2000) Appl. Phys. Lett., 77, 4049–4051.

41. Groma, G.I., Hebling, J., Kozma, I.Z., Varo, G.J., Hauer, J., Kuhl, J., and Riedle, E. (2008) Proc. Natl. Acad. Sci. U.S.A., 105, 6888–6893.

42. Leitner, D.M., Havenith, M., and Gruebele, M. (2006) Int. Rev. Phys. Chem., 25, 553–582.

43. Xu, J., Plaxco, K.W., and Allen, S.J. (2006a) J. Phys. Chem. B., 110, 24255–24259.

23

Single-Molecule Fluorescence Spectroscopy: The Ultimate Limit of Analytical Chemistry in the Condensed Phase

Dirk-Peter Herten, Arina Rybina, Jessica Balbo, and Gregor Jung

■ **Method Summary**

Acronyms, Synonyms
- Single-molecule fluorescence spectroscopy (SMFS)
- Single-molecule spectroscopy (SMS)
- Single-molecule detection (SMD)

Benefits (Information Available)
- optical resolution far better than the far-field limit of Abbé (down to several nanometers).
- high sensitivity: a detection limit of one molecule in a femtoliter (corresponding to a concentration range of $\sim$ nM) can be easily detected; a total volume of $<1\,\mu l$ is sufficient
- resolution of molecular heterogeneities and subpopulations
- no triggering of correlated motions required
- surface sensitivity as well as three-dimensional resolution possible.
- dynamic resolution down to nanoseconds possible.

Limitations (Information Not Available)
- only fluorescence works so far as contrast mechanism in condensed phase, other molecules which should be detected have to be tagged by a fluorophor
- limitation to fluorophores with high fluorescence quantum yields and excellent photostability
- information on the vibrational and geometric structures of the dye is hardly obtainable
- strong interference by background fluorescence of impurities and Raman scattering
- investigation of metal surfaces due to fluorescence quenching hampered
- experimental approaches have to be carefully chosen as analysis can be tedious.

Methods in Physical Chemistry, First Edition. Edited by Rolf Schäfer and Peter C. Schmidt.
© 2012 Wiley-VCH Verlag GmbH & Co. KGaA. Published 2012 by Wiley-VCH Verlag GmbH & Co. KGaA.

23.1
Introduction

The description of macroscopic states of molecules and interactions of the latter are based on the notion of individual particles. This theoretical concept allows the deduction of macroscopic variables from measurable quantities. In chemistry, reactions are analyzed and developed on the behavior of individual atoms within the molecular model. The derivation of microscopic processes from the macroscopically observed behavior relies on the assumption of the ergodic theorem. This claims that the molecule ensemble, consisting of an infinite number of molecules, follows the temporal averaging of a single molecule. Individual interactions and pathways which may depend on the local environment of single molecules are covered just within the statistical distribution of the population. The question of whether each molecule has slightly different properties, that is, static heterogeneity, or whether the properties of individual molecules change over time, that is, dynamic heterogeneity, cannot be explained satisfactorily by measurements of the ensemble. The ability to visualize single molecules opens entirely unexpected new perspectives to biology, chemistry, and physics. Individual molecules can be detected, identified, and counted. In some cases even their physical, chemical, and biological properties are measured as expected from the theories of statistical physics. The possibility to observe single molecules in solution, on surfaces, in cells or during chemical reactions can show if all molecules of a species behave equally or if subpopulations exist which exhibit peculiar properties. This knowledge about the heterogeneity of molecular populations has strong implications for all fields in chemistry.

23.2
Basic Principles

23.2.1
History Survey

In 1952, Erwin Schrödinger claimed that the direct observation of single molecules would remain a theoretical construct [1], whereas a few years later, Richard Feynman envisioned the feasibility of such experiments [2]. Already in 1961, a first report, which was overseen for a long time, described the accumulated fluorescence of a fluorogenic substrate by the multiple turn-over of individual enzymes [3]. Later, in 1976, Thomas Hirschfeld managed the first detection of individual antibodies, which were labeled with 80–100 fluorescein dyes [4]. The fundamental basis for the detection of single fluorescent molecules at ambient conditions was developed by the team around R. Keller at Los Alamos, based on hydrodynamic focusing to minimize the observation volume [5, 6]. Finally, in 1990, these efforts led to the first single-molecule detection (SMD) in aqueous solution [7]. Simultaneously, the detection of single dopant molecules in host crystals was developed in continuation of spectral hole-burning experiments [8–10]. While L.

Kador and W. E. Moerner provided the first evidence of a single molecule by an absorption technique, M. Orrit used the more powerful fluorescence technique on the same system. The first images of individual immobilized fluorophores, recorded by near-field [11–14] and far-field fluorescence techniques [15], stimulated the scientific world to invent new and more sensitive detection methods. In addition to these pioneering achievements in the field of fluorescence technology, other methods were developed during the 1980s to study single molecules, such as scanning probe microscopy [16, 17] and the patch-clamp technique. Sensitivity gain in electrochemistry [18] and Raman spectroscopy [19] add novel approaches to the wealth of techniques of single-molecule detection. As the whole field is constantly growing thus making a concise overview difficult, we restrict ourselves to fluorescence techniques.

Numerous excellent review articles have been published dealing with recent developments in single-molecule spectroscopy (SMS) by means of fluorescence microscopy [20–27], low-temperature solid-state spectroscopy [28–31], and their various applications [32–37]. The present overview rather focuses on various principles and techniques of SMS based on the fluorescence spectroscopy of single molecules at ambient conditions.

23.2.2
The Fluorophor and Photophysical Processes

The typical size of a fluorescent entity is far below the optical resolution of a microscope of approximately 200 nm. Optical contrast for absorbance, that is, its absorption cross section σ, is too low for detection of nanometer-sized particles. Therefore, only repeated cycles between ground and first electronic excited states, in combination with background-free emission of fluorescence photons, enable the detection of single molecules. It is, therefore, instructive to look at the photophysical requirements for fluorophores.

By absorption of a photon, the fluorophore is electronically excited from the singlet ground state S_0 into some higher vibrational level of the first excited singlet state S_1. The molecule rapidly relaxes to the lowest vibrational level of S_1 due to vibrational relaxation (Figure 23.1a). Several pathways exist for the dissipation of the electronic energy. The spontaneous return of the fluorophore to the ground state S_0 occurs with the emission of a photon and is most important for SMD. Excellent fluorescent dyes, like xanthenes, carbocyanines, and polyaromatic dyes, convert their excess energy almost entirely into light, approaching a fluorescence quantum yield $\Phi_f \sim 1$. This process occurs with the fluorescence lifetime τ_f of a few nanoseconds characteristic for the fluorophore. Similar values for the lifetime are estimated on the basis of the Strickler–Berg relation [38]. The reciprocal value of τ_f is the maximum number of emitted photons per time of a single molecule. Theoretically, a fluorescent molecule with a fluorescence lifetime of 2 ns can be excited up to 5×10^8 times per second under saturation conditions. The excitation rate k_{exc} depends on the excitation intensity I_{exc}, the absorption cross section σ,

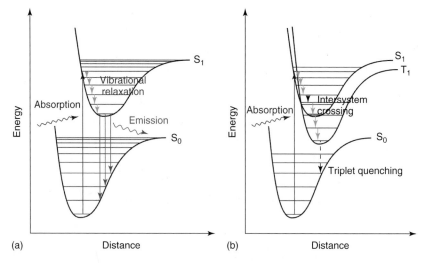

(a) Distance　　(b) Distance

Figure 23.1 (a) Jablonski diagram of fluorescence. Absorption of light stimulates a transition from the electronic ground state S_0 to the first excited electronic state S_1. After relaxation to the lowest vibronic state in S_1 the molecule can return to the electronic ground state by emission of light. The lifetime of the first excited singlet state of organic fluorophores is usually in the order of a few nanoseconds (10^{-9} s) (b) Jablonski diagram of triplet quenching. After absorption of light there is a low probability that intersystem crossing competes with fluorescence leading to a triplet state T_1. Singlet–triplet transitions are spin-forbidden by quantum mechanics. Their low probabilities give T_1 a relative long lifetime (10^{-3}–10^1 s) such that molecules in solution usually return to their groundstate S_0 by collisional quenching (dashed arrow). Reproduced from D. P. Herten, *Chemie in unserer Zeit*, 2008, **42**, 192–199.

and the saturation intensity I_S.

$$k_{exc} = \frac{I_{exc}}{h\nu} * \frac{\sigma}{1 + I_{exc}/I_S} \tag{23.1}$$

The required excitation intensity I_{exc} for saturation lies above $100\,\mathrm{kW\,cm^{-2}}$ in dependence on the absorption cross section σ of the fluorophore.

$$I_S = \frac{h\nu}{\sigma * \tau_f} \tag{23.2}$$

Hence, typical dwell times of ~0.5 ms in SMS (see below) can correspond to 2.5×10^5 excitation cycles and, correspondingly, emitted photons per fluorescence molecule.

Observed count rates, however, fall far below this value for several reasons: (i) Intersystem crossing (ISC) to the triplet state competes with the fluorescence emission, thus reducing Φ_f and dramatically enhancing saturation (Figure 23.1b). (ii) Irreversible photobleaching is even more detrimental as it limits the absolute number of photocycles which a single molecule can undergo. (iii) Inherent losses due to the imperfect detection efficiency of the optical system reduce the photon yield to only about 1% of the emitted fluorescence photons. Typically, up to 200 photons per dwell time can be detected in SMS (Figure 23.2).

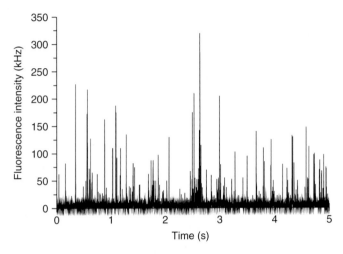

Figure 23.2 Fluorescence emission of single Cy5 molecules
(10^{-11} M) crossing the confocal detection volume due to
free diffusion. The sample was excited with a pulsed diode
laser at 635 nm and an excitation power of 100 kW cm^{-2}.
Reproduced from D. P. Herten, M. Sauer, *Bunsenmagazin*,
2003, **5**(1), 5–16.

23.2.3
Background and Noise

The above mentioned excitation conditions can be realized if several milliwatts of
a laser output are focused to the diffraction limit. Typical areas covered by laser
excitation have a size of $\sim$1 µm^2. This size exceeds the typical absorption cross
section σ of dye molecules by roughly 8 orders of magnitude. Therefore, it is
mandatory to avoid saturation of the fluorescence signal by lowering the excitation
intensity as it actually further reduces σ. It becomes obvious from the above
estimates that the ability to detect a single fluorescent molecule is less dependent
on the detection sensitivity than on the efficient elimination of background signals.
These are: (i) The elastic Rayleigh scattering, which is suppressed efficiently by
using appropriate band pass filter for fluorescence. The red-shift of the fluorescent
light is beneficial. (ii) The inelastic Raman scattering depends on the scattering cross
section of molecules in solution and is more severe in the blue region of the visible
spectrum due to the λ^{-4} dependence. As the cross section for Raman scattering is
$\sim$12 orders of magnitude smaller than the cross section for absorption, the number
of scattering molecules must be minimized by small detection volumes. (iii)
The autofluorescence of impurities, which depends considerably on the excitation
and emission wavelength due to the λ^{-3} dependence on the rate constant for
spontaneous emission. To minimize the mentioned background contributions,
excitation in the red spectral range ($>$600 nm) is convenient. For autofluorescence
only a small number of naturally abundant compounds exist, which absorb and
emit fluorescence efficiently in the spectral range $>$600 nm whereas the intrinsic

fluorescence of biological systems in the blue-green range is a major obstacle for SMD *in vivo*.

To reduce sample volume and associated Raman scattering and autofluorescence, hydrodynamic focusing, as in flow cytometry (Figure 23.3a), has been used [39, 40]. Here, the sample is injected through a thin capillary into the aqueous flow stream of a flow chamber and, due to the hydrodynamic pressure, compressed to a diameter of 5–20 μm [41]. Fluorescence of the analyte molecules is collected perpendicular to the excitation axes, as close as possible to the injection capillary. For efficient detection of all analyte molecules a comparatively large sample volume in the range of picoliters ($10 \times 10 \times 10\,\mu m^3$) is used. Due to the relatively large observation volume and concomitant strong Raman background, temporal discrimination of the delayed fluorescence against the instantaneous scattered light is necessary. In the early 1990s Rigler and coworkers established the confocal principle for the detection of individual molecules to enhance the signal-to-background (S/B) ratio [42–45]. In this method, the laser beam is focused by a microscope lens with high numerical aperture to a diffraction-limited spot at the sample (Figure 23.3b). The diffraction-limited detection volume is typically in the range of one femtoliter (10^{-15} l). It can contain only about 10^{10} solvent molecules so the absolute intensity of Raman scattering is much weaker than the fluorescence emitted by a single dye molecule. The required purity of the solvents should be better than 0.1 ppb with regard to fluorescent impurities. Observed background count rates (solvent without fluorescent dye) in the range of 1–3 kHz generate typically S/B ratios of 10–100. Photon or electronic noise is only a minor issue for the detection of single-molecules when highly sensitive photon counting detectors are used.

For the unambiguous detection of individual, freely diffusing molecules, the solution must not exceed a certain dye concentration. The concentration limit is even below the average value that only one molecule is present in the observation as estimated by the density. The number of fluorophores n in a typical confocal observation volume ($\sim$1 fl) can be described by the Poisson distribution. Assuming a concentration of 3.3×10^{-9} M we can estimate probabilities of 0.368 for the detection of a molecule, 0.184 for the simultaneous detection of two molecules, and 0.078 for the detection of more than two molecules. Therefore, the actual concentrations have to be considerably lower, in the range of $<10^{-10}$ M, which is opposed to high S/B ratios. Anyway, on the basis of confocal fluorescence detection, the way for widely-used application of SMS was quickly paved.

23.3
Methods

So far, we have presented some general requirements for SMD. The less elaborate approach is based on a confocal set-up (Figure 23.3b). Excitation as well as collection of the red-shifted fluorescence of the sample is performed by the same objective lens. Spectral separation is achieved by a semi-transparent (dichroic) beam splitter and an additional band pass filter. The confocal principle is established by a pinhole

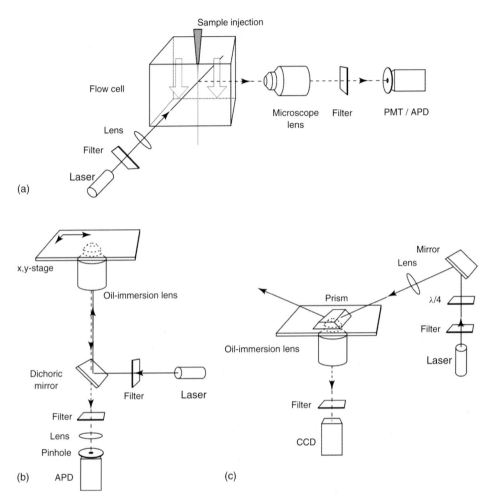

Figure 23.3 (a) Hydrodynamic focussing: A capillary injects the sample into a flow cell. The sheet flow of constantly flowing solvent (water) induces a hydrodynamic pressure compressing the sample flow to a diameter of 5–20 µm. The sample is excited by a laser beam focussing on a relatively large excitation/detection volume ($\sim 10^{-12}$ l) to ensure that every analyte molecule is detected. Emitted fluorescence is detected perpendicular to excitation. (b) Confocal microscope: spectral selection is realized by diffraction limited focussing of the excitation beam in the confocal set-up. Emitted fluorescence from the sample is separated by dichroic mirror before a pinhole performs spatial filtering by rejecting out-of-focus contributions. (c) In total internal reflection fluorescence microscopy (TIRFM) the local evanescent wave created under conditions of total internal reflection at the glass/water interface penetrates the sample only a few hundreds of nanometers. Thereby, only fluorophores at or very close to the interface are excited, thus reducing contributions from solution. Reproduced from D. P. Herten, M. Sauer, *Bunsenmagazin*, 2003, **5**(1), 5–16.

of diameter between 20 and 100 μm (in dependence on the objective magnification) before the weak fluorescence is detected by highly sensitive semiconductor devices.

This experimental arrangement is common in a wide range of SMS applications, where the sample consists of dissolved, freely mobile fluorophors. As the typical residence time of a single molecule in the detection volume is <1 ms, excitation intensities have to be as high as several 100 kW cm^{-2}. Single molecules are detected due to a burst of, typically, 10–200 photons within the dwell time (Figure 23.2). The detected fluctuations are also the basis for fluorescence correlation spectroscopy (FCS) which has developed into a widely used analytical method [46, 47]. Here, more concentrated samples can be used as only the relative intensity fluctuations are analyzed at the cost of individuality. As these relative fluctuations depend on the number of molecules in the detection volume, absolute concentrations can be determined. Other detection schemes in solution employ micro-capillaries where molecules are transported by electrokinetic forces [48]. These experiments can be considered as a continuation of the flow cytometric approach, ensuring high throughput of single molecules.

Even more information on the emission characteristics is obtained by retarding diffusion, for example, by attaching molecules to larger particles like vesicles. Ultimately, molecules can be immobilized on surfaces in thin polymer films or in solids, where photon count rates up to 100 kHz can be measured for longer time periods, up to several seconds (Figure 23.4). The applied excitation intensities are typically less than 1 kW cm^{-2}. As the sample volume can be much smaller than the detection volume in solution, due to the minute film thickness, scattering is no longer an issue. The problem of finding immobile molecules is solved by raster-scanning methods. Confocal fluorescence microscopy renders the location of single fluorophores [49, 50]. Each microscopic image consists of several thousand picture elements leading to the acquisition time for a whole image >1 s. The limitation is again the adequate photon statistics of a total of 100 photons per molecule.

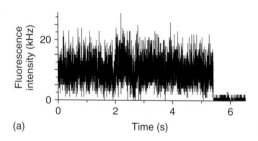

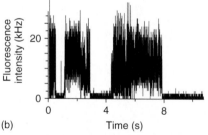

(a)   (b)

Figure 23.4 Blinking of single dye molecules immobilized on glass cover slides recorded by time-resolved confocal fluorescence microscopy. (a) Cy5 molecules show pronounced transitions to short-living triplet states ($\sim 10^{-4}$ s) while (b) MR121 molecules show significantly longer off-states (~ 1 s). The strong photon noise which exceeds the expected noise due to Poissonian statistics is the result of intersystem crossing. Both transients end abruptly by discrete photo-bleaching. Reproduced from D. P. Herten, M. Sauer, *Bunsenmagazin*, 2003, **5**(1), 5–16.

If a higher speed for full frame acquisition is mandatory, for example, for the observation of directed movement, microscopic imaging of single molecules up to video-frame rate is used. The best S/N-ratio is obtained by TIRF-imaging (total internal reflection fluorescence, Figure 23.3c). This kind of technique generates a very small excitation volume on the dielectric phase boundary. The penetration of the sample occurs by an evanescent field in the optically less dense medium, where the electromagnetic field decays exponentially. Evanescent fields next to the interfaces can be achieved when a beam used for excitation encounters a prism or microscope at an angle smaller than the critical angle as defined by Snell's law, which leads to total internal reflection on the interface. The penetration depth d of the evanescent field depends on the excitation wavelength λ_0 and the refractive indices n_1 and n_2 of the used phase (e.g., glass/water) and can be regulated by the angle of total reflection Θ. Usually, the penetration depth is between 30 and 300 nm and is largely independent of the polarization of the incident light.

$$d = \frac{\lambda}{4\pi\sqrt{n_1^2 \sin^2 \Theta - n_2^2}} \tag{23.3}$$

Emitted photons are collected with cooled emCCD or ICCD cameras on the opposite side of the interface with a lower refractive index. In microscopic applications, single fluorophores can be considered as a point source since their dimensions are much less than their mean free path. Point light sources produce in the microscopic imaging a so-called point spread function (PSF) due to the limited optical aperture. The center of the PSF can be determined with high precision (Figure 23.5). This enables, for example, the tracking of single motor proteins

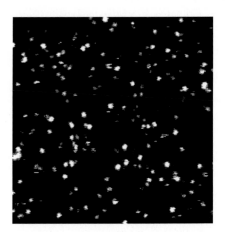

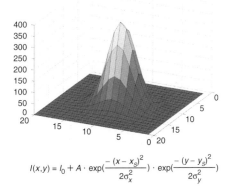

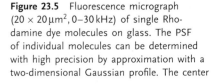

$$I(x,y) = I_0 + A \cdot \exp(\frac{-(x - x_s)^2}{2\sigma_x^2}) \cdot \exp(\frac{-(y - y_s)^2}{2\sigma_y^2})$$

Figure 23.5 Fluorescence micrograph ($20 \times 20\,\mu m^2$, 0–30 kHz) of single Rhodamine dye molecules on glass. The PSF of individual molecules can be determined with high precision by approximation with a two-dimensional Gaussian profile. The center of the profile yields the actual position of the molecule. The accuracy of this determination depends on the photon statistics. Reproduced from D. P. Herten, M. Sauer, *Bunsenmagazin*, 2003, **5**(1), 5–16.

[51] or the diffusion of single fluorescently labeled lipids in membranes [52, 53] and the study of individual molecules in gels [54]. Further experiments were done with this 2D microscopic method to study the activity and movement of myosin filaments [55, 56] and the signal transduction in membranes of living cells [57, 58]. Another alternative without reducing the sample volume provides the conventional fluorescence microscopy. Here a defocused laser excites the sample which is imaged by an emCCD or video camera [53, 59–62]. Processes in living organisms, such as the diffusion in cell membranes [62, 63], the dynamics of biomolecular machines [64], or the infection of living HeLa-cells labeled with single viruses [65] can be detected in this temporal range (see Section 23.6.2). Fluctuations in fluorescence intensity and photobleaching at the same time are limiting factors for the imaging methods. It is worth mentioning that from this area of research a whole new set of microscopy methods has been derived, exceeding the optical resolution limit by at least 1 order of magnitudes [66, 67].

Apart from the detection of fluorescence intensity more information can be extracted from the fluorescence emission. The spectral characteristics, the polarization, and the excited state lifetime of the emitting species are recorded by modifying the experimental instrumentation. Spectral data are obtained by a spectrograph or by inserting a dichroic mirror with secondary detector. Multiple fluorophores with emission spectra in different wavelength ranges can be simultaneously characterized and identified by implementation of additional laser sources [68]. In the same way, a polarization splitter in combination with two detectors allows to measure the emission polarization.

Measurement of fluorescence lifetimes requires a high time resolution ($<1\,\text{ns}$) to determine the time difference between the excitation and the emission of a fluorophore. So far, this experimental approach has only been realized in a confocal set-up with point detectors. Time-correlated single photon counting

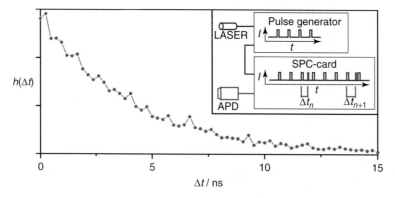

Figure 23.6 Time-correlated single-photon counting (TCSPC) is a statistic measure of photon arrival times with respect to the excitation pulse. Arrival times are collected in histograms. Time-resolution of TCSPC is of the order of $10^{-12}\,\text{s}$. Therefore, the method is well suited to studying fluorescence kinetics on the timescale of the fluorescence lifetime, which is usually of the order of $10^{-9}\,\text{s}$. Reproduced from D. P. Herten, M. Sauer, *Bunsenmagazin*, 2003, **5**(1), 5–16.

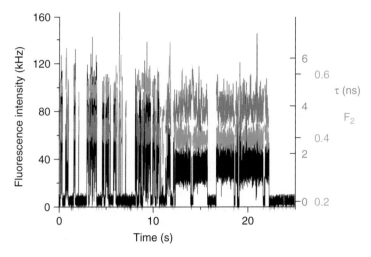

Figure 23.7 Time-resolved fluorescence intensity (black) and lifetime (red) of a single Rhodamine dye molecule immobilized on glass. Fluorescence emission was split by a dichroic mirror for simultaneous detection of two different spectral bands. F_2 denotes the fraction of the red-shifted emission. Reproduced from D.P. Herten, M. Sauer, *Bunsenmagazin*, 2003, **5**(1), 5–16.

(TCSPC) has been successfully implemented for detecting single molecules. For this purpose, the time lag between the detected photon and the previous excitation laser pulse is collected and, for statistical analysis, summarized in histograms (Figure 23.6). The introduced techniques allow simultaneous recording and analysis of fluctuations in intensity, emission spectrum, and fluorescence lifetimes of single molecules (Figure 23.7). These parameters are of crucial importance for studies of photophysical processes, such as photoelectron transfer (PET) or fluorescence resonance energy transfer (FRET), which are used for the structural analysis and study of dynamic structural changes of biologically relevant molecules. The information content, which is necessary for these complex issues, can be expanded if alternating laser excitation, switching schemes, and multiple labeling are exploited. The interplay of such spectroscopic single-molecule techniques allow exploration of the potential energy surfaces of protein folding or the action of enzymes [69].

23.4
Single Quantum Systems: Anti-Bunching, Blinking, Anisotropy

What made the discovery of SMD so exciting for many physicists and chemists was the fact that single molecules represent individual quantum systems which could now be studied at room temperature, and even under ambient conditions, with fluorescence microscopy. Besides reaching the ultimate limit of sensitivity one thereby gains access to properties and behavior of individual quantum systems, and can learn how they make up an ensemble. The distinct nature of single quantum systems also opens possibilities to prove experimentally the observation of single molecules, for example, by monitoring fluorescence fluctuations or by revealing the non-classical nature of the emitted fluorescence. The most prominent effects on fluorescence emission and polarization that can be observed in single-molecule experiments and some of their applications will be discussed briefly in the following sections.

Single-molecule fluorescence studies give access to interesting phenomena: stochastic transitions of molecular states manifested in temporal fluctuations of the measured intensity. Single-molecule fluorescence spectroscopy (SMFS) has a limited time resolution (10^{-7} to 10^{-6} s) in comparison with computer simulations of molecular dynamics (10^{-9} to 10^{-15} s). However, a wide variety of interesting aspects of the molecular dynamics of single molecules can be resolved, like fluctuations in intensity due to changes in molecular position or orientation, or spectral jumps indicating molecular interaction or transformations. Time-resolved studies on individual fluorophores can, therefore, reveal phenomena and molecular states otherwise hidden in the ensemble.

A striking effect observed in the time-resolved fluorescence emission of single molecules is photodestruction. Whereas an illuminated ensemble shows an exponential decrease in fluorescence emission with time, photo-bleaching of a single molecule occurs all of a sudden (Figure 23.4a). The more or less constant emission rate of a single fluorophore suddenly drops to background level in a single step. This phenomenon is so obviously related to single-molecule behavior that it was termed *single-step photo-bleaching* and is taken as a strong indication for the presence of a single molecule. A second indication of single-molecule observations is fluctuations in fluorescence emission. Under equilibrium conditions, time-resolved data from bulk measurements shows fluctuations only due to noise, because contributions from reversible transient changes of molecular states are cancelled out by averaging over the whole molecular ensemble. In contrast, single-molecule transients can sense individual molecular transitions if the associated change in fluorescence emission is stronger than photon shot noise. This has readily been observed for many photo-physical processes leading to spectral jumps [70–77], fluctuations in the lifetime of excited singlet- [49, 50] and triplet-states [78–80], as well as changes in orientation of the transition dipole moments [81–83]. An inevitable and well-studied example is ISC of single dye molecules into the non-fluorescent triplet state. In the bulk, individual transitions into the triplet are superimposed by the fluorescence of the remaining molecules. Upon diluting the sample an emerging number of fluctuations can be detected in the measured fluorescence intensity when only a few molecules remain in the observation volume. When only a single molecule is under observation, these fluctuations appear as discrete jumps in fluorescence intensity between distinct "on" and an "off" states. This phenomenon is called *blinking*. It is generally considered as another valid criterion for the detection of a single quantum system. Interestingly, single-molecule studies have shown that there is only a poor correlation between fluorescence intensity, radiative decay, and the emission spectrum of a single fluorophore [49, 76]. Each spectroscopic quantity appears as independent parameter for measuring the configuration or environment of a single dye. The possibility to observe such spontaneous events underscores the relevance of single-molecule experiments for kinetic and mechanistic studies of (bio-)chemical reactions (e.g., protein folding), because comparable ensemble studies demand synchronization (e.g., denatured state) of all molecules with a precision that, if at all, is very difficult to achieve. Simultaneous occurrence of phenomena like *blinking* and *photo-bleaching* indicate the presence of a single fluorophore, because it is assumed that the probability for their coincidence in more than one fluorophore is negligible.

Another opportunity to probe for the presence of single molecules which are immobilized on a surface or in a matrix (e.g., glass or polymer) is to make use of the inherent polarization properties of individual quantum systems. The transition dipole (i.e., absorption and emission dipole) of an immobilized molecule has a defined and fixed orientation in space which is then equal to the preferred polarization orientation of the absorbed/emitted light. This orientation, and whether it is well defined (for a single) or not (for multiple molecules), can easily be probed using polarized excitation in combination with a polarizing beam splitter in the fluorescence microscope [79]. Orientation imaging can be used for structural studies of macromolecules and biomolecular machines. For example, polarization of the emission light was used to identify rotationally mobile molecules like the F_0F_1-ATP synthase [84, 85]. Defocused imaging is an alternative approach to visualize the orientation of the emission dipole [86]. Recently, this technique has been used to study orientational diffusion of perylenediimide dyes in polymers [87].

23.4.1
Novel Phenomena: Non-Classical Light

A very important and exciting phenomenon is based on the fact that fluorescent dyes possess a finite excited state lifetime, the so-called fluorescence lifetime. Once excited, a single dye molecule can emit a single photon only. Consequently, the probability for detecting two consecutive photons drops to zero for time intervals shorter than the excited state lifetime (Figure 23.8).

In contrast to bulk experiments, photons now no longer appear in bunches, which is why this phenomenon is named *anti-bunching*. More simply, photon anti-bunching describes the fact that single molecules never emit two or more photons at the same time. This argument is so strong that photon-antibunching has been widely accepted as the only proof for single emitters [22, 27, 88, 89]. In SMS photon-antibunching is measured using a Hanbury-Brown, Twiss set-up in a confocal microscope where the emission is split into two detection channels using a 50 : 50 beam splitter.

The emitted photons are recorded with pico- to nanoseconds (10^{-12} to 10^{-9} s) time resolution and their autocorrelation shows a characteristic drop in the correlation amplitude around lag time zero (Figure 23.9) Interestingly, these conclusions can be carried further: The statistical analysis of coincident photons can be used to estimate the number of emitting fluorophores. First approaches to counting fluorophores made use of coincident photon pairs using pulsed laser excitation with repetition cycles significantly longer than the excited state lifetime of the

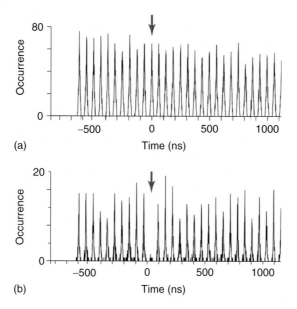

(a)

(b)

Figure 23.8 Coincidence analysis of detected fluorescence photons with pulsed excitation at 16 MHz. Owing to the dead time of TCSPC systems photon pairs are recorded by two different detectors. The excitation rate is reflected by the occurrence of photon pairs every 62.5 s. (a) An ensemble of emitters gives rise to photon pairs in the same excitation cycle at 0 ns while photon pairs at different cycles are reflected in the neighboring peaks. (b) In contrast, a single emitter shows almost no coincident photon pairs in the same excitation cycle. Reproduced from D. P. Herten, M. Sauer, *Bunsenmagazin*, 2003, **5**(1), 5–16.

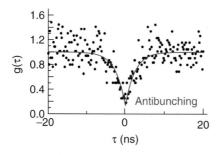

τ (ns)

Figure 23.9 Photon-pair autocorrelation function $g(\tau)$ of an immobilized molecule for investigating antibunching. In contrast to Figure 23.8, continuous-wave excitation allows for the time window below the typical time lag between repeated pulses. $g(\tau)$ can be interpreted as the probability to detect a photon at time t after the detection at $\tau = 0$. As two photons of the same molecule are delayed by the fluorescence lifetime (here: $\tau_F \sim 4$ ns), $g(\tau)$ has a dip at $\tau = 0$ and then increases exponentially by a time constant close to τ_F. The measurement was performed by means of a Hanbury-Brown and Twiss set-up, that is, a start–stop experiment like in TCSPC, thus accessing also negative values of τ, where the role of both photons is reversed.

fluorophores. Data were analyzed by either FCS or by what was called *coincidence analysis* – a simple statistical analysis of the relative occurrence of photon-pairs [89, 90]. However, neither of the two methods could resolve more than two or three molecules, most probably because the occurrence of photon-pairs rises quickly with increasing numbers of emitting molecules. Still, the statistical approach was sufficient to characterize the photo-physical dynamics of a multi-chromophoric system showing that the absorbed energy is rapidly transferred to only one of the four fluorophores which then acts as the only emitter [91, 92]. Recently, it has been shown by simulations, as well as experimentally, that the range for counting the emitting molecules in a confocal observation volume can be dramatically improved by extending the statistical analysis and by implementing four instead of only two detectors [93]. This approach uses the theoretical probability distribution for photon-pairs, triples and quadruples to estimate the number of emitting fluorophores and yields a theoretical limit of >40 fluorophores that has so far been experimentally proven for up to 15 fluorophores [94].

23.5
Conclusions and Perspectives

SMFS has become a very powerful technique in solid state physics as well as, in the past decade, in the field of biophysical research. Geometric heterogeneities and activity alterations, which are ubiquitous in the life sciences, can be resolved by techniques derived from ensemble fluorescence spectroscopy. Only recently, methods of SMD were transferred to questions in the material sciences. With the achievable optical resolution of several nanometers, the active centers in heterogeneous catalysis can be localized in an unprecedented way. We expect that more reactions on surfaces will be studied in the near future. By combination with other spectroscopic techniques, a detailed picture of ongoing transformations at these centers can be obtained. Even more chemical problems will be tackled when appropriate fluorophores with high specificity are found.

23.6
Related Fields of Research

23.6.1
Probes, Switches, and Conductors – Photophysics

A good single-molecule experiment depends on the fluorophore used and its photophysical properties. To go beyond mere single-molecule tracking, to observe and characterize a certain molecular event, like bond making or breaking, this event must be connected to changes in the spectroscopic characteristic of the fluorophore [95]. These changes can involve any spectroscopic property, such as absorption or emission spectra, fluorescence lifetime, polarization, or quantum yield, as long as it can be sensed by SMFS.

Different strategies can be classified into distance-dependent effects, such as FRET or PET, and direct changes in the chromophore structure. Some of the most important effects will be briefly discussed in the following.

23.6.1.1 Förster Resonance Energy Transfer (FRET)

FRET was first described in 1946 by Theodor Förster and has reached a significant level of importance in assay development for biological applications [96]. FRET is the non-radiative and resonant transfer of energy from an excited donor molecule to an acceptor dye molecule in close proximity (Figure 23.10).

FRET has become valuable because of the strong distance dependence in the range 1–10 nm. Figure 23.10 shows that the efficiency of FRET E_{FRET} scales with the sixth power of the distance r between donor and acceptor. It has to be mentioned, however, that the Förster distance R_0 ($E_{FRET} = 50\%$) is usually not constant but depends on the spectral overlap of donor emission with acceptor absorption J,

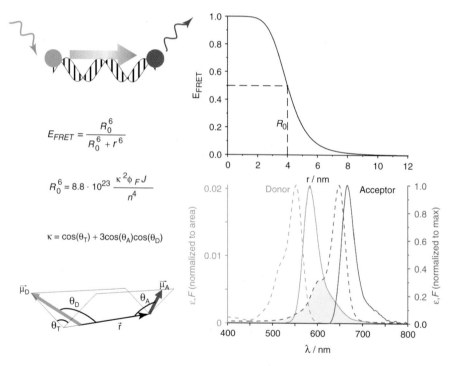

$$E_{FRET} = \frac{R_0^6}{R_0^6 + r^6}$$

$$R_0^6 = 8.8 \cdot 10^{23} \frac{\kappa^2 \phi_F J}{n^4}$$

$$\kappa = \cos(\theta_T) + 3\cos(\theta_A)\cos(\theta_D)$$

Figure 23.10 Schematic overview summarizing FRET. The non-radiative, resonant transfer from an excited donor to an acceptor dye molecule is strongly distance-dependent. FRET usually occurs in the range 1–10 nm and the FRET efficiency E_{FRET} is reciprocal to r^6. Förster distance R_0, where the FRET efficiency is 50%, is determined by the spectroscopic properties of the donor and acceptor, such as the overlap integral J and acceptor fluorescence quantum yield Φ_F. Resonant coupling is described by the overlap integral J of donor emission with acceptor absorption. Additionally, R_0 depends on the relative orientation of donor and acceptor transition dipoles expressed in κ^2.

the acceptor fluorescence quantum yield Φ_F, and the relative orientation of donor and acceptor transition dipole moments, expressed as κ^2. Usually, free rotation of donor and acceptor is assumed yielding $\kappa^2 = 2/3$. However, if only one of the two is immobilized, or their motion is correlated, interpretation of FRET experiments can become quite complicated.

FRET was first applied in single-molecule experiments by Weiss *et al.* in 1996 [97] using double-stranded DNA as scaffold to get defined distances between single pairs of donor and acceptor labels. Their first experiments gave rise to a true renaissance of FRET in single-molecule experiments in the upcoming years and spFRET (single-pair FRET) has been employed in many different experiments, like studies on protein conformation and enzymatic activity, to give but a few examples [36, 37, 98, 99]. A beautiful application of spFRET has been the study of directed energy transfer in a photonic wire consisting of multiple successive donor/acceptor pairs lined up on a DNA duplex [100]. In single-molecule experiments FRET efficiency is determined by measuring the relative contribution of acceptor emission upon donor excitation. However, the large uncertainty in κ^2 and other correction factors has limited the use of spFRET to following changes in distance rather than determining the real distance between donor and acceptor. This difficulty has recently been overcome by employing alternating laser excitation where donor and acceptor are excited in an alternating fashion. By probing the presence of the acceptor one can not only distinguish donor-only sample from zero FRET efficiency but also determine the correction factors and estimate the stoichiometry of donor/acceptor labeling in the sample [101].

It is noteworthy that the use of electronic energy transfer processes by the dipole–dipole interactions is applicable only for donor–acceptor distances above ~2 nm. Different energy transfer mechanisms, like Dexter transfer, have to be taken into account for shorter distances, complicating the situation beyond the point of suitable models for single-molecule experiments [102]. Furthermore, most spFRET applications consider only singlet–singlet energy transfer, but the possibility of resonant singlet–triplet transfer processes has also to be taken into account for precise interpretation of the data [103]. Furthermore, the so-called *energy hopping* among fluorophores of the same kind has been identified as an important process occurring in photosynthetic light harvesting complexes. This mechanism of *homo energy transfer*, also called *energy migration* between identical fluorophores was Försters initial study of FRET [96]. Such energy transfer is possible because of the strong spectral overlap of a fluorophore's absorption with its own emission.

23.6.1.2 Photo-Induced Electron Transfer

Another photophysical process exhibiting a strong distance dependence in a much shorter range <1 nm is PET [104]. In order for electron transfer (ET) to occur, contact pair formation of the two participating molecules is required. ET reactions involving excited states of a chromophore can occur in two different directions. In one, the chromophore is oxidized in its excited state, that is, ET occurs from the excited state to the lowest unoccupied molecular orbital (LUMO) of the electron acceptor. In the other, the chromophore is reduced in its excited state, that is, the electron is transferred from the electron donor's highest occupied molecular orbital (HOMO) to the lower energy orbital of the chromophore. In both cases emissive return of the excited state electron is prevented and a radical ion pair is formed which is usually thought to be kept together sufficiently long by the solvent cage to allow radical recombination to restore the ground state. The necessity of contact pair formation makes the process very sensitive to small distance changes. The process has been used in DNA probes employing fluorescence quenching by guanosine bases which have a sufficiently low oxidation potential for oxidizing certain oxazine and rhodamine derivatives [95, 105]. Tryptophan has a similarly low oxidation potential and has also been used to probe conformational changes in small proteins, thereby extending the capabilities of time-resolved single-molecule studies in proteins to very small conformational changes [106]. PET is possibly also the mechanism for quenching in some copper(II) sensing probes that were used for single-molecule studies on transition metal complex formation [107]. Here, the copper(II)-complex of a bipyridine derivative acts as an intramolecular electron acceptor that quenches the emission of tetramethylrhodamine. Important examples for the second case include work by Urano *et al.* who created a whole library of fluorescent switches by changing the electron donation potential of the phenyl side group [108].

23.6.1.3 Structural Changes

Structural changes at the chromophore can also be employed to create fluorescence switches. In principle, one can imagine direct changes in the chromophore moiety or structural changes close to the chromophoric group influencing its spectroscopic properties. One example for the first case is the reduction of the two nitrogens in the 1 and 5 positions of the flavine moiety (Figure 23.11a), which has

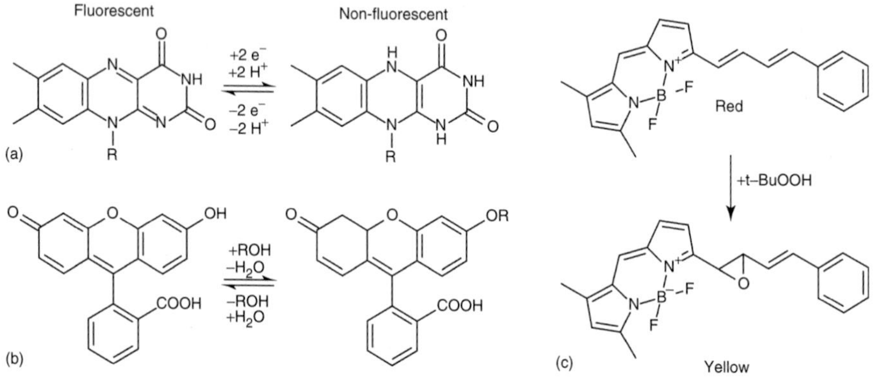

Figure 23.11 Examples of probes that have been used in single-molecule experiments to signal specific chemical reactions by undergoing structural changes influencing their spectroscopic properties. (a) Reduction of the co-factor FAD to the non-fluorescent FADH$_2$ was employed to follow catalytic turnover of a cholesteroloxidase. (b) Hydrolysis of ester-sidegroups in non-fluorescent fluorescein derivatives was used to map catalytic activity in layered double hydroxide crystals. (c) BODIPY-derivatives carrying a double bond in conjugation to the chromophoric group have been used to map individual catalytic centers in doped zeolites with nanometer resolution.

been used to study single-enzyme turn-over by monitoring the fluorescent state of the co-factor FAD/FADH$_2$ [109].

Other examples involve the hydrolysis of sidegroups from functionalized oxygen in fluorescein derivatives (Figure 23.11b) that was used to map catalytic activity in layered double hydroxide crystals [110] or oxidation of Resazurin to Resorufin by gold nanoparticles [111]. Of similar interest are chromophores with a reactive double bond in conjugation to the chromophore influencing their spectral properties as shown in Figure 23.11c [112]. Such compounds have been used successfully for mapping individual catalytic centers in doped zeolites with nanometer resolution [113]. It should be noted, however, that unexpected products have to be considered under the conditions of single-molecule chemistry [114].

23.6.2
Single-Molecule Tracking

One of the most widespread applications of single-molecule imaging is the tracing of fluorescently labeled molecules [115]. It is especially useful in life sciences to follow the active transport of molecules by motor proteins and to map the fluidity of cellular membranes [116–118]. Most of these kinds of experiments are based on an imaging set-up as depicted in Figure 23.3c. Data acquisition of an image frame can occur with frequencies of up to 0.5 kHz [119], but most often video-rate recording (or below) is established. The actual position of the molecule is analyzed by a Gaussian profile fit (Figure 23.5) where the accuracy of the center determines the localization precision. The latter is affected by the photon statistics as well as by the blurring of the signal as a result of the molecule's motion [120, 121]. The challenge here is to find a good compromise between a strong signal and a high acquisition speed to get the best spatial resolution.

Once the position of the molecules has been extracted from the image, a distinction between the different dynamical behavior can be made. The key parameter is the position change Δr of the molecule between one (ith) and the next ((i+1)th) image. As the frame rate is known, Δr can be converted into $\Delta r(t)$.

It is convenient to start with the unbiased movement which is diffusion [122, 123]. Two-dimensional diffusion obeys Fick's second law:

$$\frac{dc}{dt} = D\frac{d^2c}{d\Delta r^2} \tag{23.4}$$

A solution to this law is the concentration $c(\Delta r, t)$ which is a concentration amplitude c_0 multiplied by the probability density function $p(\Delta r, t)$ to find a particle at a distance Δr from the original point.

$$(\Delta r, t) = \frac{c_0}{4\pi Dt}\exp\left(-\frac{\Delta r^2}{4Dt}\right) = c_0^* p(\Delta r, t) \tag{23.5}$$

This description applies to most imaging experiments, even if the random walk also takes place in the third dimension along the optical axis of the experiment. The dislocations in the different directions are independent of each other, and, thus, only the projection onto the detector plane is evaluated. There are two ways of analyzing the data: averaging of the step length Δr and the cumulative probability approach. The first methods averages Δr of many images (the actual position in image i is the origin for the calculation) or, if all molecules behave similarly, over many molecules. The mean squared displacement, $\langle \Delta r^2 \rangle$, is linearly proportional to the time lag t between both images according to the integration

$$\langle \Delta r^2 \rangle = \int_0^\infty \Delta r^2 * p\left(\Delta r, t\right) * 2\pi \, \Delta r \mathrm{d}\Delta r = 4Dt \tag{23.6}$$

Here, Δr is weighted by the perimeter as there is more room for larger Δr. The proportionality constant is the diffusion coefficient D. However, it is impossible to distinguish between the different dynamical behavior only on the basis of Δr between the images i and $i + 1$. Therefore, Δr is analyzed between the image i and $i + j$, where j is an integer, adding thus more data points to a plot of $\langle \Delta r^2 \rangle$ versus t. In this plot, a steeper curvature than linear indicates an active transport, whereas leveling is interpreted as hindered or corralled diffusion (Figure 23.12).

In the cumulative probability approach, all $(= N)$ Δr are ordered according to increasing values of Δr. The smallest rank $= 1$ is obtained for the smallest Δr, and the largest step length Δr has the highest rank $= N$. The probability of finding a molecule within a certain distance Δr_{arb}, $P(\Delta r \leq \Delta r_{\mathrm{arb}})$, is just the rank number, divided by N. On the other hand, an analytical expression for $P(\Delta r \leq \Delta r_{\mathrm{arb}})$ can be derived from $p(\Delta r, t)$ by integration

$$P\left(\Delta r \leq \Delta r_{\mathrm{arb}}\right) = \int_0^{\Delta r_{\mathrm{arb}}} p\left(\Delta r, t\right) * 2\pi \, \Delta r \mathrm{d}\Delta r = \left. -\exp\left(-\frac{\Delta r^2}{4Dt}\right) \right|_0^{\Delta r_{\mathrm{arb}}}$$

$$= 1 - \exp\left(-\frac{\Delta r_{\mathrm{arb}}^2}{\langle \Delta r^2 \rangle}\right) \tag{23.7}$$

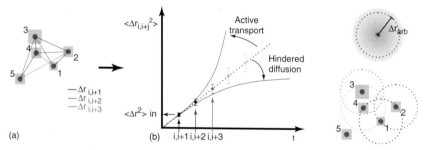

Figure 23.12 Tracking of single molecules. (a) An individual particle (filled circle) can be localized in two dimensions according to the fluorescence signal noise and its mobility (squares; see Figure 23.5). The step length Δr between the position in the ith image (here 1–5) and in consecutive images is squared and plotted against the time lag. Deviations from linearity indicate other transport mechanisms than diffusion. (b) The likelihood to find the molecule in the next image within a radius Δr_{arb} depends on $\langle \Delta r^2 \rangle$, shown in (a) and as described by Equation 23.7).

Divergence from an exponential increase is an indication that the dynamical behavior deviates from a purely random walk.

Besides the characterization of the dynamical behavior and its heterogeneity, the motion of single molecules can be used to visualize the pathways of cellular transportation. In the cases where an active transport is not realized, the random walk of single molecules can be exploited. In an intriguing example, single-molecule diffusion was used to map the channel orientation of mesoporous materials (Figure 23.13) [124]. Due to the single-molecule nature of the emitter, an optical resolution down to 10 nm was established and allowed assignment of the geometry

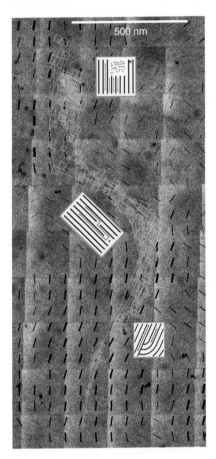

Figure 23.13 The random walk of single molecules in a mesoporous material. The bidirectional diffusion is exploited to map the channel topology including dead ends, defects, and branching points in an unprecedented way. The orientation is verified by high resolution TEM images (small squared regions) as indicated by the bars; their thickness is a measure of the local crystal quality. Due to the single-molecule nature of the tracer, its actual position (cyan squares) can be determined far beyond the diffraction-limited emission profile of ~250 nm. Courtesy of the Bräuchle group, LMU Munich.

of an ensemble of a few channels. A correlation with high resolution transmission electron micrographs verified that single-molecule microscopy revealed the nano-sized local structure. It turns out that this novel tool in material science is appropriate to characterize defects and heterogeneities of real materials in an unprecedented way [125]. Applications in heterogeneous catalysis are foreseen.

Note

Quite recently, an experimental verification of the ergodic theorem (see Section 23.1) has been achieved by comparison of single-molecule tracking data and pulsed-field gradient NMR data (Feil, F. *et al.* (2012) *Angew. Chem.*, 124, 1308, DOI: 10.1002/anie.201105388).

References

1. Schrödinger, E. (1952) *Br. J. Phil. Sci.*, 3, 233–242.
2. Feynman, R.P. (1960) *Eng. Sci.*, 23, 22–36.
3. Rotman, B. (1961) *Proc. Natl. Acad. Sci. U.S.A.*, 47, 1981–1991.
4. Hirschfeld, T. (1976) *Appl. Opt.*, 15, 3135–3139.
5. Dovichi, N.J., Martin, J.C., Jett, J.H., Trkula, M., and Keller, R.A. (1984) *Anal. Chem.*, 56, 348–354.
6. Dovichi, N.J., Martin, J.C., Jett, J.H., and Keller, R.A. (1983) *Science*, 219, 845–847.
7. Shera, E.B., Seitzinger, N.K., Davis, L.M., Keller, R.A., and Soper, S.A. (1990) *Chem. Phys. Lett.*, 174, 553–557.
8. Moerner, W.E. and Kador, L. (1989) *Anal. Chem.*, 61, 1217A–1223A.
9. Kador, L., Horne, D.E., and Moerner, W.E. (1990) *J. Phys. Chem.*, 94, 1237–1248.
10. Orrit, M. and Bernard, J. (1990) *Phys. Rev. Lett.*, 65, 2716–2719.
11. Betzig, E. and Chichester, R.J. (1993) *Science*, 262, 1422–1425.
12. Trautman, J.K., Macklin, J.J., Brus, L.E., and Betzig, E. (1994) *Nature*, 369, 40–42.
13. Xie, X.S. and Dunn, R.C. (1994) *Science*, 265, 361–364.
14. Ambrose, W.P., Goodwin, P.M., Martin, J.C., and Keller, R.A. (1994) *Phys. Rev. Lett.*, 72, 160–163.
15. Macklin, J.J., Trautman, J.K., Harris, T.D., and Brus, L.E. (1996) *Science*, 272, 5259.
16. Hecht, B., Sick, B., Wild, U., Deckert, V., Zenobi, R., Martin, O., and Pohl, D. (2000) *J. Chem. Phys.*, 112, 7761–7774.
17. Gimzewski, J.K. and Joachim, C. (1999) *Science*, 283, 1683–1688.
18. Fan, F.R.F. and Bard, A.J. (1995) *Science*, 267, 871–874.
19. Kneipp, K., Wang, Y., Kneipp, H., Perelman, L.T., Itzkan, I., Dasari, R., and Feld, M.S. (1997) *Phys. Rev. Lett.*, 78, 1667–1670.
20. Eigen, M. and Rigler, R. (1994) *Proc. Natl. Acad. Sci. U.S.A.*, 91, 5740–5747.
21. Moerner, W.E. (2002) *J. Phys. Chem. B*, 106, 910–927.
22. Tamarat, P., Maali, A., Lounis, B., and Orrit, M. (2000) *J. Phys. Chem. A*, 104, 1–16.
23. Xie, X.S. and Trautman, J.K. (1998) *Annu. Rev. Phys. Chem.*, 49, 441–480.
24. Nie, S. and Zare, R.N. (1997) *Annu. Rev. Biophys. Biomol. Struct.*, 25, 567–596.
25. Wu, M., Goodwin, P.M., Ambrose, W.P., and Keller, R.A. (1996) *J. Phys. Chem. A*, 100, 17406–17409.
26. Barnes, M.D., Whitten, W.B., and Ramsey, J.M. (1995) *Anal. Chem.*, 67, 418A–423A.

27. Basché, T., Moerner, W.E., Orrit, M., and Talon, H. (1992) *Phys. Rev. Lett.*, **69**, 1516–1519.

28. Moerner, W.E. and Orrit, M. (1999) *Science*, **283**, 1670–1676.

29. Plakhotnik, T., Donley, E., and Wild, U. (1997) *Annu. Rev. Phys. Chem.*, **48**, 181–212.

30. Basché, T. and Bräuchle, C. (1996) *Ber. Bunsen-Ges. Phys. Chem.*, **100**, 1269–1279.

31. Moerner, W.E. (1996) *Acc. Chem. Res.*, **29**, 563–571.

32. Ambrose, W.P., Goodwin, P.M., Jett, J.H., van Orden, A., Werner, J.H., and Keller, R.A. (1999) *Chem. Rev.*, **99**, 2929–2956.

33. Keller, R.A., Ambrose, W.P., Arias, A.A., Cai, H., Emory, S.R., Goodwin, P.M., and Jett, J.H. (2002) *Anal. Chem.*, **74**, 316A–324A.

34. Forkey, J.N., Quinlan, M.E., and Goldman, Y.E. (2000) *Prog. Biophys. Mol. Biol.*, **74**, 1–35.

35. Ishii, Y. and Yanagida, T. (2000) *Single Mol.*, **1**, 5–13.

36. Weiss, S. (1999) *Science*, **283**, 1676–1683.

37. Weiss, S. (2000) *Nat. Struct. Biol.*, **7**, 724–729.

38. Strickler, S.J. and Berg, R.A. (1962) *J. Chem. Phys.*, **37**, 814–822.

39. van Orden, A., Keller, R.A., and Ambrose, W.P. (2000) *Anal. Chem.*, **72**, 37–41.

40. Goodwin, P.M., Cai, H., Jett, J.H., Ishaug-Riley, S.L., Machara, N.P., Semin, D.J., van Orden, A., and Keller, R.A. (1997) *Nucleosides Nucleotides*, **16**, 543–550.

41. Goodwin, P.M., Ambrose, W.P., and Keller, R.A. (1996) *Acc. Chem. Res.*, **29**, 607–613.

42. Rigler, R. and Widengren, J. (1990) *BioScience*, **3**, 180–183.

43. Rigler, R., Mets, Ü., Widengren, J., and Kask, P. (1993) *Eur. Biophys. J.*, **22**, 169–175.

44. Mets, Ü. and Rigler, R. (1994) *J. Fluoresc.*, **4**, 259–264.

45. Nie, S., Chiu, D.T., and Zare, R.N. (1994) *Science*, **266**, 1018–1021.

46. Haustein, E. and Schwille, P. (2004) *Curr. Opin. Struct. Biol.*, **14**, 531–540.

47. Enderlein, J., Gregor, I., Patra, D., and Fitter, J. (2004) *Curr. Pharm. Biotechnol.*, **5**, 155–161.

48. Sauer, M., Angerer, B., Han, K., and Zander, C. (1999) *Phys. Chem. Chem. Phys.*, **1**, 2471–2477.

49. Tinnefeld, P., Buschmann, V., Herten, D.-P., Han, K.-T., and Sauer, M. (2000) *Single Mol.*, **3**, 215–223.

50. Tinnefeld, P., Herten, D.-P., and Sauer, M. (2001) *J. Phys. Chem. A*, **105**, 7989–8003.

51. Sase, I., Miyata, H., Corrie, J.E.T., Craik, J.S., and Kinosita, K. (1995) *Biophys. J.*, **69**, 323–328.

52. Schmidt, T., Schütz, G.J., Baumgartner, W., Gruber, H.J., and Schindler, H. (1995) *J. Phys. Chem.*, **99**, 17662–17668.

53. Schmidt, T., Schutz, G.J., Baumgartner, W., Gruber, H.J., and Schindler, H. (1996) *Proc. Natl. Acad. Sci. U.S.A.*, **93**, 2926–2929.

54. Dickson, R.M., Norris, D.J., Tzeng, Y.L., and Moerner, W.E. (1996) *Science*, **274**, 966–969.

55. Funatsu, T., Harada, Y., Tokunaga, M., Saito, K., and Yanagida, T. (1995) *Nature*, **374**, 555–559.

56. Wazawa, T., Ishii, Y., Funatsu, T., and Yanagida, T. (2000) *Biophys. J.*, **78**, 1561–1569.

57. Sako, Y., Hibino, K., Miyauchi, T., Miyamoto, Y., Ueda, M., and Yanagida, T. (2000) *Single Mol.*, **1**, 159–163.

58. Sako, Y., Minoguchi, S., and Yanagida, T. (2000) *Nat. Cell Biol.*, **2**, 168–172.

59. Harms, G., Sonnleitner, M., Schütz, G., Gruber, H., and Schmidt, T. (1999) *Biophys. J.*, **77**, 2864–2870.

60. Cognet, L., Harms, G.S., Blab, G.A., Lommerse, P.H.M., and Schmidt, T. (2000) *Appl. Phys. Lett.*, **77**, 4052–4054.

61. Adachi, K., Kinosita, K. Jr., and Ando, T. (1999) *J. Microsc.*, **195** (Pt 2), 125–132.

62. Schütz, G.J., Schindler, H., and Schmidt, T. (1997) *Biophys. J.*, **73**, 1073–1080.

63. Sonnleitner, A., Schutz, G.J., and Schmidt, T. (1999) *Biophys. J.*, **77**, 2638–2642.

64. Adachi, K., Yasuda, R., Noji, H., Itoh, H., Harada, Y., Yoshida, M., and

Kinosita, K. (2000) *Proc. Natl. Acad. Sci. U.S.A.*, **97**, 7243–7247.

65. Seisenberger, G., Ried, M.U., Endress, T., Büning, H., Hallek, M., and Bräuchle, C. (2001) *Science*, **294**, 1929–1932.

66. Hell, S.W. (2009) *Nat. Methods*, **6**, 24–32.

67. Huang, B., Bates, M., and Zhuang, X. (2009) *Annu. Rev. Biochem.*, **78**, 993–1016.

68. Prummer, M., Hubner, C.G., Sick, B., Hecht, B., Renn, A., and Wild, U.P. (2000) *Anal. Chem.*, **72**, 443–447.

69. Schuler, B., Lipman, E.A., and Eaton, W.A. (2002) *Nature*, **419**, 743–747.

70. Ambrose, W.P. and Moerner, W.E. (1991) *Nature*, **349**, 225–227.

71. Lu, H. and Xie, X. (1997) *Nature*, **385**, 143–146.

72. Ha, T., Enderle, T., Selvin, P., Chemla, D., and Weiss, S. (1997) *Chem. Phys. Lett.*, **271**, 1–5.

73. Ying, L. and Xie, X.S. (1998) *J. Phys. Chem. B*, **102**, 10399–10409.

74. Yip, W.-T., Hu, D., Yu, J., Vanden Bout, D.A., and Barbara, P.F. (1998) *J. Phys. Chem. A*, **102**, 7564–7575.

75. Weston, K.D. and Buratto, S.K. (1998) *J. Phys. Chem. A*, **102**, 3635–3641.

76. Weston, K.D., Carson, P.J., Metiu, H., and Buratto, S.K. (1998) *J. Chem. Phys.*, **109**, 7474–7484.

77. Köhn, F., Hofkens, J., and de Schryver, F.C. (2000) *Chem. Phys. Lett.*, **321**, 372–378.

78. Veerman, J.A., Garcia-Parajo, M.F., Kuipers, L., and van Hulst, N.F. (1999) *Phys. Rev. Lett.*, **83**, 2155–2158.

79. Weston, K. and Goldner, L. (2001) *J. Phys. Chem. B*, **105**, 3453–3462.

80. English, D.S., Furube, A., and Barbara, P.F. (2000) *Chem. Phys. Lett.*, **324**, 15–19.

81. Ha, T., Enderle, T., Chemla, D.S., Selvin, P.R., and Weiss, S. (1996) *Phys. Rev. Lett.*, **77**, 3979–3982.

82. Ruiter, A.G.T., Veerman, J.A., Garcia-Parajo, M.F., and van Hulst, N.F. (1997) *J. Phys. Chem. A*, **101**, 7318–7323.

83. Bartko, A.P. and Dickson, R.M. (1999) *J. Phys. Chem. B*, **103**, 11237–11241.

84. Diez, M., Zimmermann, B., Börsch, M., König, M., Schweinberger, E., Steigmiller, S., Reuter, R., Felekyan, S., Kudryavtsev, V., Seidel, C.A.M., and Gräber, P. (2004) *Nat. Struct. Mol. Biol.*, **11**, 135–141.

85. Kaim, G., Prummer, M., Sick, B., Zumofen, G., Renn, A., Wild, U.P., and Dimroth, P. (2002) *FEBS Lett.*, **525**, 156–163.

86. Patra, D., Gregor, I., and Enderlein, J. (2004) *J. Phys. Chem. A*, **108**, 6836–6841.

87. Uji-I, H., Melnikov, S.M., Deres, A., Bergamini, G., de Schryver, F., Herrmann, A., Müllen, K., Enderlein, J., and Hofkens, J. (2006) *Polymer*, **47**, 2511–2518.

88. Tinnefeld, P., Müller, C., and Sauer, M. (2001) *Chem. Phys. Lett.*, **345**, 252–258.

89. Weston, K.D., Dyck, M., Tinnefeld, P., Müller, C., Herten, D.P., and Sauer, M. (2002) *Anal. Chem.*, **74**, 5342–5349.

90. Sýkora, J., Kaiser, K., Gregor, I., Bönigk, W., Schmalzing, G., and Enderlein, J. (2007) *Anal. Chem.*, **79**, 4040–4049.

91. Hübner, C.G., Zumofen, G., Renn, A., Herrmann, A., Müllen, K., and Basché, T. (2003) *Phys. Rev. Lett*, **91**, 093903, P-03.157.

92. Tinnefeld, P., Hofkens, J., Herten, D.-P., Masuo, S., Vosch, T., Cotlet, M., Habuchi, S., Müllen, K., de Schryver, F.C., and Sauer, M. (2004) *ChemPhysChem*, **5**, 1786–1790.

93. Ta, H., Wolfrum, J., and Herten, D.-P. (2009) *Laser Phys.*, **20**, 119–124.

94. Ta, H., Kiel, A., Wahl, M., and Herten, D.-P. (2010) *Phys. Chem. Chem. Phys.*, **12**, 10295–10300.

95. Lymperopoulos, K., Kiel, A., Seefeld, A., Stöhr, K., and Herten, D.-P. (2010) *ChemPhysChem*, **11**, 43–53.

96. Förster, T.H. (1948) *Ann. Phys.*, **2**, 55–75.

97. Ha, T., Enderle, T., Ogletree, D.F., Chemla, D.S., Selvin, P.R., and Weiss, S. (1996) *Proc. Natl. Acad. Sci. U.S.A.*, **93**, 6264–6268.

98. Selvin, P.R. (2000) *Nat. Struct. Biol.*, **7**, 730–734.

99. Ha, T., Ting, A.Y., Liang, J., Caldwell, W.B., Deniz, A.A., Chemla, D.S.,

Schultz, P.G., and Weiss, S. (1999) *Proc. Natl. Acad. Sci. U.S.A.*, **96**, 893–898.

100. Heilemann, M., Tinnefeld, P., Sanchez Mosteiro, G., Garcia Parajo, M., van Hulst, N.F., and Sauer, M. (2004) *J. Am. Chem. Soc.*, **126**, 6514–6515.

101. Kapanidis, A.N., Lee, N.K., Laurence, T.A., Doose, S., Margeat, E., and Weiss, S. (2004) *Proc. Natl. Acad. Sci. U.S.A.*, **101**, 8936–8941.

102. Dexter, D.L. (1953) *J. Chem. Phys.*, **21**, 836–850.

103. Hofkens, J., Maus, M., Gensch, T., Vosch, T., Cotlet, M., Köhn, F., Herrmann, A., Müllen, K., and de Schryver, F. (2000) *J. Am. Chem. Soc.*, **122**, 9278–9288.

104. Sauer, M. (2003) *Angew. Chem. Int. Ed.*, **42**, 1790–1793.

105. Seidel, C.A.M., Schulz, A., and Sauer, M.H.M. (1996) *J. Phys. Chem.*, **100**, 5541–5553.

106. Neuweiler, H., Doose, S., and Sauer, M. (2005) *Proc. Natl. Acad. Sci. U.S.A.*, **102**, 16650–16655.

107. Kiel, A., Kovacs, J., Mokhir, A., Krämer, R., and Herten, D.-P. (2007) *Angew. Chem. Int. Ed.*, **46**, 3363–3366.

108. Kamiya, M., Urano, Y., Ebata, N., Yamamoto, M., Kosuge, J., and Nagano, T. (2005) *Angew. Chem. Int. Ed.*, **44**, 5439–5441.

109. Xie, X.S. and Lu, H.P. (1999) *J. Biol. Chem.*, **274**, 15967–15970.

110. Roeffaers, M.B.J., Sels, B.F., Uji-I, H., de Schryver, F.C., Jacobs, P.A., de Vos, D.E., and Hofkens, J. (2006) *Nature*, **439**, 572–575.

111. Xu, W., Kong, J.S., Yeh, Y.-T.E., and Chen, P. (2008) *Nat. Mater.*, **7**, 992–996.

112. Jung, G., Schmitt, A., Jacob, M., and Hinkeldey, B. (2008) *Ann. N.Y. Acad. Sci.*, **1130**, 131–137.

113. de Cremer, G., Roeffaers, M.B.J., Bartholomeeusen, E., Lin, K., Dedecker, P., Pescarmona, P.P., Jacobs, P.A., de Vos, D.E., Hofkens, J., and Sels, B.F. (2010) *Angew. Chem. Int. Ed.*, **49**, 908–911.

114. Schmitt, A., Hinkeldey, B., Hötzer, B., and Jung, G. (2009) *J. Phys. Org. Chem.*, **22**, 1233–1238.

115. Bräuchle, C., Lamb, D.D., and Michaelis, J. (eds) (2009) *Single Particle Tracking and Single Molecule Energy Transfer*, Wiley-VCH Verlag GmbH, Weinheim.

116. Levi, V. and Gratton, E. (2007) *Cell Biochem. Biophys.*, **48**, 1–15.

117. Ritchie, K. and Spector, J. (2007) *Biopolymers*, **87**, 95–101.

118. Peterman, E.J.G., Sosa, H., and Moerner, W.E. (2004) *Annu. Rev. Phys. Chem.*, **55**, 79–96.

119. Ritter, J.G., Veith, R., Veenendaal, A., Siebrasse, J.P., and Kubitscheck, U. (2010) *PLoS One*, **5**, e11639.

120. Thompson, R., Larson, D., and Webb, W. (2002) *Biophys. J.*, **82**, 2775–2783.

121. Kubitscheck, U., Kuckmann, O., Kues, T., and Peters, R. (2000) *Biophys. J.*, **78**, 2170–2179.

122. Kirstein, J. (2007) Diffusion of single molecules in nanoporous mesostructured materials. Doctoral thesis. Ludwig-Maximilians-University, Munich.

123. Kirstein, J., Platschek, B., Jung, C., Brown, R., Bein, T., and Bräuchle, C. (2007) *Nat. Mater.*, **6**, 303–310.

124. Zuerner, A., Kirstein, J., Doeblinger, M., Braeuchle, C., and Bein, T. (2007) *Nature*, **450**, 705–708.

125. Michaelis, J. and Bräuchle, C. (2010) *Chem. Soc. Rev.*, **39**, 4731–4740.

24
Scanning Probe Methods: From Microscopy to Sensing

Rüdiger Berger, Jochen Gutmann, and Rolf Schäfer

■ Method Summary

Acronyms, Synonyms
- Scanning probe microscopy (SPM)
- Scanning force microscopy (SFM)
- Atomic force microscopy (AFM)
- Molecular force probe (MFP)
- Differential thermogravimetry (DTG)
- Micro-electromechanical system (MEMS)
- Scanning near-field optical microscopy (SNOM).

Benefits (Information Available)
- Imaging: topography with atomic or molecular resolution (ultra-high vacuum, air, and liquids)
- Force probe: investigation of unbinding forces of single molecules, kinetic parameters of complex formation at a single molecule level
- Gravimetry: mass resolution of femtograms for biological species and of zeptograms in an ultrahigh vacuum environment
- Biosensors: detection limit in the nanomolar concentration range
- Calorimetry: heat resolution of nanojoule under ambient conditions and of picojoule in a vacuum environment, nanogram sample amounts can be measured.

Limitations (Information Not Available)
- Imaging: chemical contrast is often hardly observed
- Force probe: needs a microscopic model for the analysis of the experimental data
- Gravimetry: sample preparation is demanding
- Biosensors: transduction mechanisms not clear-cut
- Calorimetry: quantitative evaluation is challenging.

Methods in Physical Chemistry, First Edition. Edited by Rolf Schäfer and Peter C. Schmidt.
© 2012 Wiley-VCH Verlag GmbH & Co. KGaA. Published 2012 by Wiley-VCH Verlag GmbH & Co. KGaA.

24.1
Introduction

In daily life we obtain information on material properties by touching surfaces with our fingers. We can feel the roughness of surfaces. In addition, we are able to sense elasticity, friction, hardness, and temperature of surfaces and objects. In order to obtain a profile of these properties we approach our fingers toward the object of interest and move the finger along the object at a desired velocity and pressure. Actually we adjust our mode of touching surfaces for specific purposes. For example, we knock with our fingers against an object to obtain information about a specific type of material, for example, wood or steel; or we press our fingers into an area of interest to find out if the object is hard or soft. Sometimes we touch objects to feel their temperature. In order to avoid burning our fingers, we touch surfaces for a very short time. All those surface properties – and many more – can be analyzed on a million times smaller lateral scale by means of a micromechanical cantilever. Instead of using a finger, an atomically sharp tip is scanned over a sample and the interaction strength, that is, the force between the tip and the sample, is measured by determination of the deflection of a micromechanical cantilever spring in a scanning probe microscope. The fundamental principle of this method is a measurement of forces thus the method is called atomic force microscopy (AFM) or scanning force microscopy (SFM) as a synonym. The SFM method allows one to study the surface topography with atomic or molecular resolution [1–3]. However, specific probes and operation modes can be used to probe lateral variations of frictional [4], adhesion [5], thermal [6, 7], electrical [8], and magnetic properties [9]. Therefore here we use the term scanning probe microscopy (SPM) as a generic term for all modes.

The family of SPM methods has been established as a powerful technique in research, even in the analysis of industrial electronic products [8], food research [10], and cosmetics [11]. Besides the static characterization of surface properties, dynamic processes on surfaces can also be studied. Examples are investigations of changes in the wetting behavior of surfaces with changing temperature [12], the motion of nanometer scale objects induced by solvent exposure [13] and surface topography changes in cells that have been exposed to different chemicals [14]. In particular, SPM plays a major role in the identification of molecular arrangement and morphology at length scales from nanometers to micrometers. The characterization of surfaces at these length scales is mandatory for a rational design of functional materials. Besides the possibility of recording the surface topography, the micromechanical cantilever technique allows investigation of molecular interactions at a single molecule level [15, 16]. This single molecular force probe (MFP) spectroscopy not only gives insight into the specific forces involved during the formation of single (bio)molecular complexes but also allows one to study the mechanical properties of individual complexes [17, 18].

However, micromechanical cantilevers are not only very sensitive force sensors but they can also be used as an almost universal sensor. Micromechanical cantilevers can be used as a microbalance [19], a microcalorimeter [20, 21], or a molecular stress sensor [22]. Owing to the small size of the cantilever sensors they can be easily fabricated in arrays, thus allowing parallel measurements [23, 24].

The aim of this chapter is not to cover fully all of the possible application areas of SPM and micromechanical cantilevers. Instead we would like to introduce the basic working principles of SPM and illustrate the specific advantages of the method, for example, the possibility to use liquid environments. Furthermore, we present examples of the unusual properties of micromechanical cantilevers as sensors. The applications of such sensors range from materials analysis to biomedical detection of single molecules.

24.2
Basic Principles and Experimental Technqiues

24.2.1
Working Principle of Scanning Probe Microscopy

The basic components of a scanning probe microscope are given here. A sharp tip is situated at the end of a cantilever spring (see scanning electron microscopy image in Figure 24.1). Approaching the tip towards the sample leads to a deflection of the cantilever spring owing to additional tip–surface forces: depending on the nature of these forces the deflection can be away from the surface (e.g., repulsive electrostatic forces) or towards the surface (e.g., Van der Waals forces, capillary forces) [5]. The deflection of the cantilever is proportional to the magnitude of the sum of all acting

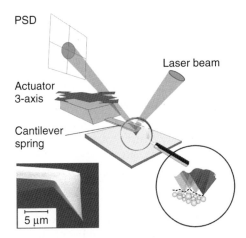

Figure 24.1 The basic components of a scanning probe microscope. A 3-axis actuator moves the cantilever spring. The deflection of the cantilever spring is measured with a laser beam reflected from the cantilever's end and a position sensitive detector (PSD). Magnifying glass: schematic constant force profile (dotted line) of the tip. Inset: a scanning electron microscopy image of a micromachined Silicon tip.

forces and can be measured by an optical beam deflection principle. A laser is focused on the end of the cantilever spring from where it is reflected and falls on a PSD. This detector generates electrical signals corresponding to the light intensities of each of the illuminated quadrants, respectively. Using this method, deflection changes smaller than 1 nm can be measured routinely. Thus by controlling the position and the tip–sample distance with a three-axis actuator the force between tip and sample can be adjusted and kept constant using a feedback loop. Upon scanning the tip laterally line profiles and, finally, a map of the sample topography are obtained.

24.2.2
Scanning Probe Microscopy

During the last 25 years, the family of SPM methods has been extended by numerous operating principles. In the following, we just illustrate two basic operating methods: (i) contact and (ii) intermittent contact mode, which are both used for topography mapping.

In contact mode the sample is scanned by a tip, typically fabricated onto a soft cantilever spring. With respect to its mechanical properties a cantilever is a simple elastic beam fixed at one end. Its bending behavior follows from Hook's law

$$F = -Kz \tag{24.1}$$

where a force applied to the cantilever leads to a deflection z, with the spring constant K as a proportionality constant. Contact mode, intermittent mode, and force spectroscopy measurements require micromechanical cantilevers having different spring constants, which can be addressed by the geometry of the cantilever. For a simple rectangular-shaped cantilever, the spring constant K is obtained by $K = Ewt^3/4L^3$ where E, is the Young's modulus of the cantilever material, w is the width, t the thickness, and L the length of the cantilever. The first harmonic resonance frequency f_0 of the micromechanical cantilever is then given by

$$f_0 = \frac{1}{2\pi}\sqrt{\frac{K}{0.24m}} \tag{24.2}$$

where m is the mass of the cantilever. For example, micromechanical cantilevers can be fabricated from silicon which feature a first harmonic resonance frequency of the order of 1 MHz having a spring constant of the order of 7 N m^{-1} ($L = 40\,\mu m$, $w = 10\,\mu m$, $t = 1\,\mu m$, $E_{Si} = 169\,GPa$, $\rho_{Si} = 2.33\,g\,cm^{-3}$).

While scanning the surface with the tip, the deflection of the cantilever is kept constant by controlling the tip–sample distance by using an electronic feedback loop. A topographic map of the sample is obtained by plotting the electronic feedback signal versus the position of the tip. The obtained lateral resolution depends on the properties of the sample, the geometry of the apex of the SPM-tip and the interaction between the two. For some sample surfaces an atomic lattice resolution can be obtained [25–27]. However, in contact mode, soft sample surfaces like polymers are often altered or scratched by the tip. Thus an intermittent contact mode is more widely used for imaging soft surfaces. In this mode, the cantilever is vibrated by an additional piezoelectric actuator element. Stiffer cantilevers are typically used (K ranges from 2 to $50\,N\,m^{-1}$), with resonance frequencies in the range 70–1000 kHz. When the oscillating SPM-tip is brought into close contact with a sample surface, additional tip–sample forces reduce the amplitude of the vibrating cantilever. While scanning the SPM-tip along the surface the electronic feedback maintains a constant amplitude of vibration by adjusting the tip sample distance. In analogy to the contact mode, the signal from the electronic feedback is used to represent the sample topography. Furthermore, in the intermittent contact mode the energy dissipation during the short contact time can be mapped by analyzing the phase relation between the piezoelectric actuator element and the actual movement of the cantilever [28]. Changes in this phase relation are often used as a measure for mechanical or adhesion properties of sample surfaces.

In order to record surfaces at an atomic resolution, a precise feedback loop is required which is able to track minute changes in tip–sample forces. In addition, very clean and flat surfaces must be prepared. Therefore ultra-high vacuum conditions are often used [29]. Furthermore, precise knowledge of the tip's apex chemistry helps to unravel internal molecular structures [30]. During the last five years, the noise of electronic components has been significantly decreased, allowing real atomic resolution of single atomic defects in liquids [31, 32]. In order to map molecules in the vicinity of surfaces, the standard two-dimensional imaging techniques are extended by simultaneously acquiring distance-dependent parameters. For example, Fukuma and coworkers have investigated the atomic scale distribution of water molecules at the mica interface [33].

Besides the application of atomic resolution imaging of surfaces, SPM methods are routinely applied in ambient air conditions. In the latter case a lateral resolution of the order of 3–4 nm can be achieved while the vertical (i.e., height) resolution is below 1 nm. One of the reasons for the lower lateral resolution in air is the presence of water at the sample surface which increases tip–sample forces [34].

24.2.3
Force Spectroscopy

The family of SPM methods is not only a powerful tool to map surface properties but also allows one to study interactions between molecules. For that purpose, one of the molecules involved (ligand) is immobilized at a surface, while the other one (receptor) is fixed on the tip of a cantilever. On bringing the tip into contact with the surface a complex formation between molecules might take place. By separating the cantilever from the surface, the ligand–receptor complex will be exposed to an external loading force until it dissociates. The so-called unbinding forces obtained from these force-separation experiments are a measure of the strength of the ligand–receptor complex. In Figure 24.2a the principle of the MFP experiment using complementary DNA single strands as model ligand–receptor pairs is shown: during contact the formation of a DNA complex, consisting of

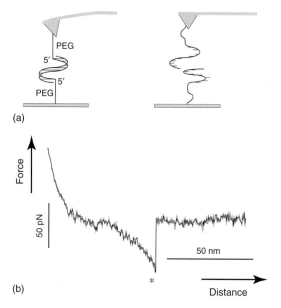

(a)

(b)

Figure 24.2 Molecular force probe experiment with double-stranded DNA, which consists of 30 complementary base pairs. The complementary DNA strands are fixed at the 5' end via polyethylene glycol (PEG) spacers [35]. (a) Principle of the experiment. (b) Exemplary force–distance curve with an unbinding force of about 55 pN.

two complementary DNA strands each containing 30 bases, takes place. When the cantilever is retracted an additional bending of the cantilever is observed until the complex dissociates. The corresponding force–distance measurement is displayed in Figure 24.2b, revealing an unbinding force of about 55 pN.

The absolute value of the obtained unbinding forces could be quantitatively analyzed in terms of molecular parameters of the complex formation, if force–distance experiments with different retract velocities are performed [36].

SFM-based force spectroscopy has emerged as a powerful tool to study the rupture of molecular bonds, ranging from non-covalent and covalent bonds [16, 37] to the unfolding of proteins [38] and nucleic acids [15]. An MFP can also be used to investigate the structure [39] and dissociation constant [40] of supramolecular assemblies and to reveal and discriminate multiple unbinding events [41]. Force spectroscopy measurements of ligand binding, antibody–antigen interactions, and protein unfolding are well established nowadays and MFP instruments are now routinely applied to probe the affinity, mechanics, and recognition properties of (bio)molecular interactions [5, 17, 42]. An interesting new development is the combination of SFM imaging with force mapping and spectroscopy. This technique allows the investigator to scan the surface and subsequently target specific topological structures for stretching measurements with sub-nanometer accuracy. In one example, purple membrane patches were imaged with the SFM to find the location of an individual bacteriorhodopsin molecule, which was then extracted with the same SFM tip so that the unbinding forces and force spectrum of removal could be measured [43]. These experiments highlight the unique capability of SFM instruments to both image the surface topology at high resolution and measure the unbinding or unfolding forces at well-defined locations.

24.2.4
Mass Sensing

The resonance frequency of micromechanical cantilevers depends on their mass. Therefore, changes in the mass loading can be detected by measuring resonance frequency changes. In the most simple case of a micromechanical cantilever that is end-loaded by a mass m_1, the resonance frequency f_1 is lowered by

$$f_1 = \frac{1}{2\pi} \sqrt{\frac{K}{0.24m + m_1}} \qquad (24.3)$$

Thus the frequency shift Δf becomes proportional to the additional mass loading

$$\Delta f = f_1 - f_0 \approx -\frac{1}{2} \frac{m_1}{0.24m} \cdot f_0 \qquad (24.4)$$

In order to estimate the minimum detectable mass, the smallest measurable frequency shift has to be defined. For this purpose the quality factor of the resonating cantilever beam must be considered. It can be approximated by $Q = \frac{f_0}{\Delta f_{FWHM}}$, where Δf_{FWHM} is the full width at half maximum of the resonance peak. Large values of Q are characterized by a small value of Δf_{FWHM}. It is obvious that the sharper the peak the better resonance frequencies can be detected. Thus the smallest measurable frequency shift is inversely proportional to the Q-factor: $\Delta f \approx \frac{f_0}{Q}$. Consequently, the smallest measurable mass changes can be estimated by $m_1 \approx 2 \frac{0.24m}{Q}$. In other words, small cantilever masses have to be used and high quality factors have to be realized in order to achieve a high mass sensitivity. Taking a standard SPM cantilever made of silicon ($L = 150\,\mu m$, $w = 40\,\mu m$, $t = 2\,\mu m$) in air ($Q \approx 400$) we obtain a mass of the cantilever of $\approx 28\,ng$ and a mass resolution in air of the order of 34 pg. However, the mass sensitivity decreases when the cantilever sensor is operated in viscous media, such as water, owing to damping, which reduces Q significantly.

24.2.5
Surface Stress

Besides measuring the dynamic properties of micromechanical cantilevers, it is also possible to investigate its time-averaged static profile, that is, its curvature, as a measurement signal [44–46]. Here, it is noteworthy that the bending of micromechanical cantilevers owing to gravity is negligible. However, any asymmetry of surface properties results in an imbalance in surface tension. This imbalance induces a surface stress σ in the micromechanical cantilever. Consequently, the micromechanical cantilever responds to such a surface stress via a bending until the stress and elastic bending energy are in equilibrium. Very often stresses acting on a micromechanical cantilever are the result of a second coating layer applied to only one side of a cantilever. If the thickness of this second layer is significantly less than the thickness of the cantilever (the so-called thin film approximation), the contribution of the thin layer to the elastic bending energy can be ignored and one obtains the following relation proposed by Stoney [47–49]

$$\sigma = \frac{Et^2}{6t_c} C, \tag{24.5}$$

where E is the Young's modulus, t the cantilever thickness, t_c the coating thickness, and C the curvature of the micromechanical cantilever. If the coating thickness is less than 1% of the cantilever thickness, the overall error resulting from the thin film approximation is typically less than 5%. However, for thicker coatings Stoney's formula needs to be corrected, and numerous correction schemes have been proposed [49–52].

Measuring the deflection change Δz of a cantilever, rather than its curvature, is often more convenient. Assuming a constant curvature and a Poisson ratio v of the micromechanical cantilever then Stoney's formula can be rewritten as

$$\Delta z = \frac{4L^2 (1 - v)}{Et^2} \Delta\sigma. \tag{24.6}$$

In any case, the relation between bending and changes in surface stress serves as a junction point linking a thermodynamic description of the surface stress to a macroscopic quantity. The modeling of surface stresses can either start at the Shuttleworth equation linking surface stress and interfacial tension. For example, if one considers the adsorption of antibodies to receptor molecules covalently bound to a single cantilever surface, the adsorption can be described in analogy to a classical van't Hoff isotherm [53]. Since the Gibbs energy G of the micromechanical cantilever system has to include contributions from surface area changes, the change in Gibbs energy leads to a modified van't Hoff isotherm

$$\Delta_r G^0 = -\frac{\Delta\sigma}{\Gamma} - RT \ln K \tag{24.7}$$

with Γ denoting the interfacial excess of bound species and K the equilibrium constant of the adsorption process. Investigating the hybridization of a DNA 12mer, a Gibbs energy of the reaction of $\Delta_r G^0 = -55.6 \, \text{kJ mol}^{-1}$ was determined from a single bending measurement using a silicon cantilever with a Poisson ratio of 0.25 at a temperature of 295 K.

24.2.6
Calorimetry

Calorimetry is the science of measuring the heat of chemical reactions or physical changes. The change of heat in a system is given by $\Delta Q = mc\Delta T$, where m is the mass, c is the specific heat capacity, and ΔT is the temperature change. Bimaterial micromechanical cantilevers (material 1 and material 2) bend with minute changes in temperature. The temperature-induced deflection z of micromechanical cantilevers is given by Bradley *et al.* [54] as:

$$\Delta z = \frac{L^2}{2} \frac{6 E_1 E_2 t_1 t_2 (t_1 + t_2) \cdot \Delta \alpha \cdot \Delta T}{\left(E_1 t_1^2\right)^2 + \left(E_2 t_2^2\right)^2 + 2 t_1 t_2 E_1 E_2 \left(2 t_1^2 + 3 t_1 t_2 + 2 t_2^2\right)} \tag{24.8}$$

where $\Delta \alpha$ corresponds to the difference in the expansion coefficient of the two materials, L is the cantilever length, t_1, t_2 are the thicknesses of the two layers, and E_1, E_2 are the Young's moduli of the materials. Barnes and coworkers estimated heat sensitivity down to $10\,\text{pW}$. The thermal response time of micromechanical cantilevers is $<1\,\text{ms}$ resulting in a minimum detectable energy of the sensor of the order of $20\,\text{fJ}$ [21]. Hereby, heat losses into the environment also have to be considered [55].

24.3
Applications

24.3.1
Imaging in Solution

One of the main advantages of SFM is that samples can be investigated in various environments ranging from ultra-high vacuum via ambient air to liquids. In particular, the latter environment is essential for many investigations in biotechnology and material sciences. Imaging in liquids at molecular resolution makes the SFM method unique.

In 1992 Hansma and coworkers demonstrated the imaging of plasmid DNA on mica [56]. As one example we show here the SFM investigation of a hybrid material containing a nucleic acid segment (DNA) and an organic polymer (polypropylene-oxide) (PPO) [57]. These amphiphilic single-stranded DNA block copolymers form micelles of spherical shape in aqueous solution [58]. Single-stranded DNA-*b*-PPO molecules, drawn schematically in Figure 24.3a, can be hybridized with complementary DNA strands. Upon using a DNA strand having five times the complementary sequence (Figure 24.3b) a very specific rod-like DNA-*b*-PPO architecture can be obtained, represented by Figure 24.3c. This specific architecture was visualized by SFM on a mica surface and four representative images are shown in Figure 24.3.

Recently, Rothemund reported that long, single-stranded DNA molecules can be folded into arbitrary two-dimensional shapes. SFM images in buffer solution revealed self-assembled DNA structures in the shape of stars, triangles, and 'smileys' on a $100\,\text{nm}$ scale [59]. In conclusion, SFM operation in liquid has become a valuable tool for investigating molecular architectures mediated by biomolecular

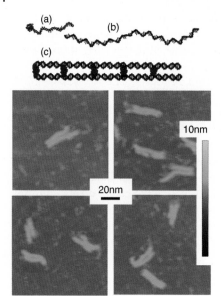

Figure 24.3 For the DNA-b-PPO-T110 hybridization product rod-like structures were observed. T110 indicates the template consisting of 110 nucleotides. The DNA-b-PPOs are organized in a linear fashion along the template molecule and the hydrophobic PPOs interact with other aggregates to form twins. Four representative SFM images of the aggregate are shown here. The images were recorded in buffer solution using intermittent contact mode [57].

and hydrophobic interaction. In addition to imaging, atomic force spectroscopy methods can be applied in liquids, as will be discussed in the following.

24.3.2
Binding Forces and Off-Rates

As a consequence of the high lateral resolution, individual antibody fragments could be addressed. The measured unbinding forces of mutated antibodies show a strong correlation with the change in the Gibbs energy for the binding of the antigen [60]. However, the absolute value of the obtained unbinding forces could only be quantitatively analyzed in terms of molecular parameters of the complex formation, if MFP experiments with different retract velocities are performed.

The rate constant for a unimolecular dissociation reaction is given by Eyring theory as [36]:

$$k_d(0) = \frac{k_B T}{h} \exp\left(-\frac{\Delta^{\ddagger} G(0)}{k_B T}\right) \tag{24.9}$$

An exponential increase in the rate constant $k(F)$

$$k_d(F) = k_d(0) \exp\left(\frac{F x_0}{k_B T}\right) \tag{24.10}$$

is now characteristic for a sharp energy barrier where the transition state is located at a distance x_0, projected along the direction of applied force F, to the energy minimum, because the Gibbs energy for dissociation decreases linearly with applied force, that is, $\Delta^{\ddagger}G(F) = \Delta^{\ddagger}G(0) - Fx_0$. This is schematically shown in Figure 24.4.

The behavior of this model in the context of an unbinding force measurement, where the force increases until the complex unbinds, has been discussed in detail by Evans and Ritchie [36]. For the sake of simplicity we assume here that the force F on the complex increases with a constant loading rate $r = K_{\text{eff}}v$, where K_{eff} is an effective spring constant of the cantilever–molecule system and v is the retract velocity or pulling speed. Because the measurement is done with a soft spring, the ligand and receptor are further separated after crossing the transition state and therefore rebinding will be neglected. The stochastic nature of the unbinding events is captured by solving a master equation [61] for the probability $N(t)$ to be in the bound state under an increasing force $F = rt$

$$\frac{dN(t)}{dt} = -k_{\text{d}}(rt)N(t) \tag{24.11}$$

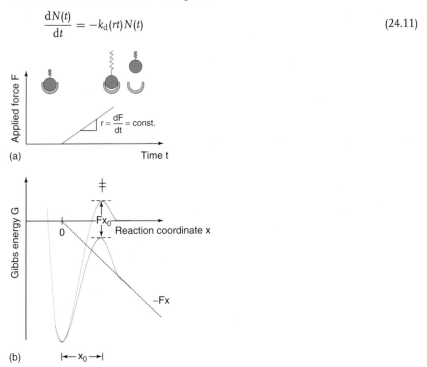

(a)

(b)

Figure 24.4 (a) Variation in the applied force F with time t for a constant pulling speed v. The force on a single complex increases linearly until it dissociates. The dissociation is monitored by an abrupt re-laxation of the macroscopic spring of a force probe. (b) Influence of an applied force F on the Gibbs energy profile for a unimolecular dissociation, if only a single energy barrier is existing (blue curve: without external load; dashed line: contribution from the applied load red curve: with applied load). The dissociation over the sharp barrier is characterized by a linear decrease in the barrier height with applied force, because the spring constant of the cantilever is much smaller than the curvature of the molecular potential well.

With $N(0) = 1$ the probability to be in the bound state is therefore given by

$$N(t) = \exp\left(-\frac{k_d(0)k_B T}{rx_0}\left(\exp\left(\frac{x_0 rt}{k_B T}\right) - 1\right)\right) \tag{24.12}$$

This results in a probability density of unbinding forces $p(F) = k_d(F)N(F/r)/r$ for a given loading rate r, which leads, with the exponential increase in the rate constant $k_d(F)$ to

$$p(F) = \frac{k_d(0)}{r}\exp\left(\frac{Fx_0}{k_B T} + \frac{k_d(0)k_B T}{rx_0}\left(1 - \exp\left(\frac{Fx_0}{k_B T}\right)\right)\right) \tag{24.13}$$

The most probable unbinding force F_{max}, the maximum of the probability density, is then determined by

$$F_{max} = \frac{k_B T}{x_0}\ln\left(\frac{K_{eff}vx_0}{k_d(0)k_B T}\right) \tag{24.14}$$

Therefore, in the simplest case, the parameters governing the dissociation kinetics under an applied force, the thermal lifetime of the complex $\tau = 1/k_d(0)$ and the characteristic length x_0, can be determined directly from a plot of the most probable unbinding force F_{max} versus the logarithm of the loading rate $r = K_{eff}v$.

This is exemplary shown for a double-stranded DNA complex in Figure 24.5, which consists of 10 complementary base pairs connected via a polyethylene glycol (PEG) linker to the cantilever and, respectively, to the substrate. Thereby, histograms of unbinding forces are presented for two different retract velocities v. It becomes obvious that the most probable unbinding force F_{max} grows with increasing retract velocity. A plot of F_{max} versus v indeed gives a straight line, which opens the possibility to determine the characteristic length x_0 from the slope, and the life-time τ from the zero crossing. However, to calculate the values for the thermal life-time, the effective spring constant of the cantilever–PEG linker system had to be determined from the force–displacement curves: although the effective spring constant (force increase per piezo displacement) K_{eff} is a nonlinear function of the force, attributable to the nonlinear entropic elasticity of the PEG linker, Equation 24.14 is still a good approximation if the effective spring constant is chosen to be the elasticity at the most probable unbinding force [36]. This elasticity or effective spring constant is given by the slope of the force–distance curve before the unbinding event occurs, and values from 1.5 to 3.0 pN nm^{-1} for unbinding forces from 20 to 50 pN have been found. The moderate increase in the effective spring constant with the unbinding forces preserves the linear scaling of the unbinding forces with the logarithm of the retract velocity. Therefore, a constant value of $K_{eff} = 2$ pN nm^{-1} for the effective spring constant has been adopted to calculate a numeric value of the thermal life-time. For the DNA complex with 10 base pairs this yields values of 8.5 Å for x_0 and 121 s for τ. The length x_0 thereby increases almost linearly with the number of base pairs while the life-time grows exponentially. This is in full accordance with ensemble methods to determine the kinetic parameters of the DNA complex formation and points to a cooperative dissociation mechanism.

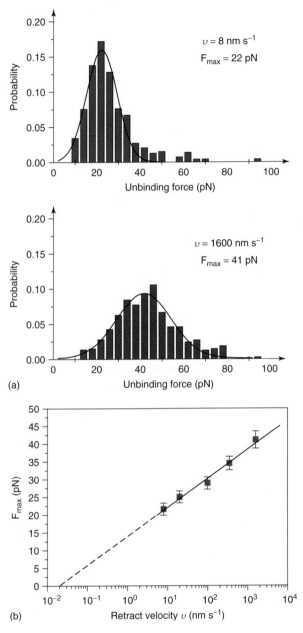

Figure 24.5 (a) Histograms of unbinding forces for two different retract velocities. (b) Variation of the most probable unbinding force F_{max} with the retract velocity. The observed logarithmic dependence points to a cooperative dissociation mechanism, which takes place over a single energy barrier [35]. In order to calculate the thermal life-time of the complex, an effective spring constant K_{eff} of $2\,pN\,nm^{-1}$ has to be taken into account.

24.3.3
Gravimetry

The characterization of small quantities in the picogram regime is very important in analytical chemistry, physics, and forensic analysis. Standard gravimetric tools allow the measurement of mass loadings down to the microgram regime. Using micro-electro-mechanical systems (MEMS), e.g. micromechancial cantilevers, sample masses down to the picogram regime can be investigated.

For example, a cantilever-based microbalance has been implemented in a molecular beam apparatus in order to investigate the physical and chemical properties of small clusters. Clusters, which are generated with a pulsed laser vaporization source, are deposited in the experiments from the molecular beam at a Si cantilever, which is used as a film thickness monitor [62]. The frequency shift of the cantilever is detected from the thermally excited eigenmodes by Fourier analysis of the signal of a position sensitive detector (PSD). A shift of 3.3 Hz from the deposition of 5×10^{11} tin atoms is shown in Figure 24.6, which corresponds to a film thickness of 6.4 Å. The ultimate detectable frequency variation with this kind of cantilever amounts to 0.2 Hz, which is equal to an ultimate mass and amount of substance sensitivity of 6.0×10^{-12} g and 5.0×10^{-14} mol, respectively. The use of cantilever arrays allows one, for example, to measure the total molecular beam profile, which has interesting applications for electric and magnetic molecular beam deflection experiments.

Gravimetry using micromechanical cantilever sensors in air is now done routinely [19, 63–65]. Even a mass resolution in the realm of zeptograms (10^{-21} g) has been reported in a cryogenically cooled, ultra-high vacuum environment [66]. As mentioned before, liquids that surround the cantilever lead to a significant damping of the cantilever oscillation, thus decreasing the Q-factor significantly. Burg and coworkers reporteded mass measurements with a cantilever containing

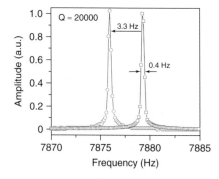

Figure 24.6 Si cantilever used as a thin film rate monitor for the high vacuum deposition of metal clusters. The quality factor Q of the cantilever is given by the value of the eigenfrequency divided by the full width at half maximum of the resonance [62].

a microchannel. Using this approach they demonstrated a mass resolution of femtograms of biologically relevant species [67].

With the progress of microfabrication methods more advanced probes can be fabricated that include additional functional elements. Such functional elements could be of a passive nature, such as reservoirs and microchannels for writing surfaces with specific molecules [68] – later called dip pen lithography [69]; or electronic components can be integrated, such as strain sensors or heater elements. IBM pioneered the integration of heater elements into scanning probe microcomponents to realize an advanced scanning probe-based memory device [70] with the potential to be used also as a patterning tool [71]. Besides the latter application, which is strongly related to SPM, the integrated resistive heater can be powered by an applied voltage allowing control of the temperature of the cantilever. Thermal response times of micro-devices are typically in the millisecond range, indicating that a thermal steady state can be achieved on a similar timescale. Any additional mass that is placed on top of the micromechanical cantilever will be heated. This approach enables the analysis of temperature-dependent processes in the samples, in particular in tiny amounts of samples [72, 73].

Thermal analytical techniques such as thermogravimetry (TG) are routinely used for forensic and material analysis. For micromechanical thermogravimetric analysis the temperature-dependent frequency variation is given by [72]:

$$\frac{1}{f}\frac{df}{dT} = \frac{1}{2K}\frac{dK}{dT} - \frac{1}{2m}\frac{dm}{dT} \tag{24.15}$$

Consequently, tracking the resonance frequency as a function of T permits determination of changes in the total sensor mass. Such changes may arise from a sample mass m_l, that is loaded onto the micromechanical cantilever. The zeolite sample shown in Figure 24.7a had a mass of approximately 150 ng and was loaded with p-nitroanaline, a yellow dye molecule [74, 75]. Upon heating the micromechanical cantilever sensor to a temperature of 300 °C a mass loss of $\approx$12 ng was measured at a temperature of around 200 °C (Figure 24.7b). This mass loss corresponded to the loss of the p-nitroanaline from the zeolite.

However, the temperature dependence of K also leads to changes in the resonance frequency. This change can be corrected by recording the resonance frequency during both the heating and the cooling of the sample-cantilever sensor [72]. In this case there is a negligible mass change during cooling, the difference between the curves allows one to calculate the pure mass change in the heating cycle.

Micromechanical thermal analysis allows one to measure individual substances such as a chlorobenzene-containing polyurethane microcapsule [76]. The stepwise increase in frequency at two temperatures indicated rupture of the capsule and the thermal degradation of the polyurethane shell. The advantage of analyzing single material units is that a much sharper signal can be measured in comparison to analysis of conglomerates. The analysis of the latter was observed to exhibit a broader distribution in the reaction due to a different mass or shell thickness.

There are several occasions where the sample masses are restricted to <1 μg: a large quantity might not be available owing to price, synthetic constrictions, or

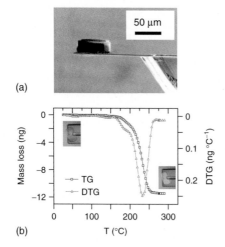

(a)

(b)

Figure 24.7 (a) Scanning electron micrograph of a zeolite single crystal on a micromechanical cantilever in side view. (b) Heating the zeolite resulted in a mass loss of ≈12 ng corresponding to the amount of p-nitroanaline molecules that was captured in the zeolite crystal. Here also the differential mass loss (DTG) is plotted. The inset shows the zeolite before (left upper corner) and after heating (lower right corner). After heating, the zeolite crystal appears transparent.

availability. The latter restriction is highly relevant in forensic analysis [77]. Here, the use of MEMS fabricated analytical tools such as micromechanical cantilever sensors is beneficial. Furthermore, screening of a large parameter space in material composition and properties can be achieved because micromechanical cantilevers can be mass fabricated even in one-dimensional or two-dimensional arrays [78, 79]. Finally, the mass of the entire sensing unit is crucial for analysis of soil material or micro-objects in outer space.

24.3.4
Biosensors (Nose)

The extremely high mass sensitivity alone is not an outstanding attribute for micromechanical cantilevers. One prerequisite for their use in analytic applications is a high selectivity of the binding events. Selectivity can be achieved by chemical functionalization of surfaces. In biotechnology, one of the most prominent examples is the detection of hybridization of DNA [80], which is shown in Figure 24.8. Here the advantages of a micromechanical cantilever system as a detector platform are label-free detection and a detection limit in the nanomolar concentration range. Given the high sensitivity of the available detection systems, it has been possible to detect single base mismatches during hybridization [80, 81].

Similar examples for ultra-sensitive detection exist for the discrimination of 3′ and 5′ overhangs [82] and hybridization induced changes in hydration [83] have been successfully demonstrated. Extending the use of the microcantilever as an

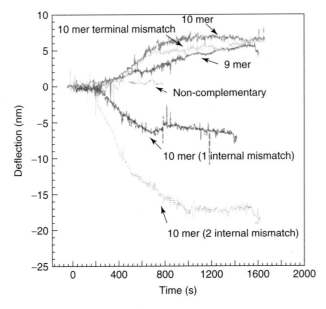

Figure 24.8 Cantilever deflection for hybridized oligonucleotides. Cantilever surfaces functionalized with a thiolated
20-mer oligonucleotide and challenged with complementary,
mismatch, and noncomplementary target oligonucleotides.

analytic tool in bioscience, Braun *et al.* [84] have used ink-jet deposited layers of
bacteriorhodopsin, in order to measure the mechanostress induced upon light
exposure.

In the broader field of soft matter science, much work has been carried out on
the use of polymers as active layers in sensing applications. Initial efforts used
physisorbed layers applied via spray coating or ink-jet printing [85, 86]. Using
micromechanical cantilever arrays in connection with machine assisted learning
techniques Baller *et al.* have successfully demonstrated an artificial nose design for
the detection of gas phase analytes [87]. The use of physisorbed polymer layers limits
cantilever sensors to the detection of gaseous analytes as immersion in liquids
would dissolve the active layer material. Using covalently grafted polymer layers
extends the application of micromechanical cantilever arrays to liquid environments
[88]. With these covalently bound brushes the stress generation during exchange
of solvents [89] and of solvent quality have been monitored. The change in solvent
quality can either be induced via temperature changes, initiating a lower critical
solution temperature phase transition of the polymer brush, or by changes in pH.
In all cases the reversible volume transition upon swelling and collapse of the
polymeric brushes leads to a reversible bending motion of the micromechanical
cantilever. This allows the use of micromechanical cantilever arrays as analytic
tools to study these systems. In a detailed analysis of these systems, great care has
to be exercised in order to separate individual stress components, leading to an

overall bending [90]. In the case of thermal responsive brushes, the temperature bending behavior shows two clear contributions from chain collapse and unspecific thermal expansion [54]. In more complex systems, such as polyelectrolyte brushes [91] or mixed polymer brushes [92], care has to be taken to separate energetic from structural changes as both result in a stress acting on the cantilever [93].

24.3.5
Thermochemistry

With a microcalorimeter based on a bimetallic cantilever, the thermochemistry of small metal and semiconductor clusters has been investigated. For that purpose, the heat of condensation is measured after cluster deposition at the bimetallic cantilever. An example of this is shown in Figure 24.9. The cluster deposition leads to the growth of a thin metal film at the cantilever. Thermodynamically, the condensation of the deposited material corresponds, therefore, to the formation of the bulk metal from gas phase clusters. A heat of condensation of 26 pJ is released for the deposition of 2.3×10^8 cluster atoms per pulse for a mean cluster size of 150 atoms, that is, the condensation of only about 1 million clusters could be detected.

For the experiment shown in Figure 24.9 the calorimeter signal consists of two parts, the released heat due to the difference in energy of tin atoms in isolated clusters and in the bulk and the contribution of the kinetic energy of the clusters in the molecular beam, which amounts to about 50% of the totally measured heat of condensation. Hence, with an experimental determination of the initial kinetic energy the formation energies per atom $\Delta_f u$ of isolated clusters, that is,

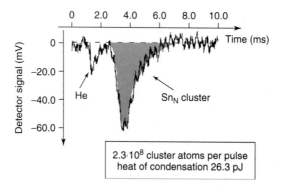

Figure 24.9 Calorimetric measurements of the released heat due to the deposition of Sn clusters on a gold-coated Si_3N_4 cantilever. The starting point for the bending measurements is the opening time of the He valve. The measured bending curves are averaged over 50 cluster pulses and obtained after the deposition of a few nanometers of Sn clusters. The first peak in the enlarged bending curve is due to the momentum transfer of the He carrier gas. The second peak is a consequence of the momentum transfer and the released heat during the cluster deposition. The area under the second peak corresponds to a released heat of 26.3 pJ [94].

the difference in binding energies, are accessible for a tin atom in the bulk and in isolated clusters. However, one has to take into account that the deflection of the bimetallic cantilever depends not only on the released heat but also on the momentum transfer during the cluster deposition process. This effect can also be seen in Figure 24.9, because the first signal is due to the elastic scattering of the inert gas He, which is needed for the generation of the cluster in the laser vaporization source. Figure 24.10 illustrates how the formation energies depend on the mean cluster size. Herein experiments with a pyro-electric calorimeter have also been taken into account. Within the experimental accuracy the values of $\Delta_f u$ are the same for both microcalorimeters used.

In the investigated cluster size regime of $N = 100–1000$ atoms per cluster, a characteristic size dependence of the formation energies is observed. This becomes particularly obvious if the values of $\Delta_f u$ are plotted against $N^{-1/3}$ because, for a small spherical particle, the ratio of the number of atoms at the surface with respect to the total number of atoms scales with $N^{-1/3}$. In this so-called droplet model the proportionality factor depends on the particle number density and the surface energy. Taking values of bulk metallic tin into account, the observed behavior could be quantitatively explained within the experimental uncertainties, that is, tin clusters with 100–1000 atoms exhibit properties similar to small but still macroscopic spherical particles [95].

In addition to the investigation of deposition processes in high vacuum phase transitions in air have also been studied with the cantilever calorimeter [73]. In Figure 24.11 temperature variations during solid–solid phase transitions of a long chain alkane (tricosane, $C_{23}H_{48}$) are shown. The small dimensions of the bimetallic cantilever allow one not only to investigate ultra-small quantities but also to detect temperature changes on a millisecond timescale and, therefore, to follow the kinetics of phase transformations. Enthalpy changes below 1 nJ on a time scale of

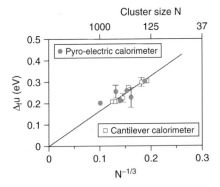

Figure 24.10 Variation in the formation energies per atom $\Delta_f u$ of tin clusters with size. The formation energies have been obtained with a bimetallic cantilever from the measured heats of condensation. In comparison the values of $\Delta_f u$, which have been recorded with a pyro-electric calorimeter [95], are displayed.

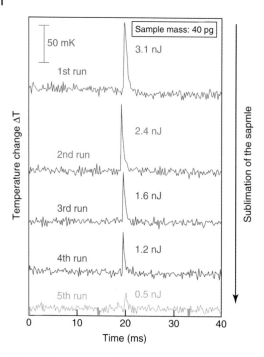

Figure 24.11 Temperature variations during solid–solid phase transitions of a long chain alkane (tricosane, $C_{23}H_{48}$) measured with a cantilever calorimeter in air. The reduction in the temperature increase at the phase transition for subsequent runs is due to a loss of material by sublimation. Initially 40 pg of Tricosane was attached to the cantilever. The loaded mass was determined from the decrease in the bimetallic cantilevers' eigenfrequency. The transition enthalpies are obtained by integration over the temperature variation taking the calorimeter constant into account. The transition enthalpies thereby decrease from 3.1 to 0.5 nJ [96].

less than 1 ms have been detected for the solid–solid phase transitions of tricosane in air. This thermal analysis corresponds to a quantity of only 7 pg or an amount of substance of 20 fmol.

24.4
Summary and Conclusion

The described SPM methods in this chapter cover only a small area of the research fields in which it is used. Topography recording by SFM was reported for the first time in 1986 [2]. While in the first years after development only topographic images were recorded, it is now possible to investigate surfaces with physical and chemical specificity. Thus the SPM methods are now used in physical/chemical science as well as in engineering and in industry. The H-index [97] can be considered as one indicator for the relevance of a scientific topic. The enormous contribution and acceptance of SPM in science for the last

25 years is reflected by an H-index >250. The fabrication of micromechanical cantilevers with custom tailored properties is a crucial point if one wants to increase the sensitivity and speed of the different operation modes. The integration of a cantilever into an optical microscope has already been realized in scanning near-field optical microscopy (SNOM) [98]. This method allows one to investigate the optical behavior of single molecules. The extension of this method into the near-infrared [99] spectral range or tip-enhanced Raman spectroscopy [100] aims to measure the vibrational spectra of single molecules. Interestingly, no stand alone international conference exists which covers recent aspects in SPM. Nowadays SPM is integrated as one of the major techniques in surface analytic sessions, for example, at the MRS and ACS meetings. Furthermore, the field has become so diverse in applications that individual conferences with specific topics are organized nowadays (e.g., International Conference on Non-Contact Atomic Force; *http://www.afm.eei.eng.osaka-u.ac.jp/ncafm2010/*).

The situation is different in the field of micromechanical cantilever sensors which emerged from the field of SPM around 1994 [20, 21, 101]. The field of micromechanical cantilever sensors has an H-index of around 40. We interpret that this H-index value indicates that the research field of micromechanical cantilever sensors currently has still a strong methodological background and has not reached significant acceptance in broader research fields. Currently, the number of researchers in the field of micromechanical cantilever sensors is of the order of 150. Researchers meet annually at a workshop on nanomechanical cantilever sensors (*http://www.nmc2011.org/*). However, the transition between cantilever-based techniques and other MEMS is fluent. One of the main issues that needs to be addressed for micromechanical cantilever sensors is the control of the surface chemistry. In particular, methods that are able to control the surface functionalization on a micrometer scale are needed. Otherwise molecules might interact in different ways and, consequently, follow different transduction mechanisms. Furthermore, methods need to be developed to deposit small amounts of samples at desired positions on the micromechanical cantilever sensor. For thermal applications such as calorimetry the heat management in micromechanical cantilevers needs to be investigated in detail, in order to determine reaction heats and absolute values of transition enthalpies. Particularly important for chemistry is the functionalization of cantilevers in order to study the electric [102] and magnetic [9] properties of samples with the smallest dimensions. For example, based on a microfabricated cantilever, some research groups try to detect an electron paramagnetic [103] or a nuclear magnetic [104–106] resonance signal of a single molecule.

For both the SPM and the micromechanical cantilever method there is already a huge number of different operation modes. However, there is plenty of scope for the development of new applications to characterize samples on a nanometer scale. Even more than 25 years after development of SPM, we are sure that the field is still expanding.

Acknowledgment

We thank J.-H. Fabian and L. Scandella for providing data on the micromechanical cantilever thermal analysis.

References

1. Binnig, G., Rohrer, H., Gerber, C., and Weibel, E. (1982) *Appl. Phys. Lett.*, **40**, 178–180.

2. Binnig, G., Quate, C.F., and Gerber, C. (1986) *Phys. Rev. Lett.*, **56**, 930–933.

3. Gould, S., Marti, O., Drake, B., Hellemans, L., Bracker, C.E., Hansma, P.K., Keder, N.L., Eddy, M.M., and Stucky, G.D., (1986) *Nature*, **332**, 332–324.

4. Fompeyrine, J., Berger, R., Lang, H.P., Perret, J., Machler, E., Gerber, C., and Locquet, J.P. (1998) *Appl. Phys. Lett.*, **72**, 1697–1699.

5. Burnham, N.A., Dominguez, D.D., Mowery, R.L., and Colton, R.J., (1990) *Physical Review Letters*, **64**, 1931–1934.

6. Cretin B., Gomes S., and Trannoy N. et al. (2007) *Microscale and Nanoscale Heat Transfer*, Topics in Applied Physics **107**, 181–238.

7. Sadat, S., Tan, A., Chua, Y.J., and Reddy, P. *Nano Lett.*, **10**, 2613–2617.

8. Berger, R., Butt, H.J., Retschke, M.B., and Weber, S.A.L. (2009) *Macromol. Rapid Commun.*, **30**, 1167–1178.

9. Hug, H.J., Stiefel, B., van Schendel, P.J.A., Moser, A., Hofer, R., Martin, S., Güntherodt, H.J., Porthun, S., Abelmann, L., Lodder, J.C., Bochi, G., and O'Handley, R.C. (1998) *J. Appl. Phys.*, **83**, 5609–5620.

10. Funami, T. (2010) *Food Sci. Technol. Res.*, **16**, 1–12.

11. Bhushan, B. (2008) *Prog. Mater. Sci.*, **53**, 585–710.

12. Junk, M.J.N., Berger, R., and Jonas, U. (2010) *Langmuir*, **26**, 7262–7269.

13. Santer, S., Kopyshev, A., Donges, J., Yang, H.K., and Ruhe, J. (2006) *Adv. Mater.*, **18**, 2359–2362.

14. Camesano, T.A., Natan, M.J., and Logan, B.E. (2000) *Langmuir*, **16**, 4563–4572.

15. Lee, G.U., Chrisey, L.A., and Colton, R.J. (1994) *Science*, **266**, 771–773.

16. Grandbois, M., Beyer, M., Rief, M., Clausen-Schaumann, H., and Gaub, H.E. (1999) *Science*, **283**, 1727–1730.

17. Neuman, K.C. and Nagy, A. (2008) *Nat. Methods*, **5**, 491–505.

18. Butt, H.J., Cappella, B., and Kappl, M. (2005) *Surf. Sci. Rep.*, **59**, 1–152.

19. Berger, R., Gerber, C., Lang, H.P., and Gimzewski, J.K. (1997) *Microelectron. Eng.*, **35**, 373–379.

20. Gimzewski, J.K., Gerber, C., Meyer, E., and Schlittler, R.R. (1994) *Chem. Phys. Lett.*, **217**, 589–594.

21. Barnes, J.R., Stephenson, R.J., Woodburn, C.N., Oshea, S.J., Welland, M.E., Rayment, T., Gimzewski, J.K., and Gerber, C. (1994) *Rev. Sci. Instrum.*, **65**, 3793–3798.

22. Chen, G.Y., Thundat, T., Wachter, E.A., and Warmack, R.J. (1995) *J. Appl. Phys.*, **77**, 3618–3622.

23. Itoh, T., Ohashi, T., and Suga, T. (1996), Ninth Annual International Workshop on Micro Electro Mechanical Systems, IEEE Proceedings – an Investigation of Micro Structures, Sensors, Actuators, Machines and Systems, pp. 451–455.

24. Lang, H.P., Berger, R., Battiston, F., Ramseyer, J.P., Meyer, E., Andreoli, C., Brugger, J., Vettiger, P., Despont, M., Mezzacasa, T., Scandella, L., Güntherodt, H.J., Gerber, C., and Gimzewski, J.K. (1998) *Appl. Phys. A-Mater. Sci. Process.*, **66**, S61–S64.

25. Binnig, G., Gerber, C., Stoll, E., Albrecht, T.R., and Quate, C.F. (1987) *Europhys. Lett.*, **3**, 1281–1286.

26. Marti, O., Ribi, H.O., Drake, B., Albrecht, T.R., Quate, C.F., and Hansma, P.K. (1988) *Science*, **239**, 50–52.

27. Ohnesorge, F. and Binnig, G. (1993) *Science*, **260**, 1451–1456.

28. Magonov, S.N. (2000) in *Encyclopaedia of Analytical Chemistry* (ed. R.A. Meyers), John Wiley & Sons, Ltd, Chichester, p. 7432.

29. Giessibl, F.J. (2003) *Rev. Mod. Phys.*, **75**, 949–983.

30. Gross, L., Mohn, F., Moll, N., Liljeroth, P., and Meyer, G. (2009) *Science*, **325**, 1110–1114.

31. Fukuma, T., Kobayashi, K., Matsushige, K., and Yamada, H. (2005) *Appl. Phys. Lett.*, **87**, 034101.

32. Fukuma, T. (2009) *Jpn. J. Appl. Phys.*, **48**, 08JA01.

33. Fukuma, T., Ueda, Y., Yoshioka, S., and Asakawa, H. (2010) *Phys. Rev. Lett.*, **104**, 016101.

34. Burnham, N.A., Colton, R.J., and Pollock, H.M. (1991) *J. Vac. Sci. Technol. A-Vac. Surf. Films*, **9**, 2548–2556.

35. Strunz, T., Oroszlan, K., Schäfer, R., and Güntherodt, H.J. (1999) *Proc. Natl. Acad. Sci. U.S.A.*, **96**, 11277–11282.

36. Evans, E. and Ritchie, K. (1997) *Biophys. J.*, **72**, 1541–1555.

37. Lee, H., Lee, B.P., and Messersmith, P.B. (2007) *Nature*, **448**, 338–341.

38. Fernandez, J.M. and Li, H.B. (2004) *Science*, **303**, 1674–1678.

39. Brown, A.E.X., Litvinov, R.I., Discher, D.E., and Weisel, J.W. (2007) *Biophys. J.*, **92**, L39–L41.

40. Nguyen, T.-H., Steinbock, L.J., Butt, H.-J., Helm, M., and Berger, R. (2011) *J. Am. Chem. Soc.*, **133**, 2025–2027.

41. Rief, M., Gautel, M., Oesterhelt, F., Fernandez, J.M., and Gaub, H.E. (1997) *Science*, **276**, 1109–1112.

42. Puchner, E.M. and Gaub, H.E. (2009) *Curr. Opin. Struct. Biol.*, **19**, 605–614.

43. Oesterhelt, F., Oesterhelt, D., Pfeiffer, M., Engel, A., Gaub, H.E., and Müller, D.J. (2000) *Science*, **288**, 143–146.

44. Alvarez, M. and Lechuga, L.M. (2010) *Analyst*, **135**, 827–836.

45. Battiston, F.M., Ramseyer, J.P., Lang, H.P., Baller, M.K., Gerber, C., Gimzewski, J.K., Meyer, E., and Güntherodt, H.J. (2001) *Sens. Actuators B-Chem.*, **77**, 122–131.

46. Fritz, J., Baller, M.K., Lang, H.P., Strunz, T., Meyer, E., Güntherodt, H.J., Delamarche, E., Gerber, C., and Gimzewski, J.K. (2000) *Langmuir*, **16**, 9694–9696.

47. Stoney, G.G. (1909) *Proc. R. Soc. Lond. Ser. A*, **82**, 172–175.

48. Butt, H.-J. (1996) *J. Colloid Interface Sci.*, **180**, 251–260.

49. Klein, C.A. (2000) *J. Appl. Phys.*, **88**, 5487–5489.

50. Pureza, J.M., Lacerda, M.M., De Oliveira, A.L., Fragalli, J.F., and Zanon, R.A.S. (2009) *Appl. Surf. Sci.*, **255**, 6426–6428.

51. Sushko, M.L., Harding, J.H., Shluger, A.L., McKendry, R.A., and Watari, M. (2008) *Adv. Mater.*, **20**, 3848–3853.

52. Zhang, Y. and Zhao, Y.P. (2006) *J. Appl. Phys.*, **99**, 053513.

53. Bergese, P., Oliviero, G., Alessandri, I., and Depero, L.E. (2007) *J. Colloid Interface Sci.*, **316**, 1017–1022.

54. Bradley, C., Jalili, N., Nett, S.K., Chu, L.Q., Forch, R., Gutmann, J.S., and Berger, R. (2009) *Macromol. Chem. Phys.*, **210**, 1339–1345.

55. Toda, M., Ono, T., Liu, F., and Voiculescu, I. (2010) *Rev. Sci. Instrum.*, **81**, 055104.

56. Hansma, H.G., Vesenka, J., Siegerist, C., Kelderman, G., Morrett, H., Sinsheimer, R.L., Elings, V., Bustamante, C., and Hansma, P.K. (1992) *Science*, **256**, 1180–1184.

57. Ding, K., Alemdaroglu, F.E., Boersch, M., Berger, R., and Herrmann, A. (2007) *Angew. Chem. Int. Ed.*, **46**, 1172–1175.

58. Wang, J., Alemdaroglu, F.E., Prusty, D.K., Herrmann, A., and Berger, R. (2008) *Macromolecules*, **41**, 2914–2919.

59. Rothemund, P.W.K. (2006) *Nature*, **440**, 297–302.

60. Ros, R., Schwesinger, F., Anselmetti, D., Kubon, M., Schäfer, R., Plückthun, A., and Tiefenauer, L. (1998) *Proc. Natl. Acad. Sci. U.S.A.*, **95**, 7402–7405.

61. Strunz, T., Oroszlan, K., Schumakovitch, I., Güntherodt, H.J., and Hegner, M. (2000) *Biophys. J.*, **79**, 1206–1212.

62. Bachels, T. and Schäfer, R. (1998) *Rev. Sci. Instrum.*, **69**, 3794–3797.

63. Ono, T., Li, X.X., Miyashita, H., and Esashi, M. (2003) *Rev. Sci. Instrum.*, **74**, 1240–1243.

64. Ekinci, K.L., Huang, X.M.H., and Roukes, M.L. (2004) *Appl. Phys. Lett.*, **84**, 4469–4471.

65. Dohn, S., Sandberg, R., Svendsen, W., and Boisen, A. (2005) *Appl. Phys. Lett.*, **86**, 233501.

66. Yang, Y.T., Callegari, C., Feng, X.L., Ekinci, K.L., and Roukes, M.L. (2006) *Nano Lett.*, **6**, 583–586.

67. Burg, T.P., Godin, M., Knudsen, S.M., Shen, W., Carlson, G., Foster, J.S., Babcock, K., and Manalis, S.R. (2007) *Nature*, **446**, 1066–1069.

68. Jaschke, M. and Butt, H.J. (1995) *Langmuir*, **11**, 1061–1064.

69. Piner, R.D., Zhu, J., Xu, F., Hong, S.H., and Mirkin, C.A. (1999) *Science*, **283**, 661–663.

70. Vettiger, P., Despont, M., Drechsler, U., Durig, U., Haberle, W., Lutwyche, M.I., Rothuizen, H.E., Stutz, R., Widmer, R., and Binnig, G.K. (2000) *IBM J. Res. Dev.*, **44**, 323–340.

71. Pires, D., Hedrick, J.L., De Silva, A., Frommer, J., Gotsmann, B., Wolf, H., Despont, M., Duerig, U., and Knoll, A.W. (2010) *Science*, **328**, 732–735.

72. Berger, R., Lang, H.P., Gerber, C., Gimzewski, J.K., Fabian, J.H., Scandella, L., Meyer, E., and Güntherodt, H.J. (1998) *Chem. Phys. Lett.*, **294**, 363–369.

73. Berger, R., Gerber, C., Gimzewski, J.K., Meyer, E., and Güntherodt, H.J. (1996) *Appl. Phys. Lett.*, **69**, 40–42.

74. Scandella, L., Fabian, J.H., Von Scala, C., Berger, R., Lang, H.P., Gerber, C., Gimzewski, J.K., and Meyer, E. (1998) *Helv. Phys. Acta*, **71**, 3–4.

75. Fabian, J.H., Berger, R., and Lang, H.P. *et al.* (2000) *Micromechanical Thermogravimetry on single Zeolite Crystals*, in Micro Total Analysis Systems '98: 3rd International Symposium on Micro-Total Analysis Systems (mu-TAS'98), Banff, Canada, October 13–16, 1998, Mesa Monographs (eds D.J. Harrison A. VanDenBerg), Klumer Academic Publishers, pp. 117–120.

76. Lee, D., Park, Y., Cho, S.H., Yoo, M., Jung, N., Yun, M., Ko, W., and Jeon, S. (2010) *Anal. Chem.*, **82**, 5815–5818.

77. Causin, V. (2010) *Anal. Methods*, **2**, 792–804.

78. Hamann, S., Ehmann, M., Thienhaus, S., Savan, A., and Ludwig, A. (2008) *Sens. Actuators A-Phys.*, **147**, 576–582.

79. Ludwig, A., Zarnetta, R., Hamann, S., Savan, A., and Thienhaus, S. (2008) *Int. J. Mater. Res.*, **99**, 1144–1149.

80. Fritz, J., Baller, M.K., Lang, H.P., Rothuizen, H., Vettiger, P., Meyer, E., Güntherodt, H.J., Gerber, C., and Gimzewski, J.K. (2000) *Science*, **288**, 316–318.

81. Hansen, K.M., Ji, H.F., Wu, G.H., Datar, R., Cote, R., Majumdar, A., and Thundat, T. (2001) *Anal. Chem.*, **73**, 1567–1571.

82. McKendry, R., Zhang, J.Y., Arntz, Y., Strunz, T., Hegner, M., Lang, H.P., Baller, M.K., Certa, U., Meyer, E., Güntherodt, H.J., and Gerber, C. (2002) *Proc. Natl. Acad. Sci. U.S.A.*, **99**, 9783–9788.

83. Mertens, J., Rogero, C., Calleja, M., Ramos, D., Martin-Gago, J.A., Briones, C., and Tamayo, J. (2008) *Nat. Nanotechnol.*, **3**, 301–307.

84. Braun, T., Backmann, N., Vogtli, M., Bietsch, A., Engel, A., Lang, H.P., Gerber, C., and Hegner, M. (2006) *Biophys. J.*, **90**, 2970–2977.

85. Battiston, F.M., Ramseyer, J.P., Lang, H.P., Baller, M.K., Gerber, C., Gimzewski, J.K., Meyer, E., and Guntherodt, H.J. (2001) *Sens. Actuators B-Chem.*, **77**, 122–131.

86. Bietsch, A., Zhang, J.Y., Hegner, M., Lang, H.P., and Gerber, C. (2004) *Nanotechnology*, **15**, 873–880.

87. Baller, M.K., Lang, H.P., Fritz, J., Gerber, C., Gimzewski, J.K., Drechsler, U., Rothuizen, H., Despont, M., Vettiger, P., Battiston, F.M., Ramseyer, J.P., Fornaro, P., Meyer, E., and Güntherodt, H.J. (2000) *Ultramicroscopy*, **82**, 1–9.

88. Bumbu, G.G., Kircher, G., Wolkenhauer, M., Berger, R., and Gutmann, J.S. (2004) *Macromol. Chem. Phys.*, **205**, 1713–1720.

89. Bumbu, G.G., Wolkenhauer, M., Kircher, G., Gutmann, J.S., and Berger, R. (2007) *Langmuir*, **23**, 2203–2207.

90. Zhao, J., Berger, R., and Gutmann, J.S. (2006) *Appl. Phys. Lett.*, **89**, 033110.

91. Zhou, F., Biesheuvel, P.M., Chol, E.Y., Shu, W., Poetes, R., Steiner, U., and Huck, W.T.S. (2008) *Nano Lett.*, **8**, 725–730.

92. Lenz, S., Nett, S.K., Memesa, M., Roskamp, R.F., Timmann, A., Roth, S.V., Berger, R., and Gutmann, J.S. *Macromolecules*, **43**, 1108–1116.

93. Wolkenhauer, M., Bumbu, G.G., Cheng, Y., Roth, S.V., and Gutmann, J.S. (2006) *Appl. Phys. Lett.*, **89**, 054101.

94. Bachels, T. and Schäfer, R. (1999) *Chem. Phys. Lett.*, **300**, 177–182.

95. Bachels, T., Schäfer, R., and Güntherodt, H.J. (2000) *Phys. Rev. Lett.*, **84**, 4890–4893.

96. Nakagawa, Y. and Schäfer, R. (1999) *Angew. Chem. Int. Ed.*, **38**, 1083–1085.

97. Hirsch, J.E. (2005) *Proc. Natl. Acad. Sci. U.S.A.*, **102**, 16569–16572.

98. Kingsley, J.W., Ray, S.K., Adawi, A.M., Leggett, G.J., and Lidzey, D.G. (2008) *Appl. Phys. Lett.*, **93**, 213103.

99. Kjoller, K., Felts, J.R., Cook, D., Prater, C.B., and King, W.P. (2010) *Nanotechnology*, **21**, 185705.

100. Hayazawa, N., Inouye, Y., Sekkat, Z., and Kawata, S. (2000) *Opt. Commun.*, **183**, 333–336.

101. Thundat, T., Warmack, R.J., Chen, G.Y., and Allison, D.P. (1994) *Appl. Phys. Lett.*, **64**, 2894–2896.

102. Lu, J., Delamarche, E., Eng, L., Bennewitz, R., Meyer, E., and Güntherodt, H.J. (1999) *Langmuir*, **15**, 8184–8188.

103. Rugar, D., Budakian, R., Mamin, H.J., and Chui, B.W. (2004) *Nature*, **430**, 329–332.

104. Degen, C.L., Lin, Q., Hunkeler, A., Meier, U., Tomaselli, M., and Meier, B.H. (2005) *Phys. Rev. Lett.*, **94**, 207601.

105. Eberhardt, K.W., Hunkeler, A., Meier, U., Tharian, J., Mouaziz, S., Boero, G., Brugger, J., and Meier, B.H., (2008) *Phys. Rev. B.*, **78** (21), 214401, doi: 10.1103/PhysRevB.78.214401

106. Nestle, N., Schaff, A., and Veeman, W.S. (2001) *Prog. Nucl. Magn. Reson. Spectrosc.*, **38**, 1–35.

25
Superconducting Quantum Interference Device Magnetometry

Heiko Lueken

■ **Method Summary**

Acronyms, Synonyms
- Superconducting quantum interference device (SQUID)
- Direct current (DC)
- Alternating current (AC).

Benefits (Information Available)
- materials: solids, liquids (e.g., intermetallics), gaseous elements (e.g., oxygen), and compounds
- electronic structure and oxidation state
- spin–orbit coupling and ligand field effect
- magnetic anisotropy (technique: direction-dependent measurements using single-crystalline specimen)
- interatomic spin–spin exchange coupling
- magnetic ordering (ferro-, ferri-, antiferromagnetism)
- spin glass behavior
- superparamagnetism
- single molecule magnets.

Limitations (Information Not Available)
- crystal structure and magnetic structure.

25.1
Introduction

In today's magnetic-materials sciences automated measurements of high accuracy are state of the art. Superconducting quantum interference device (SQUID) magnetometers [1] are extensively used to fully characterize magnetic properties in the temperature range 2–800 K, with applied magnetic (direct current (DC)) fields up to 7 T and a large frequency range ($10^{-3} - 1.5 \times 10^3$ Hz, extensible

Methods in Physical Chemistry, First Edition. Edited by Rolf Schäfer and Peter C. Schmidt.
© 2012 Wiley-VCH Verlag GmbH & Co. KGaA. Published 2012 by Wiley-VCH Verlag GmbH & Co. KGaA.

down to 10^{-4} Hz and up to 10^8 Hz [2]) in the alternating current (AC) option. The measurements serve to determine electronic configurations of magnetically active ions, interatomic exchange interactions, diamagnetic contributions, metallic character, magnetic anisotropy, superconductivity, characteristics of spin glasses, superparamagnetism of nanoparticles, single molecule magnetism, delocalization of electrons in mixed-valence systems, multiferroic behavior, and so on. For a deeper insight into the magnetic behavior of antiferromagnets and frustrated spin systems, measurements at high magnetic fields up to 40 T (continuous fields) and 100 T (pulsed fields) have appeared as favorable [3].

Systems with localized valence electrons have the advantage that the free ion behavior is a convenient starting point to clarify the magnetism of the condensed phase. Systems with delocalized valence electrons, for example, 3d, 4d, 5d inter-metallics, defy this magnetic treatment on account of band magnetism with a non-integral number of magnetic electrons per atom. For intermetallics of the lanthanide (actinide) series the magnetic behavior is generally characterized by delocalized 5d (6d) and 6s (7s) electrons of the outer subshells as well as localized 4f (5f) electrons of the inner subshell.

To take full advantage of experimental data, special attention should be given to the purity of the samples, measurement conditions, graphical presentation of the results, and adequate models. In practical operation, equations for the magnetic quantities are applied that contain, as a function of temperature (T) and applied magnetic field (B), the magnetically relevant parameters.

Two principal systems of electromagnetic units are utilized in magnetometry which can be called "*rational*" (the SI system) and "*irrational*" (the CGS-emu system) [4]. In the present article SI units are applied throughout. Conversion factors are collected in Table 25.1.

25.2
Basic Principles

25.2.1
Magnetic Quantities and Units [5, 6]

25.2.1.1 Magnetic Dipole Moment m

The orbital motion of an electron with charge $-e$ and mass m_e has a magnetic dipole moment m which is connected with an angular momentum l:

$$m = \gamma_e l \quad \text{where} \quad \gamma_e = -\frac{e}{2m_e} \qquad (\gamma_e : \textit{magnetogyric ratio}) \qquad (25.1)$$

In an atom m is specified by quantum numbers l and m_l and the component of orbital angular momentum along the z axis is $m_l \hbar$ and the magnitude $\sqrt{l(l+1)}\hbar$. In units of Bohr magnetons, $\mu_B = e\hbar/(2m_e)$, the component of the magnetic moment along z is $-m_l \mu_B$ and the magnitude of the orbital magnetic dipole moment is $\sqrt{l(l+1)}\mu_B$. In addition to an orbital contribution an electron has a spin, characterized by $s = 1/2$ and $m_s = \pm 1/2$. The magnitude of the spin magnetic

Table 25.1 Magnetic quantities; definitions, units, and conversion factors in the SI and in the CGS-emu system [4].

Quantity		SI	CGS-emu	Factor[a]	
μ_0	Permeability of vacuum	$\dfrac{4\pi}{10^7}$ H m^{-1b}	1		
B	Magnetic induction	T	G	10^{-4}	$\dfrac{\text{T}}{\text{G}}$
H	Magnetic field strength	A m^{-1}	Oe	$\dfrac{10^3}{4\pi}$	$\dfrac{\text{Am}^{-1}}{\text{Oe}}$
M	Magnetization	A m^{-1}	G	10^3	$\dfrac{\text{Am}^{-1}}{\text{G}}$
M_m	Molar magnetization	A m^2 mol^{-1}	G cm^3 mol^{-1}	10^{-3}	$\dfrac{\text{Am}^2}{\text{Gcm}^3}$
μ_a^c	Atomic magnetic dipole moment	$\mu_\text{a} = M_\text{m}/N_\text{A}$ A m^2	$\mu_\text{a} = M_\text{m}/N_\text{A}$ G cm^3		
m	Magnetic dipole moment	A m$^2 \equiv$ J T^{-1}	Gcm$^3 \equiv$ erg G^{-1}	10^{-3}	$\dfrac{\text{Am}^2}{\text{Gcm}^3}$
μ_B	Bohr magneton	A m^2	Gcm^3	10^{-3}	$\dfrac{\text{Am}^2}{\text{Gcm}^3}$
χ	Magnetic volume susceptibility	$M = \chi H$ $\quad$ 1	$M = \chi^{(\text{ir})} H^{(\text{ir})}$ $\quad$ 1	4π	
χ_g	Magnetic mass susceptibility	$\chi_\text{g} = \chi/\rho$ m^3/kg	$\chi_\text{g}^{(\text{ir})} = \chi^{(\text{ir})}/\rho$cm^3/g	$\dfrac{4\pi}{10^3}$	$\dfrac{\text{m}^3/\text{kg}}{\text{cm}^3/\text{g}}$
χ_m	Molar magnetic susceptibilty	m^3 mol^{-1}	cm^3 mol^{-1}	$\dfrac{4\pi}{10^6}$	$\dfrac{\text{m}^3}{\text{cm}^3}$
μ_eff	Effective Bohr magneton number [8]	1^d	1^e	1	
$N_\text{A}\mu_\text{B}^f$		5.58494 Am^2 mol^{-1}	5.58494 × 10^3 erg G^{-1} mol^{-1}		
$\dfrac{\mu_\text{B}}{k_\text{B}}$		0.67171 KT^{-1}	0.67171 × 10^{-4} KG^{-1}		
$\mu_0\dfrac{N_\text{A}\mu_\text{B}^2}{3k_\text{B}}$		1.57141 × 10^{-6} m^3 Kmol^{-1}	1.25049 × 10^{-1} cm^3 Kmol^{-1}		

[a] Factor applied to the value in CGS-emu units to obtain the value in SI units.
[b] H = Henry; H/m = Vs/Am.
[c] $\mu_\text{a}/\mu_\text{B}$ gives the atomic magnetization in Bohr magnetons, independent of the unit systems SI and CGS-emu.
[d] $\mu_\text{eff} = \sqrt{3k_\text{B}/\mu_0 N_\text{A}\mu_\text{B}^2}\sqrt{\chi_\text{m}T} = 797.74\sqrt{\chi_\text{m}T}$.
[e] $\mu_\text{eff} = \sqrt{3k_\text{B}/N_\text{A}\mu_\text{B}^2}\sqrt{\chi_\text{m}T} = 2.8279\sqrt{\chi_\text{m}T}$.
[f] $\mu_0 N_\text{A}\mu_\text{B} = 7.018\,24 \times 10^{-6}$ T m^3 mol^{-1}.

dipole moment is $\sqrt{s(s+1)}\hbar = \sqrt{3}\hbar/2$. The spin angular momentum is associated with a magnetic moment which can have a component along a particular axis equal to $-g\mu_\text{B}m_s$ and a magnitude equal to $\sqrt{3}g\mu_\text{B}/2$; the g-factor takes a value of approximately 2, so that the component of the magnetic spin moment along the z axis is $\mp \mu_B$. In the classical picture the energy of a magnetic dipole in a magnetic field is $E = -\boldsymbol{m} \cdot \boldsymbol{B}$. Due to space quantization the energy for a single electron reads $E = g \mu_B m_s B$ and the energy difference between the two states is $\Delta E = g \mu_B B$. In general, electrons in atoms may have both orbital and spin angular momentum which can couple (called *spin–orbit coupling*). The g-factor can therefore take different values depending on the contributions of spin and orbital angular momenta.

25.2.1.2 Magnetic Field, Magnetization, Magnetic Susceptibility

A magnetic field can be described by the vector fields $\boldsymbol{B}$ (measured in Tesla (T)) and $\boldsymbol{H}$ (measured in amperes per meter (A m^{-1})). In free space (vacuum) they are related by $\boldsymbol{B} = \mu_0 \boldsymbol{H}$ where $\boldsymbol{B}$ is the magnetic induction and $\boldsymbol{H}$ is the magnetic field strength; the factor $\mu_0 = 4\pi \times 10^{-7}$ H m^{-1} is the permeability of free space (Table 25.1). Moving charges are the *origin* of $\boldsymbol{H}$ while $\boldsymbol{B}$ describes the *effect* of the field upon materials.

In a magnetic solid the vector relationship is $\boldsymbol{B} = \mu_0(\boldsymbol{H} + \boldsymbol{M})$ where the magnetization is defined as $\boldsymbol{M} = \boldsymbol{m} \, V^{-1}$ (magnetic dipole moment per volume). In the case of weak magnetism $\boldsymbol{M}$ and $\boldsymbol{H}$ are linearly related: $\boldsymbol{M} = \chi \boldsymbol{H}$, where χ is the dimensionless magnetic volume susceptibility. In this case also $\boldsymbol{B}$ and $\boldsymbol{H}$ are linearly related: $\boldsymbol{B} = \mu_0(1 + \chi)\boldsymbol{H} = \mu_0 \mu_r \boldsymbol{H}$, where $\mu_r = 1 + \chi$ is the relative permeability. Many materials have μ_r values that deviate only a little from 1, that is, $|\chi| \ll 1$, and χ serves to classify materials: paramagnets $\chi > 0$, vacuum $\chi = 0$, diamagnets $\chi < 0$. In practice, the molar susceptibility $\chi_m = \chi M / \rho$ (M: molar mass; ρ: density) is used.

Diamagnetism In the presence of an external magnetic field there is a contribution to the magnetic moment, called *diamagnetism*, arising from the change in orbital angular momentum due to the applied field. In the one-electron approximation the diamagnetic susceptibility per mole of an atom is given by $\chi_m = -(\mu_0 N_A e^2 / 6 m_e) \sum_{i=1}^{n} < r_i^2 >$, where the radial integrals $< r_i^2 >$ describe the extension of the orbitals occupied by the n electrons and N_A, e, and m_e are the Avogadro constant, elementary charge, and mass of the electron, respectively. For closed shell atoms the diamagnetic susceptibility is independent of both T and B. Generally, diamagnetic contributions are necessary to correct experimental susceptibility data of compounds with open shell ions. For uses of diamagnetic materials see Section 25.4.1.

Paramagnetism The precondition for paramagnetism is the presence of ions with unpaired electrons (open shell atoms). Net spins and orbital angular momenta are polarized in the direction of the applied field. Paramagnetism is observed in various forms, differing in magnitude and dependence on T and $\boldsymbol{B}$: (i) *Curie paramagnetism*: inverse dependence of χ_m on T, independence of χ_m on $\boldsymbol{B}$ at weak applied fields, but inverse dependence at strong fields. While an increase in $\boldsymbol{B}$ will tend to line up the magnetic dipoles, an increase in T will disorder them. Therefore, the magnetization of a Curie paramagnet depends on the ratio $\boldsymbol{B}/T$.

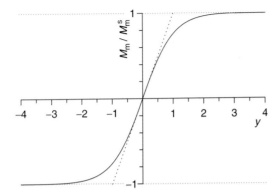

Figure 25.1 Brillouin function $B_J(y)$ for $J = \frac{1}{2}$ and $g = 2$, ($y = \mu_B B/k_B T$). The dotted line passing the origin corresponds to Curie law behavior (see text).

(ii) *Temperature-independent paramagnetism* (TIP) called *Van Vleck paramagnetism* (second order effect that involves mixing of the ground state with excited multiplets by the applied field), and *Pauli paramagnetism* (of conduction electrons) observed for metallic systems.

25.2.1.3 Brillouin Function, Curie Law

The magnetic moment of a free ion is characterized by its total angular momentum composed of the orbital and the spin angular momentum: $J = L + S$, measured in units of $\hbar$. The total angular momentum quantum number J can take an integer or a half-integer value according to an even or odd number, respectively, of unpaired electrons. In Equation 25.2 the magnetization as a function of B and T is given by the Brillouin function $B_J(y) = M_m/M_m^s$ where M_m^s is the molar saturation magnetization at $T = 0$ and $y = g_J J \mu_B B/(k_B T)$.[1]

$$B_J(y) = \frac{M_m}{M_m^s} = \frac{2J+1}{2J} \coth\left[\left(\frac{2J+1}{2J}\right)y\right] - \frac{1}{2J}\coth\left(\frac{y}{2J}\right) \tag{25.2}$$

where

$$y = \frac{g_J J \mu_B B}{k_B T} \quad \text{and} \quad M_m^s = N_A g_J J \mu_B$$

For $J = \frac{1}{2}$ and $g = 2$ the Brillouin function reduces to $M_m/M_m^s = \tanh y$, outlined in Figure 25.1. Figure 25.2 displays the atomic magnetic dipole moment divided by the Bohr magneton, μ_a/μ_B, as a function of B/T for compounds[2] of Cr^{3+} ($S = \frac{3}{2}$), Mn^{2+} ($S = \frac{5}{2}$), and Gd^{3+} ($S = \frac{7}{2}$) [8].

For $y \ll 1$ the Brillouin function approximates to $B_J(y \ll 1) = (J+1)y/(3J)$ (see dotted line in Figure 25.1); magnetization and magnetic susceptibility read

1) For a pure spin system the argument of the Brillouin function $B_s(y)$ reads $y = g S \mu_B B/(k_B T)$.

2) $KCr(SO_4)_2 \cdot 12H_2O$, $NH_4Fe(SO_4)_2 \cdot 12H_2O$, $Gd_2(SO_4)_3 \cdot 8H_2O$.

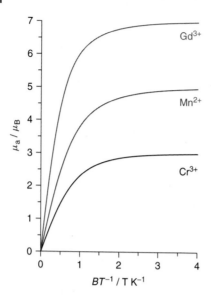

Figure 25.2 Variation μ_a vs. B/T for $S = \frac{3}{2}$ (Cr^{3+}), $\frac{5}{2}$ (Mn^{2+}), $\frac{7}{2}$ (Gd^{3+}). (Adapted from Ref. [7].)

$M_m = (N_A \mu_B^2 B/3k_B T)g_J^2 J(J+1)$ and $\chi_m = \mu_0(N_A \mu_B^2/3k_B T)g_J^2 J(J+1)$, respectively. Apart from measurements at low T and large B, the *Curie law* is obeyed:

$$\chi_m = \frac{C}{T} \quad \text{where} \quad C = \mu_0 \frac{N_A \mu_B^2}{3k_B} \underbrace{g_J^2 J(J+1)}_{\mu_{eff}^2}, g_J = \frac{3}{2} + \frac{S(S+1) - L(L+1)}{2J(J+1)} \quad (25.3)$$

Written as *term symbol* $^{2S+1}L_J$ the quantum numbers S, L, and J specify the ground state of the free ion and are determined using Hund's rules;[3] μ_{eff} is the *effective Bohr magneton number* [8] while $\mu_a^s = M_m^s/N_A = g_J J \mu_B$ is the *atomic magnetic saturation moment* measured at low T and very large $\boldsymbol{B}$.

25.2.2
Classification of Magnetic Materials

25.2.2.1 Magnetically Dilute Systems
In insulators the electronic states of a magnetically isolated d^N or f^N metal ion, surrounded by ligands, undergo energy splittings by interelectronic repulsion (H_{ee}), spin–orbit coupling (H_{so}), the electrostatic field produced by the ligands

---

3) Hund's rules [6]: (1) In consideration of Pauli's exclusion principle arrange the spins of the electrons in a way that $S = |m_{s_1} + m_{s_2} \cdots + m_{s_N}| = |M_S|$ maximizes. (2) Given the first rule, maximize L by summing up the m_l values. (3) J is obtained using $J = |L - S|$ if the shell is less than half full and $J = L + S$ if it is more than half full.

(H_{lf}), and the applied magnetic field, H_{mag} (Zeeman effect). Depending on the relative strengths of H_{ee}, H_{so}, and H_{lf} the following situations can be distinguished for $3d^N$ [9] and $4f^N$ [10] ions:

$3d^N$	H_{ee}	>	H_{lf}	>	H_{so}		Weak ligand field
	H_{lf}	>	H_{ee}	>	H_{so}		Strong ligand field
	H_{lf}	$\approx$	H_{ee}	>	H_{so}		Intermediate ligand field
$4f^N$	H_{ee}	>	H_{so}	>	H_{lf}		Ln strong field
	H_{ee}	>	H_{so}	>>	H_{lf}		Ln weak field

Energy splittings of d and f electron states caused by H_{ee} are around 10^4 cm^{-1} [4] and splittings by H_{so} up to 10^3 cm^{-1}. The effect of H_{lf} on d ions extends to 2×10^4 cm^{-1} [5] and is much weaker on 4f ions (10^2 cm^{-1}) and 5f ions (10^3 cm^{-1}). Referring to interatomic exchange interactions (H_{ex}), effects up to 10^2 cm^{-1} have been observed for nd systems. For 4f ions only very small splittings < 1 cm^{-1} have been detected, comparable to H_{mag} which corresponds to an energy equivalent of ≈ 0.5 cm^{-1} for $B = 1$ T. [6] Magnets generating high continuous fields up to 40 T and pulsed fields up to 100 T are available for specific new problems (see Sections 25.3.3, 25.4.6).

The magnetic properties of d and f metal compounds arise from the ground state of the metal ion as well as from higher, thermally populated states. A special electronic situation arises for $3d^5$ and $4f^7$ ions with ground states $^6S_{5/2}$ (Mn^{2+}, Fe^{3+}) and $^8S_{7/2}$ (Eu^{2+}, Gd^{3+}), respectively. Practically, H_{lf} and H_{so} have no effect because orbital contributions are negligible, so that spin-only paramagnetism according to $S = 5/2$ and $7/2$, respectively, is observed. [7] Therefore, compounds like $NH_4Fe(SO_4)_2 \cdot 12H_2O$ and $Gd_2(SO_4)_3 \cdot 8H_2O$ obey the Curie law (Equation 25.3) with *effective Bohr magneton number* $\mu_{eff} = 2\sqrt{S(S+1)} = \mu_{s.o.}$ (spin-only formula), that is, $\mu_{eff} = 5.9$ and 7.9 for Fe^{3+} and Gd^{3+}, respectively. Another special case is the orbital singlet ground state $^1A_1$ of $3d^6$ low-spin systems in a strong octahedral ligand field leading to $\chi_m =$ const. (TIP), corresponding to a parabolic variation $\mu_{eff} \sim \sqrt{T}$ (cf. footnotes d and e in Table 25.1).

In the case of a non-half-filled 3d subshell, the ground state can be A, E, or T [14]. For the former two, χ_m is often given as a sum of a temperature-dependent

---

4) Instead of cm^{-1} other energy equivalents are in use, for example, K, kJ mol^{-1}, s^{-1} = Hz [11].

5) On the basis of the *spectrochemical series* for ligands (L) and metals (M) the splitting factor $\Delta = f(L) \cdot g(M)$ for octahedral complexes [ML$_6$] is predictable to a good approximation [12].

6) For a reliable detection of such weak exchange interactions H_{mag} must be several orders of magnitude weaker than 1 T.

7) Spectroscopic studies have clarified that the ground term of the Gd^{3+} free ion is composed of 97% $^8S_{7/2}$ and 3% $^6P_{7/2}$, $^6P_{5/2}$, $^6P_{3/2}$ corresponding to $g = 1.993(2)$ [13] instead of $g = 2.00$.

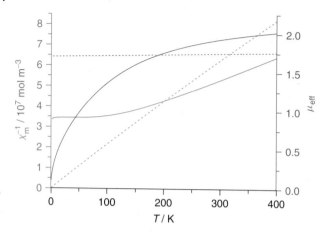

Figure 25.3 Calculated variation μ_{eff} vs. T and χ_m^{-1} vs. T of a $3d^1$ ion in octahedral (———, ———) and orthorhombic ligand field (- - - -, - - - -).

Curie term C/T and a TIP contribution χ_0 where μ_{eff} is close to the spin-only value and varies only little with temperature. The μ_{eff} behavior of systems with T ground terms, however, is strongly temperature dependent [15, 16]. As an example, the variation μ_{eff} vs. T and χ_m^{-1} vs. T of a $3d^1$ ion, for example, Ti^{3+}, V^{4+}, is exhibited in Figure 25.3 where the ligand field has octahedral and orthorhombic symmetry, respectively. For the octahedral system (ground term $^2T_2$), the unquenched orbital contribution to the magnetic moment and the spin part produce, via H_{so}, a nonmagnetic ground state ($\mu_{eff} \to 0$ for $T \to 0$) [15]. The variation χ_m^{-1} vs. T shows another feature in that for $T > 150$ K a linear increase with temperature is observed. Reducing further the ligand field symmetry, for example to C_{2v}, the quenching of orbital contributions to μ_{eff} is (nearly) complete so that H_{so} has no effect. Pure spin magnetism results with a temperature-independent $\mu_{eff} = 1.73$.

In 4d and 5d electron systems H_{so} increases and H_{ee} decreases compared to the 3d case. No simple formula is available to describe the variation μ_{eff} vs. T. As a rule, the μ_{eff} values are lower than those of comparable 3d systems.

The magnetic behavior of 4f systems (insulating and metallic as well[8]) is accurately predictable with programs [17, 18]. Neglecting the influence of the ligand field, which means temperatures well above 100 K, and having a magnetic ground state of specific J, Equation 25.3 applies. Because of Kramers' rule [19] for centers with an odd number of 4f electrons, a degenerate ground state is always observed, leading to Curie behavior at temperatures below 10–20 K (in the absence of magnetic cooperative effects). In the case of an even number of f electrons, degenerate as well as nondegenerate ground states can be found, leading to Curie behavior and TIP, respectively, at low temperature. Going to actinides, H_{so} and

---

8) Exceptions are Ln intermetallics with intermediate valence, for example, $Ce^{III/IV}$, $Eu^{II/III}$, $Yb^{II/III}$ [6].

H_{lf} increase according to the larger effective nuclear charge and the fact that 5f electrons are more accessible for ligands, respectively. No simple approximation can be made. So, the only possibility to predict the magnetic behavior for these ions is computational methods.[9]

25.2.2.2 Local Magnetic Ordering within a Dinuclear Molecule

On the way from dilute to concentrated magnetic systems the simplest one is a dinuclear molecule where each metal ion has a single electron ($S_1 = S_2 = \frac{1}{2}$), corresponding to a dinuclear Cu(II) complex with negligible intermolecular inter-actions. Switching on the isotropic spin–spin coupling, described by the operator $\hat{H}_{ex} = -2J_{ex}\hat{S}_1 \cdot \hat{S}_2$,[10] the system undergoes a coupling of antiferromagnetic na-ture ($J_{ex} < 0$, ground state $S' = S_1 - S_2 = 0$) or of ferromagnetic nature ($J_{ex} > 0$, ground state $S' = S_1 + S_2 = 1$). The magnetic behavior, outlined by means of the Bleaney–Bowers equation [27]

$$\chi_m = \mu_0 \frac{N_A \mu_B^2}{3k_B T} g^2 \left[1 + \frac{1}{3}\exp\left(\frac{-2J_{ex}}{k_B T}\right)\right]^{-1}$$

for $g = 2$ and various J_{ex} values between -50 and $+50\,\text{cm}^{-1}$, is shown in Figure 25.4, presented as variation χ_m vs. T, χ_m^{-1} vs. T, and μ_{eff} vs. T. The $\chi_m - T$ plot is well suited for antiferromagnetic spin–spin couplings ($J_{ex} < 0$) showing the typical maximum of χ_m, whereas for a coupling of ferromagnetic nature the $\mu_{eff} - T$ plot is suitable.[11] The $\chi_m^{-1} - T$ plot exhibits that for $J_{ex} = 0$ the Curie law is obeyed while otherwise straight lines are not observed before high temperature is achieved (see next section and Equation 25.4).

25.2.2.3 Magnetically Condensed Systems

Materials with high magnetic dipole concentration tend to order magnetically on cooling below the Curie temperature T_C (ferro- and ferrimagnets) and below the Néel temperature T_N (antiferromagnets), respectively. When the magnetism of the metal ion is only caused by the spin, the magnetic behavior in the paramagnetic regime is described by the *Curie–Weiss law*

$$\chi_m = \frac{C}{T - \theta} \quad \text{with} \quad \theta = \frac{2S(S+1)}{3k_B}\sum_i^n z_i J_{ex,i}, \tag{25.4}$$

9) Computer programs are available to treat single ion effects and, in addition, interatomic exchange interactions: (i) CONDON [17, 18] computes the magnetic susceptibility of d and f systems and interatomic spin–spin coupling under varying T and B [20], (ii) the high-temperature series expansion (HTSE) method [21–23] is applicable to isotropic interionic exchange interactions in extended spin lattices, (iii) MAGPACK [24] solves the exchange problem in molecular assemblies, formed by a finite number of exchange-coupled magnetic d ions, and (iv) MVPACK [25], applicable to d systems, determines the magnetic properties of polynuclear *mixed valence* molecular systems.

10) In the literature, different definitions of J_{ex} exist. Instead of the original version $\hat{H}_{ex} = -2J_{ex}\hat{S}_1 \cdot \hat{S}_2$, preferred here, one finds $\hat{H}_{ex} = -J_{ex}\hat{S}_1 \cdot \hat{S}_2$ and $\hat{H}_{ex} = J_{ex}\hat{S}_1 \cdot \hat{S}_2$ [26].

11) As an alternative to the $\mu_{eff} - T$ plot the $\chi_m T - T$ plot is used, where $\chi_m T \propto \mu_{eff}^2$.

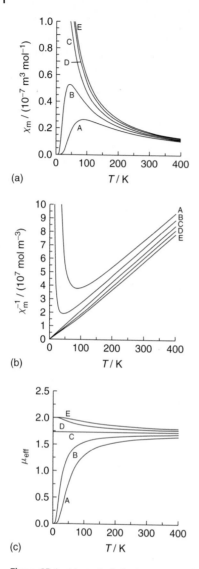

Figure 25.4 Magnetic behavior per metal ion of an exchange coupled dinuclear system with $S_1 = S_2 = \frac{1}{2}$; (a) χ_m vs. T, (b) χ_m^{-1} vs. T, and (c) μ_{eff} vs. T; J_{ex}/cm^{-1}: (A) −50, (B) −25, (C) 0, (D +25, and (E) +50.

where θ is the Weiss constant, z_i the number of ith nearest magnetic neighbors of a given magnetic center, $J_{ex,i}$ stands for the exchange interaction between the ith neighbors, and n is the number of sets of neighbors for which $J_{ex,i} \neq 0$ [28]. Positive and negative θ values refer to predominating ferro- and antiferromagnetic interactions, respectively. For non-half-filled subshells where both ligand field and exchange effects play a major role, and where the single ion behavior deviates from Curie law, the variation of the magnetic susceptibility with temperature above the ordering temperature is described by the molecular field (mf) model.

$$\frac{1}{\chi_m} = \frac{1}{\chi_m'} - \lambda_{mf} \quad \text{where} \quad \lambda_{mf} = \frac{2\sum\limits_{i}^{n} z_i J_{ex,i}}{\mu_0 N_A \mu_B^2 g^2} \tag{25.5}$$

χ_m' refers to the single ion susceptibility and the mf parameter λ_{mf} produces a parallel shift of the $(\chi_m')^{-1}$–T curve, similar to θ for the Curie–Weiss law; g may deviate more or less from 2, depending on orbital contributions.

25.2.2.4 Ferromagnetism

Ferromagnets exhibit spontaneous magnetization below T_C. The magnetic dipoles tend to align parallel to each other on account of interatomic spin–spin exchange couplings leading to long-range magnetic order. The most important properties of ferromagnets are a high relative permeability, $\mu_r = 1 + \chi$, and the saturation magnetization, M^s, as well as the hysteresis loop, recorded in Figure 25.5 as variation of M vs. H [29]. A ferromagnet in its initial state is not magnetized, indicated by O. Domains exist within each of which the local magnetization reaches the saturation value M^s, and the direction of magnetization of the various domains are such that the crystal as a whole has no net magnetization. After

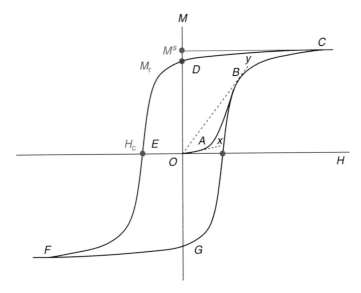

Figure 25.5 Magnetization curve (*OABC*) and hysteresis loop (*CDEFGC*) of a typical ferromagnetic material; M^s: saturation magnetization, M_r: remanent magnetization, H_c: coercive field; *Ox, Oy*: see text. (Adapted from Ref. [29].)

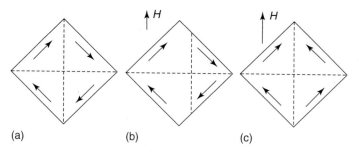

Figure 25.6 Schematic representation of domains in a ferromagnetic crystal; (a) demagnetized state; (b) magnetization through movement of domain walls; and (c) magnetization by rotation of the magnetization of whole domains. (Adapted from Ref. [30].)

saturation to M^s along the *magnetization curve OABC*, a decrease in H to zero reduces M to M^r (point D). The reversed field required to reduce M to zero is called coercivity H_c. Increasing the field in the reverse direction, decreasing it to zero, and then increasing it to the original value, the hysteresis loop $CDEFGC$ is obtained (symmetric with respect to O). The initial χ is the slope of the tangent of M^s at O (slope Ox), while the maximum of χ (slope Oy) normally occurs near the knee of the M vs. H curve (point B). The movement of domain walls during a magnetization process is schematized in Figure 25.6. Domains oriented parallel to H grow at the expense of antiparallel domains through movement of domain walls (b) and by rotation (c). The domains remain magnetized along a preferred (*easy*) direction. Stronger applied fields are necessary to force the magnetization parallel to the applied field.

Carefully grown single crystals of ultra-pure iron (bcc structure) are magnetically soft, that is, they have an exceptionally small area of the hysteresis loop and a small remanence. The domain walls are easily movable by applied fields and the crystals are magnetically anisotropic: saturation is most easily obtained for applied fields parallel to the unit cell edges [100] (Figure 25.7).

Ferromagnets are broadly divided into magnetically soft materials (high permeability, easily magnetized, and demagnetized on account of low coercive force; used e.g., in transformers), and magnetically hard materials (low permeability, difficult to magnetize and demagnetize on account of a high coercive force; used in permanent magnets). Systems exhibiting both magnetic and electric order (so-called *multiferroics*) are interesting materials for future information-technology devices in which data can be written to magnetic memory elements by applied magnetic fields [32, 33].

25.2.2.5 Antiferromagnetism
Antiferromagnetic ordering is described by the concept of a *sublattice*, that is, a unit of the lattice with parallel magnetic dipoles [34, 35]. MnF_2, a typical

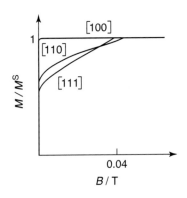

Figure 25.7 Initial magnetization curves for a single crystal of iron. (Adapted from Ref. [31].)

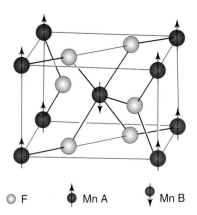

Figure 25.8 Spin structure of MnF_2.

○ F   Mn A   Mn B

antiferromagnet, crystallizes in a body-centered tetragonal structure (rutile-type, space group $P4_1/nmm$; $Mn^{2+}[3d^5]$, $S = 5/2$) and can be regarded as composed of two primitive sublattices A and B (Figure 25.8). Below the Néel temperature $T_N = 74$ K [36] A and B ions are antiferromagnetically coupled, where each magnetic dipole is collinear to the fourfold axis and oriented antiparallel to the dipoles of its eight nearest neighbors.

At $T > T_N$ the thermomagnetic behavior of MnF_2 obeys the Curie–Weiss law with $\theta = -113$ K. Susceptibility measurements with a weak applied field at $T < T_N$ parallel and perpendicular to the fourfold axis revealed that $\chi_\parallel$ decreases with decreasing T, while $\chi_\perp$ remains constant (Figure 25.9)[12].

12) At $T = 0$ both sublattices are saturated. Thermal agitation effects of the dipoles can be ignored. Therefore, a small applied field collinear to the magnetization directions has no effect ($\chi_\parallel = 0$) while an applied field perpendicular to the magnetization vectors deflects both sublattices in the field direction and yields a component along the applied field ($\chi_\perp \neq 0$). Increasing the temperature to T_N yields an increase in $\chi_\parallel$ on account of increasing thermal fluctuations, while $\chi_\perp$ remains constant [5, 6].

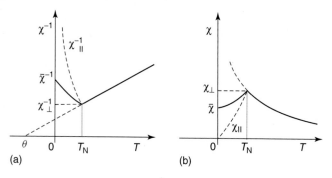

Figure 25.9 $\chi^{-1} - T$ (a) and $\chi - T$ plot (b) of an anti-ferromagnet with single-axis anisotropy. $\chi_\parallel$, $\chi_\perp$, and $\bar\chi = (\chi_\parallel + 2\chi_\perp)/3$ of a single crystal are measured with a weak applied field parallel and perpendicular to the unique axis, while $\bar\chi$ refers to the polycrystalline sample.

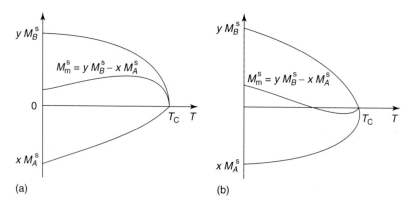

Figure 25.10 Special variations M_m^s vs. T of a ferrimagnet; (a) maximum of the magnetization and (b) magnetization reverse.

25.2.2.6 Ferrimagnetism

In ferrimagnets the sublattices A and B are structurally non-equivalent and can be occupied by different magnetic atoms and by different numbers of atoms, so that the antiparallel spin arrangement results in an uncompensated spontaneous magnetization below T_C. On the basis of mf theory the M_m^s vs. T behavior can be simulated and compared with the measured curves [31, 34]. In Figure 25.10 two typical examples are represented showing a maximum in M_m^s (a) and a magnetization reverse (b)[13].

Figure 25.11 displays the variation χ_m^{-1} vs. T above T_C in the form of a hyperbola with asymptotes $\chi_m^{-1} = T/C + \chi_0^{-1}$ and $T = \Theta$. The intersection point of the former

13) A Prussian blue analog, $(Ni_{0.22}Mn_{0.60}Fe_{0.18})_{1.5}[Cr^{III}(CN)_6]\cdot 7.6H_2O$, is a ferrimagnet exhibiting two compensation temperatures, that is, two reversals at 35 and 53 K [37, 38].

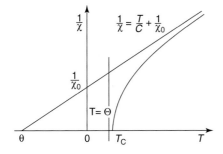

Figure 25.11 Variation χ_m^{-1} vs. T of a simple ferrimagnet.

asymptote with the temperature axis at $\theta = -C/\chi_0$ is called the *asymptotic Curie temperature*.

An important class of ferrimagnetic substances adopts the spinel structure (space group $Fd\bar{3}m$), among them the normal spinel Zn^{II} $[Fe_2^{III}]O_4$ and the inverse spinels Fe^{III} $[Mn^{II}Fe^{III}]O_4$ as well as $Fe_3O_4 \equiv Fe^{III}[Fe^{II}Fe^{III}]O_4$ (magnetite). The octahedral site, generally assigned B, is marked with square brackets. The tetrahedral site is assigned A. In this class of cubic ferrites all interactions are antiferromagnetic, but the A–B interaction is considerably stronger than A–A and B–B. Thus, so long as all cations in the spinel structures are magnetically active, the dominating A–B interaction makes the spins within each sublattice parallel. This is supported by the fact that the normal spinel $Zn[Fe_2]O_4$ has no net saturation moment on account of diamagnetic Zn^{2+} and antiferromagnetically coupled Fe^{3+}. By mixing either two inverse base ferrites or a normal and an inverse ferrite, the fine-tuning of magnetic properties is possible and of technological importance [39].

25.3
Experimental Magnetometrical Methods

SQUID magnetometry is a highly sensitive way of measuring magnetic properties. A SQUID is capable of amplifying very small changes in magnetic field (produced by a magnetized sample) into a large electrical signal, even in the presence of large static magnetic fields (see Refs [1, 40, 41] for further details). Based on an inductive detection of the magnetic moment, the sensitivity of a SQUID magnetometer can be up to $10^{-11}\,J\,T^{-1}$. To investigate individual magnetic nanoparticles a micro-SQUID magnetometer has been developed operating at 0.04 K $\leq T \leq 7$ K [42–44]. Moreover, fabrication improvements have led to nano-SQUIDs [45].

25.3.1
Direct Current Method

In DC magnetometry the paramagnetic sample is placed in a static magnetic field H, and the static magnetic volume susceptibility is derived by dividing the

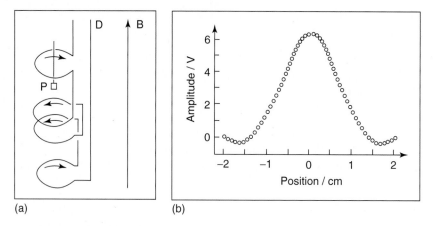

(a)   (b)

Figure 25.12 (a) Pick-up coil (P: magnetic dipole, D: superconducting wire, B: applied magnetic field). (b) SQUID
response of P driven through the detection coil in a series
of steps.

magnetization M by H, $\chi = M/H$. Only equilibrium states are considered; the
time required to establish equilibrium when H changes is irrelevant.

In the experiments the sample is moving relative to a set of superconducting
pick-up coils (see Figure 25.12a) which are coupled to a SQUID to measure the
current induced in the superconducting pick-up coils. Figure 25.12b displays the
output of the SQUID as the magnetized sample is moved through the pick-up coil
[40, 41]. Conventional temperature and applied field ranges of operation extend to
2–400 K and up to ±7 T, respectively.

25.3.2
Alternating Current Method

In AC magnetometry the oscillating field with varying frequency ω (as a rule
$10^{-3} - 1.5 \times 10^3$ Hz) furnishes information about the *magneto-dynamics* of the
system. At low frequency one can expect that χ will be identical to the DC result
while for an AC magnetic field of sufficiently high frequency the magnetization of
the material is unable to follow the change in the field, but instead lags in phase
[29]. If the applied field is of the form $H(t) = H + H_1 \cos \omega t$, the magnetization
may be represented by

$$M(t) = M_0 + M_1 \cos(\omega t - \phi)$$
$$= M_0 + M_1 \cos \omega t \cos \phi + M_1 \sin \omega t \sin \phi \qquad (25.6)$$

where H is the static field, $M_0 = \chi_0 H$ is the equilibrium value of the magnetization
in this static field, $H_1 \cos \omega t$ is the oscillating field, and ϕ is the phase angle

by which the magnetization lags the field.[14] Defining $\chi' = M_1 \cos\phi / H_1$ and $\chi' = M_1 \sin\phi / H_1$, Equation 25.6 becomes

$$M(t) = \chi_0 H + \chi' H_1 \cos\omega t + \chi'' H_1 \sin\omega t \tag{25.7}$$

To develop the corresponding susceptibility equations it is convenient to use complex notation (cf. Section 25.5.1):

$$H(t) = H + H_1 e^{i\omega t} \quad and \quad M(t) = M_0 + \chi H_1 e^{i\omega t} \tag{25.8}$$

where $e^{i\omega t} = \cos\omega t + i\sin\omega t$ and $\chi = \chi' - i\chi''$. Both χ' and χ'' depend on the frequency ω as well as the magnitude of the static field H: χ' is called *the paramagnetic dispersion; χ''* is proportional to the energy absorbed by the substance from the high frequency field. At low frequencies $\chi' = \chi_0$ and $\chi'' = 0$, that is, M and H are in phase.

The complex susceptibility can provide information on the *relaxation* process (i.e., the return to equilibrium), energy exchange between magnetic dipoles and the lattice, structural details of the materials, electrical conductivity by induced currents, and so on. The continual redistribution of the magnetic dipoles over the energy levels (e.g., two in number for $J = \frac{1}{2}$, $M_J = \pm\frac{1}{2}$) proceeds via the *spin–lattice* relaxation process. The process has a characteristic time constant τ and is caused by the coupling of the spin system to the environment via the spin–phonon interaction originating from the perturbation of the ligand field by lattice vibrations.[15]

To obtain information about the relaxation time τ, the behavior of χ with respect to ω has to be investigated. If ω of the AC field is low, $\omega\tau << 1$, the measured susceptibility is the isothermal one, $\chi = \chi_T$. In the limit $\omega\tau >> 1$, however, the system has no time to exchange energy with the environment and the adiabatic susceptibility $\chi = \chi_S$ is measured. With regard to the intermediate range the susceptibility is given by

$$\chi(\omega) = \chi_S + \frac{\chi_T - \chi_S}{1 + i\omega\tau} \tag{25.9}$$

The real and imaginary components of the susceptibility $\chi(\omega) = \chi' - i\chi''$[16] [46–48] are given by

$$\chi' = \chi_S + \frac{\chi_T - \chi_S}{1 + \omega^2\tau^2} \quad and \quad \chi'' = \frac{(\chi_T - \chi_S)\omega\tau}{1 + \omega^2\tau^2} \tag{25.10}$$

According to Equation 25.10, the frequency dependence of χ' and χ'' is drawn schematically in Figure 25.13. The two susceptibility limits display the in-phase

---

14) The equations are written in scalar form, since the constant H is usually parallel to the oscillating one.

15) With increasing frequency of the oscillatory magnetic field a spin–spin absorption additional to the spin–lattice absorption occurs. The spin–spin relaxation time $\tau_{ss} \approx 10^{-10}$s is much shorter than τ and is temperature independent. For this reason, the

16) two relaxation processes can usually be studied independently.

The relation between $\chi(\omega)$, χ', and χ'' can be verified by showing that $\underbrace{\chi_S + \frac{\chi_T - \chi_S}{1 + i\omega\tau}}_{\chi(\omega)}$

$$= \underbrace{\chi_S + \frac{\chi_T - \chi_S}{1 + \omega^2\tau^2}}_{\chi'} - i\underbrace{\frac{(\chi_T - \chi_S)\omega\tau}{1 + \omega^2\tau^2}}_{\chi''}.$$

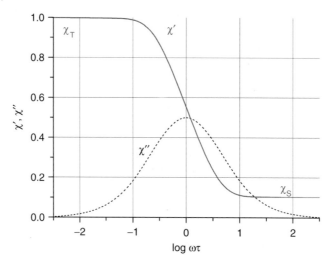

Figure 25.13 Frequency dependence of the real and imaginary components χ' and χ'', respectively, of the magnetic susceptibilities. (Adapted from Ref. [46].)

component χ' ($\chi' = \chi_T$ for $\omega \ll \tau^{-1}$ and $\chi' = \chi_S$ for $\omega \gg \tau^{-1}$), while the out-of-phase component χ'' approaches zero at these limits, but shows a maximum around the frequency $\omega = \tau^{-1}$, the height of the maximum being $\frac{1}{2}(\chi_T - \chi_S)$. Another possibility to present the results of AC measurements is the Argand diagram (Cole–Cole plot [49, 50]). For a given temperature the susceptibility that follows Equation 25.10 can be represented as a semicircle in the complex χ-plane with the midpoint on the real axis (χ' axis, see Figure 25.14). The semicircle intersects the real axis at χ_T on the low frequency side and at χ_S at high frequencies, while $\omega^{-1} = \tau$ is found at the top of the semicircle, so that the relaxation time can be easily extracted.

Equations 25.9 and 25.10 are restricted to a single relaxation process characterized by a single τ. For a distribution of relaxation times, on account of several relaxation processes, the empirical law $\chi(\omega) = \chi_S + (\chi_T - \chi_S)/[1 + (i\omega\tau)]^{1-\alpha}$ applies, that is, the wider the distribution in relaxation times the larger is α [46, 49b]; α is available by the Argand plot which becomes an arc of a circle with the center shifted downwards. The angle that subtends the arc is given by $\pi(1 - \alpha)$ (see Figure 25.14).

25.3.3
Magnetometry at High Fields

Continuous magnetic fields up to 40 T [3] and pulsed magnetic fields up to 100 T [51] (with nearly sinusoidal shape as a function of time and pulse durations of 10–1000 ms) are available to elucidate, for example, the strength of antiferromagnetic exchange couplings and the spin dynamics in polynuclear 3d systems (cf. Section 25.4.6).

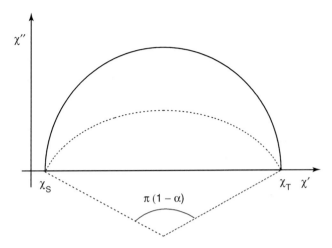

Figure 25.14 Variation of χ' vs. χ'' (Argand plot) at a given temperature; solid line: single relaxation time, broken line: distribution of relaxation times; see text. (Adapted from Ref. [46].)

25.4
Selected Applications

25.4.1
Diamagnetic Materials

With regard to applications important materials are (i) superconductors, on account of their transition to an ideal diamagnet below the critical temperature T_C, (ii) diamagnetic metals alloyed with paramagnets to exactly cancel out the magnetic susceptibility at a distinct temperature, and (iii) diamagnetically anisotropic liquid crystals that are aligned by a strong magnetic field [39, 52].

25.4.2
Metamagnetic $Sc_2MnRh_5B_2$

Metamagnetism is often observed for materials that have an antiferromagnetic ground state, dominating ferromagnetic interactions ($\theta > 0$) and crystal anisotropy. Compounds of this class exhibit an unusual applied field dependence of the magnetization[17]. Metamagnetic behavior has been established in the intermetallic

17) A typical example is $FeCl_2$ [53] ($CdCl_2$-type layer structure) with spin–spin coupling of ferromagnetic nature within each layer of Fe^{2+} ions and a weak coupling of antiferromagnetic nature between the layers. The compound orders antiferromagnetically below $T_N = 24$ K and shows Curie–Weiss behavior at high temperature with $\theta = +48$ K. In applied fields $B \geq 1$ T below T_N a parallel orientation of the spins is induced corresponding to paramagnetic saturation.

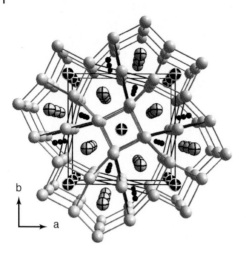

b

a

Figure 25.15 Perspective view of the crystal structure of $Sc_2MnRh_5B_2$ along the [001] direction. Classification of the spheres: Rh (gray, at $z = 1/2$, all other atoms at $z = 0$), Sc (gray, black cross), Mn (black, white cross), and B (black) (Adapted from Ref. [56d].)

$Sc_2MnRh_5B_2$. The phase is an ordered substitutional variant of the $Ti_3Co_5B_2$-type structure [54] ($P4/mbm$) in which the magnetically active Mn atoms are arranged in chains along [001] with intrachain and interchain Mn–Mn distances of 3.058 and 6.606 Å, respectively [55] (see Figure 25.15). DC susceptibility measurements exhibit Curie–Weiss behavior above 300 K, where $\theta = 217$ K and $\mu = 4.2\mu_B$, calculated from the slope of the Curie–Weiss straight line (insert of Figure 25.16). At $T_N = 130$ K the onset of an antiferromagnetic ordering is observed where an intermediate maximum in the atomic magnetic dipole moment μ_a is detected, so long as the applied field does not exceed $B = 1.7$ T. Increasing B shifts the maximum of μ_a to lower temperature. Above $B \approx 2.5$ T this maximum disappears and μ_a is enhanced strongly. At 2 K and 5 T $\mu_a = 1.8$ μ_B is achieved without reaching saturation (see Figure 25.17).

Obviously the magnetic behavior is dominated by ferromagnetic interactions within the chains while weaker antiferromagnetic interactions exist between the chains. Higher fields destroy the antiferromagnetic interchain coupling and force the transition to the paramagnetically saturated state.

In the meantime, numerous ordered substitutional variants of the $Ti_3Co_5B_2$-type structure with the general composition $A_2MM'_5B_2$ (A: Mg, Sc; M: Mn, Fe, Co, Ni; M': Ru, Rh, Ir) have been intensively investigated both experimentally and theoretically, in particular with respect to itinerant magnetism [56]. A clear trend from antiferro- to ferromagnetism is observed by increasing the valence electron concentration. For example, replacing Mn in $Sc_2MnRh_5B_2$ by Fe a ferromagnetic phase is obtained that orders at $T_C = 450$ K.

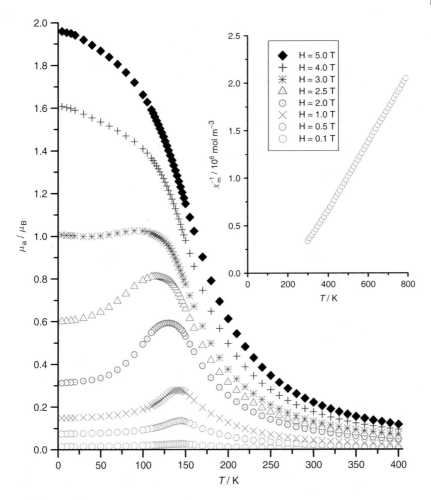

Figure 25.16 $Sc_2MnRh_5B_2$: Variation μ_a vs. T at various applied fields. Insert: $\chi_m^{-1}-T$ plot. (Adapted from Ref. [55].)

25.4.3
Superparamagnets

Nanosized ferromagnetic particles of single-domain structure[18] behave like small permanent magnets. The direction of the particles' magnetization is determined by the applied field and by internal forces, that is, shape anisotropy and magnetocrystalline anisotropy. At sufficiently high temperature the magnetization vector

18) Domain walls become energetically unfa-
vorable when a certain size of the particle
falls short of [5].

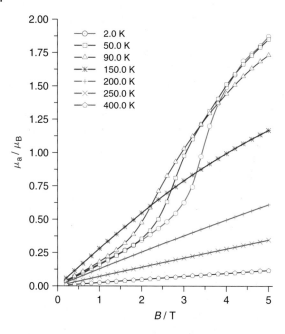

Figure 25.17 Sc$_2$MnRh$_5$B$_2$: Variation μ_a vs. B at various temperatures. (Adapted from Ref. [55].)

begins to flip from one easy direction to the opposite easy direction, thereby overcoming an energy barrier ΔE [5, 57, 58]. For $k_B T \gg \Delta E$ and in the absence of interparticle interactions the system behaves like a paramagnet in which the independent moments are huge compared to atomic moments, for example, $\approx 10^4 \mu_B$ for a 50 Å particle. To characterize such *superparamagnetic* particles, AC magnetometry is applied. The relaxation time τ of the particles' moment is given by $\tau = \tau_0 \exp[\Delta E/(k_B T)]$, where τ_0 is $\approx 10^{-9}$ s. With decreasing T, τ increases and the system becomes static when τ exceeds the measuring time t, which for the AC susceptibility is in the range 10^{-1}–10^{-5} s. The temperature below which the particles are locked, is the blocking temperature T_B.

$$T_B = \frac{\Delta E}{\ln(\tau/\tau_0)k_B} = \frac{\Delta E}{\ln(\gamma t/\tau_0)k_B}, \qquad \tau = \gamma t, \quad \gamma = 100$$

where τ refers to a much longer relaxation time compared to the measuring time t[19]. As superparamagnets are technologically important in recording material science [39] and biology, as well as biomedicine [59, 60] the magnetic characterization comprises several aspects: (i) the magnetic interactions between the particles, (ii) the size of the particles, and (iii) the variation of T_B as a function of frequency ω.

19) "Much longer" is defined to be $\tau > \gamma t$
where $\gamma = 100$ [5].

As an example the results of AC magnetometrical investigations into a sample of ferritin[20] are presented [62]. The in-phase χ' and out-of-phase component χ'' are plotted in Figure 25.18 as a function of temperature. In the temperature range 30–300 K, χ' follows the Curie law, while $\chi'' \approx 0$ (cf. Section 25.3.2). This result indicates superparamagnetism in the timescale of the experiments and, with regard to item (i), to the absence of magnetic interactions between the particles. For item (ii): from the slope of the $\chi^{-1}-T$ straight line the magnetic dipole density and the particle size of about 7 nm has been derived. The low-temperature behavior permits the answer to item (iii): at $T \leq 30$ K the magnetic behavior is qualitatively similar to the low-temperature behavior of single-molecule magnets (cf. Figure 25.20), as the blocking of magnetic moments leads to a distinct decrease in χ' and the onset of χ''. $T_B(\chi')$ and $T_B(\chi'')$ increase as the frequency $\omega/2\pi$ of the AC field increases.

25.4.4
Spin Glasses

In the scope of spin glasses one distinguishes essentially between metallic and insulating systems. The former are alloys consisting of a nonmagnetic metal with fcc type structure (Cu, Ag, Au, Pt, Pd) and a small fraction (only an "impurity") of a magnetically active 3d (Mn, Fe) or 4f metal (Gd, Er), typified for example by CuMn ($Cu_{1-x}Mn_x$, $x << 1$ [63]). A representative of an insulating spin glass is $Sr_{1-x}Eu_xS$ ($0.1 \leq x \leq 0.5$) [64], adopting the NaCl-type structure[21].

The preconditions for magnetic spin glass behavior are site or bond *disorder* (preventing standard long-range order) and *frustration* on account of competing ferro- and antiferromagnetism spin–spin exchange couplings as a function of distances between the magnetic dipoles[22]. A spin glass is characterized by a well-defined freezing temperature T_f. Above T_f, it exhibits Curie paramagnetism on account of more or less independent spins. On cooling, the spins locally interact to form clusters which, on further cooling, couple among one another via single spins located between the clusters. At T_f, the interactions become more long range and after reaching one of its many ground states the system freezes [5, 67, 68].

Figure 25.19 exhibits the spin glass transition via the cusp in the AC susceptibility for CuMn [63] (0.94 at%), while the fine-scale of the insert reveals a series of peaks shifted downwards in temperature with decreasing frequency of the measurements and accompanied by a small decrease in T_f, a feature that is not present in

20) Ferritin is a protein consisting of a Fe[III] oxide core surrounded by a protein shell [61].

21) Further examples are: (YGd)Al$_2$, (LaGd)B$_6$ (intermetallics); Mg$_{1-x}$Fe$_x$Cl$_2$, (insulating system) [65].

22) In intermetallics the exchange interactions are of RKKY-type (Ruderman–Kittel–Kasuya–Yoshida model [66]) and are mediated by conduction electrons which become spin-polarized, and this polarization couples in turn a neighboring localized magnetic moment. The interaction is rather strong and long range and has an oscillatory dependence on the distance between the magnetic dipoles [5]. By way of contrast the Eu[II]–Eu[II] interactions in the insulating series Sr$_{1-x}$Eu$_x$S are weak and short-range mediated by the sulfide ions between nearest and next-nearest Eu neighbors.

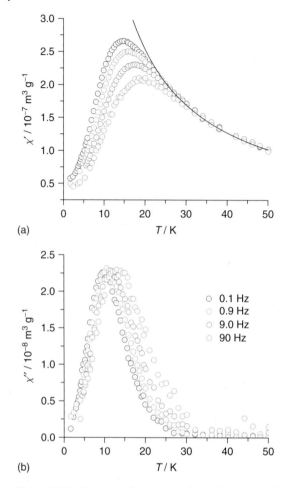

Figure 25.18 Frequency dependence of the AC susceptibility of ferritin. In-phase component: χ'_g–T plot; the continuous line is a least-squares fitting to Curie's law to data at $T \geq$ 30 K. Out-of-phase component: χ''_g–T plot (Adapted from Ref. [60].)

magnetically ordered systems and, therefore, confirms the spin glass phase. In contrast to the susceptibility behavior, the magnetic specific heat shows a very smooth temperature dependence and there is no way of determining T_f [65].

At $T < T_f$, spin glasses show anomalous dynamical characteristics: slow relaxation, irreversibility, memory effects, hysteresis, and aging. For example, nonequilibrium dynamics has been studied in $Ag_{0.89}Mn_{0.11}$ [69] by the out-of-phase component χ'' of AC susceptibility and relaxation experiments after the following sequence: The sample was cooled to 23 K, paused for 2 h at 23 K and subsequently cooled to 20 K. Upon immediate reheating, a dip in χ'' is clearly visible at 23 K,

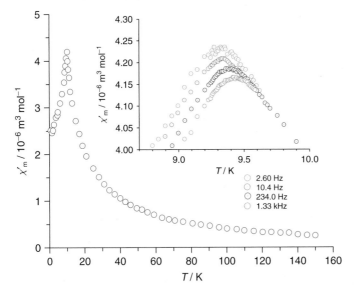

Figure 25.19 Real part, χ', of the AC susceptibility for CuMn (0.94 at.%) as a function of T (234 Hz, AC driving field $H_1 \approx 10^{-4}$ T) and, in the insert, at varying frequencies between 2.6 and 1.33 kHz. The irreversibility in spin glasses leads to a nonzero out-of-phase component, χ'', below T_f (not shown in the figure). Adapted from Ref. [63].

indicating that the system can remember the effect of the relaxation during the wait time.

In the last four decades computer-aided efforts have been undertaken to develop models that describe the thermodynamic properties and the nature of the phase transitions as well as the magnetic ground state of a geometrically frustrated spin lattice [68]. The situation is still under debate [70]. Good promise has been shown by experimental attempts to switch over to systems with spin glass properties that are easier to analyze and computerize than conventional spin glasses: (i) two-dimensional Ising systems with triangular (hexagonal) lattice symmetry and nearest neighbor antiferromagnetic interactions [67], (ii) nanosized mesoscopic systems, and (iii) transition from three-dimensional arrangements to layers and wires [71].

25.4.5
Single-Molecule Magnets

Polynuclear coordination complexes, comprising several magnetically active metal ions, display unique magnetic properties under the condition that they have a (i) large ground spin state S, (ii) large easy axis magnetic anisotropy, and (iii) negligible spin–spin coupling between the molecules. They are candidates for molecular systems with magnetic bistability, that is the ability of isolated so-called

SMMs to remain in one of two states (with opposite directions of the magnetic moment) for a long time at sufficiently low temperatures in the absence of a magnetic field [72] .[23)]

The archetype of the SMMs, $[Mn_{12}O_{12}(O_2CCH_3)_{16}(OH_2)_4]$, displayed as its core $Mn_{12}O_{12}$ fragment in the inset of Figure 25.20, is well characterized with regard to the magnetic properties (cf. Ref. [46] and references cited therein). It comprises eight Mn(III) ions ($S = 2$) and four Mn(IV) ions ($S = \frac{3}{2}$); the ground state $S = 10$ of the complex results from a strong antiparallel coupling of the total of Mn(III) spins ($8 \times 2 = 16$) and Mn(IV) spins ($4 \times \frac{3}{2} = 6$) along the lines of a molecular ferrimagnet.

On account of a uniaxial anisotropy due to H_{so} and H_{lf}, the spin multiplet $S = 10$ splits into ten doublets and one singlet. The effect is described by a zero-field splitting parameter $D < 0$ which gives rise to an energy ladder with $M_S = \pm 10$ as the ground state, followed by $|M_S| < S$ states and finished at the $M_S = 0$ state, that is, at the cusp of the barrier which lies by an amount of $|D|S^2$ above the $M_S = \pm S$ states (Figure 25.21). The axial distortion determines the Ising character of the ground doublet as its magnetic moment is aligned with the anisotropy axis. The states $M_S = \pm 10$ manifest the magnetic bistability, for its minima are separated by a barrier with a height of $|D|S^2$. The switching on of a DC field changes the symmetrical distribution of the M_S states. As the application of the ground state spin Hamiltonian $\hat{H} = D\hat{S}_z^2 - g\mu_B H\hat{S}$ shows, the energy of states with positive M_S decreases while the energy of states with negative M_S increases.

The crucial characteristic of a magnet is the critical temperature below which the magnetization persists. For conventional magnets with an infinite number of spin centers this is the Curie temperature T_C. For a SMM, comprising a limited number of spin centers, the critical temperature is determined by the boundary of the range in which the relaxation time (τ) of the molecule increases rapidly. For the temperature dependence of τ the Arrhenius law is applicable: $\tau = \tau_0 \exp(U/k_B T)$ where $U = |D|S^2$ is the barrier height. Below the blocking temperature T_B (see Section 25.4.3) τ becomes very large. With $\tau_0 = 2 \times 10^{-6}$ s and $U/k_B = 62$ K, the average relaxation time should be several months at 2 K and about 50 years at 1.5 K [78, 79].

With regard to spin–lattice relaxation processes below T_B the repopulation of higher states is hindered by the anisotropy barrier because phonon-induced transitions are limited to $\Delta M_S = \pm 1, \pm 2$, causing slow relaxation of the magnetization, as evidenced in AC susceptibility data (Figure 25.20). However, at very low temperature ($T << T_B$) when only the ground level with $M_S = \pm S$ is populated, relaxation is caused by tunneling transitions between $M_S = S$ and $M_S = -S$ states [79]. In

23) SMMs promise access to dynamic random access memory (RAM) devices for quantum computing and to ultimate high-density memory storage devices in which each bit of digital information might be stored on a single molecule [73–75].

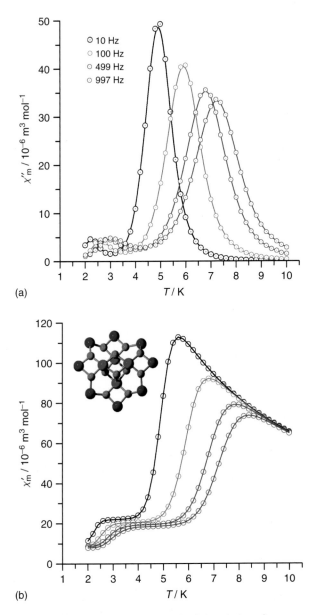

(a)

(b)

Figure 25.20 [Mn$_{12}$O$_{12}$(O$_2$CCH$_3$)$_{16}$(OH$_2$)$_4$]; variation of the real (χ'_m) and imaginary (χ''_m) component of the AC susceptibility vs. *T* measured with frequencies in the range 10–997 Hz. Inset: Mn(III) (blue), Mn(IV) (green), O (red). (Adapted from Ref. [76].)

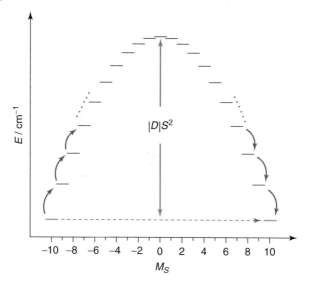

Figure 25.21 $[Mn_{12}O_{12}(O_2CCH_3)_{16}(OH_2)_4]$; Energies $-|D|S_z^2$ of the zero-field split $S = 10$ ground state as a function of S_z in zero applied field yielding an anisotropy barrier. Curved arrows indicate phonon-induced transitions and dashed arrows quantum tunneling transitions. Adapted from Ref. [77]. (See also Figure 5.7, page 174)

this region, the dependence $\tau = \tau_0 \exp(U/k_B T)$ is violated: τ ceases to depend on temperature. For the ground state $M_S = \pm S$, the tunneling rate is very low, but higher for smaller $|M_S|$. However, several quantum tunneling mechanisms exist that enable direct tunneling between positive and negative M_S states if their energies match, which happens at specific applied fields. These tunneling effects can be seen as discontinuous steps at specific fields in the hysteresis curves at low temperatures [46]. Moreover, the presence of not only an axial D term but also nonaxial magnetic anisotropy allows additional direct transitions, effectively lowering the blocking temperature.

25.4.6
Magnetometry at High Fields

25.4.6.1 Tetranuclear Mn (II) System

High field investigations into a molecular tetranuclear $Mn(II)[3d^5]$ system with cuban-like $[Mn_4O_4]$ core [80] have revealed that antiferromagnetic spin–spin exchange couplings between adjacent metal ions are controlled by electron density withdrawal through their bridging carboxylate ligands (acetate, benzoate, or trifluoroacetate). By means of μ_a vs. B plots it has been identified that the Mn_4 unit bridged by acetate shows the strongest antiferromagnetic coupling, while for the benzoate, and, even more so, the trifluoroacetate ligands the antiferromagnetic coupling is weakened in consequence of an increasing electron withdrawal by the ligands.

25.4.6.2 Triangular V(IV) Units

A magnetic complex of vanadium [81] displaying a pair of identical triangular V(IV)[3d^1] units has been investigated with regard to thermal relaxation mechanism. Each triangle consists of three spins $S = \frac{1}{2}$ that interact via isotropic antiferromagnetic exchange resulting in a ground state doublet $S = \frac{1}{2}$ of each triangle. The results of time-resolved magnetization measurements at $T = 1.7$ and 4.2 K and pulsed fields $B < 25$ T (sweep rates of order $1\,\mathrm{T\,ms^{-1}}$) have shown hysteresis loops and magnetization steps for $B \approx 0$. The steps were attributed to adiabatic Landau–Zener–Stückelberg transitions between the lowest magnetic energy levels split by an anisotropic intertriangle exchange interaction of order $-0.1\,\mathrm{cm^{-1}}$. One-phonon resonant transitions among the Zeeman-split doublet of each triangle have been identified as the dominant mechanism underlying the hysteresis behavior.

25.5
Supplementary Material

25.5.1
Complex Functions as Solutions of Differential Equations

To characterize the AC behavior of a system of dipoles the *distribution function* of the moments is necessary. For this purpose, Debye [82] defined a differential equation – originally for dielectric properties – that has the distribution function as solution. A differential equation may be easier to solve when it is comprehended as an equation with a complex variable instead of a real variable. Using a theorem and a simple example the mathematical procedure will be outlined [83]. If a linear homogeneous differential equation, for example, $ay'' + by' + cy = 0$, has a complex function $f(x) = f_1(x) + if_2(x)$ as solution, then both the real part $f_1(x)$ and the imaginary part $f_2(x)$ is a solution. The theorem is proved in such a way:

$$a[f_1 + if_2]'' + b[f_1 + if_2]' + c[f_1 + if_2] = 0 \rightarrow af_1'' + bf_1' + cf_1 + i[af_2'' + bf_2' + cf_2] = 0$$

As the real and imaginary part, $af_1'' + bf_1' + cf_1 = 0$ and $af_2'' + bf_2' + cf_2 = 0$, respectively, do not cancel out, both f_1 and f_2 are separate solutions.

25.5.2
Example

We consider the frictionless motion of a mass m, fixed at a spring, on a horizontal baseplate. In the balanced position the force on the spring is zero, while an extension in length of the spring by x produces a retroactive force, whose absolute value is proportional to x. Assigning D the constant of proportionality, the force is

$-Dx$ which must be equal to the product of mass and acceleration: $-Dx = m\ddot{x}$ and $m\ddot{x} + Dx = 0$, respectively ($\ddot{x} = \mathrm{d}^2x/\mathrm{d}t^2$). As solution of the differential equation the complex function $x = Ae^{i\omega t}$ is suited which, introduced into $m\ddot{x} + Dx = 0$ yields.

$$-mA\omega^2 e^{i\omega t} + DAe^{i\omega t} = 0$$

After reduction by $Ae^{i\omega t}$, $\omega = \pm\sqrt{D/m}$ is obtained. The solutions are the functions $e^{i\omega_0 t}$ and $e^{-i\omega_0 t}$ where $\omega_0 = \sqrt{D/m}$. On account of $e^{i\omega_0 t} = \cos\omega_0 t + i\sin\omega_0 t$ it can thus be concluded that, according to the above mentioned theorem, $\cos\omega_0 t$ and $\sin\omega_0 t$ are also solutions.

Acknowledgment

The author would like to thank Prof. Dr. Paul Kögerler and Prof. Dr. Wolfgang Stahl for helpful discussions and Dr. Manfred Speldrich for graphical settings.

References

1. For informations on SQUID magnetometers see *hyperphysics.phy-astr.gsu.edu/hbase/solids/squid.html* (accessed October 2010).

2. Dormann, J.L., Fiorani, D., and Tronc, E. (1997) in *Advances in Chemical Physics*, vol. **98** (eds I. Prigogine and S.A. Rice), John Wiley & Sons, Inc., pp. 283–494.

3. Peerenboom, J.A.A.J., Wiegers, S.A.J., Christianen, P.C.M., Zeitler, U., and Maan, J.C. (2003) *J. Low Temp. Phys.*, **133**, 181–201.

4. Mills, I., Cvita, T., Homann, K., Kallay N., and Kuchitsu, K. (eds) (1993) *Quantities, Units and Symbols in Physical Chemistry (the 'Green Book')*, 2nd edn, Blackwell Science, Oxford. (see especially Sections 2.3, 7.2, 7.3, 7.4).

5. Blundell, S. (2001) *Magnetism in Condensed Matter*, Oxford University Press, Oxford.

6. Lueken, H. (1999) *Magnetochemie*, Teubner, Stuttgart.

7. Henry, W.E. (1952) *Phys. Rev.*, **88**, 559–652.

8. Van Vleck, J.H. (1932) *The Theory of Electric and Magnetic Susceptibilities*, Oxford University Press, Oxford.

9. Williams, A.F. (1979) *A Theoretical Approach to Inorganic Chemistry*, Springer, Berlin.

10. Görller-Walrand, C. and Binnemans, K. (1996) in *Handbook on the Physics and Chemistry of Rare Earths*, vol. **23**, Chapter 155 (eds K.A. Gschneidner and L. Eyring), Elsevier, Amsterdam.

11. Hellwege, K.-H. (1988) *Einführung in die Festkörperphysik*, Springer, Berlin.

12. Jørgensen, C.K. (1962) *Absorption Spectra and Chemical Bonding in Complexes*, Pergamon Press, Oxford, London, New York, Paris.

13. Sytsma, J., Murdoch, K.M., Edelstein, N.M., Boatner, L.A., and Abraham, M.M. (1995) *Phys. Rev. B*, **52**, 12668–12676.

14. Schläfer, H.L. and Gliemann, G. (1967) *Einführung in die Ligandenfeldtheorie*, Akademische Verlagsgesellschaft, Frankfurt; English version: (1969) *Basic Principles in Ligand Field Theory*, Wiley-Interscience, New York.

15. Mabbs, F.E. and Machin, D.J. (1973) *Magnetism and Transition Metal Complexes*, Chapman and Hall, London.

16. Griffith, J.S. (1971) *The Theory of Transition-Metal Ions*, Cambridge University Press, Cambridge.

17. Schilder, H., Lueken, H. (2004) *J. Magn. Magn. Mater.*, **281**, 17–26 (PROGRAM CONDON is free software, covered by the GNU General Public Licence), *http://www.condon.fh-aachen.de* (accessed October 2010).

18. (a) Urland, W. (1976) *Chem. Phys.*, **14**, 393–401; (b) Urland, W. (1977) *Chem. Phys. Lett.*, **46**, 457–460.

19. Abragam, A. and Bleaney, B. (1970) *Electron Paramagnetic Resonance of Transition Ions*, Clarendon Press, Oxford.

20. (a) Eifert, T., Hüning, F., Lueken, H., Schmidt, P., and Thiele, G. (2002) *Chem. Phys. Lett.*, **364**, 69–74; (b) Eifert, T. (2001) Software 'HTSE Package', RWTH Aachen University,

21. Rushbrooke, G.S. and Wood, W.J. (1958) *Mol. Phys.*, **1**, 257–283.

22. Handrick, K. (1991) Modelle zur Beschreibung magnetischer Wechselwirkungen zwischen paramagnetischen Zentren in niedrigdimensionalen Systemen. Dissertation. RWTH Aachen University, Reihe Chemie D(82), Shaker.

23. Eifert, T., Handrick, K., Hüning, F., Neuhausen, U., Schilder, H., and Lueken, H. (2006) *Z. Anorg. Allg. Chem.*, **632**, 521–529.

24. (a) Borrás-Almenar, J.J., Clemente-Juan, J.M., Coronado, E., and Tsukerblat, B.S. (1999) *Inorg. Chem.*, **38**, 6081–6088; (b) Borrás-Almenar, J.J., Clements-Juan, J.M., Coronado, E., and Tsukerblat, B.S. (2001) *J. Comput. Chem.*, **22**, 985–991.

25. Borrás-Almenar, J.J., Cardona-Serra, S., Clemente-Juan, J.M., Coronado, E., Paulii, A.V., and Tsukerblat, B.S. (2010) *J. Comput. Chem.*, **31**, 1321–1332.

26. Hatscher, S., Schilder, H., Lueken, H., and Urland, W. (2005) *Pure Appl. Chem.*, **77**, 497–511; German version: (2006) *Angew. Chem.*, **118**, 8233–8240.

27. Bleaney, B. and Bowers, K.D. (1952) *Proc. R. Soc. London, Ser.A*, **214**, 451–465.

28. Smart, J.S. (1966) *Effective Field Theories of Magnetism*, Saunders, Philadelphia.

29. Morrish, A.H. (1965) *The Physical Principles of Magnetism*, John Wiley & Sons, Inc., New York, London, Sydney.

30. Bleaney, B.I. and Bleaney, B. (1994) *Electricity and Magnetism*, 3rd edn, Oxford University Press, Oxford.

31. Kneller, E. (1962) *Ferromagnetismus*, Springer, Berlin.

32. Lueken, H. (2008) *Angew. Chem. Int. Ed.*, **47**, 8562–8564.

33. Gerhard, L., Yamada, T.K., Balashov, T., Takács, A.F., Wesselink, R.J.H., Däne, M., Fechner, M., Ostanin, S., Ernst, A., Mertig, I., and Wulfhekel, W. (2010) *Nat. Nanotechnol.*, **5**, 792–797.

34. Néel, L. (1948) *Ann. Phys. (Paris)*, **3**, 137–198.

35. Authier, A. (ed.) (2003) *International Tables for Crystallography*, International Union of Crystallography, Vol. **D**, Kluwer Academic Publishers, Dordrecht, Boston, London; section 1.5 Borovik-Romanov, A.S. and Grimmer, H. *Magnetic Properties*.

36. Erickson, R.A. (1953) *Phys. Rev.*, **90**, 779–785.

37. Ohkoshi, S.-I., Abe, Y., Fujishima, A., and Hashimoto, K. (1999) *Phys. Rev. Lett.*, **82**, 1285–1288.

38. Kahn, O. (1999) *Nature*, **399**, 21–23.

39. Spaldin, N. (2003) *Magnetic Materials, Fundamentals and Device Applications*, Cambridge University Press.

40. Reiff, W.M. (1994) *Am. Lab.* **24**, 26–35.

41. Clarke, J. (1990) SQUIDs: principles, noise, and applications, in *Superconducting Devices* (eds S.T. Ruggiero and D.A. Rudman), Academic Press, San Diego, Chapter 2.

42. Chapelier, C., El Khatib, M., Perrier, P., Benoit, A., Mailly, D., Koch, H., and Lübbing, H. (eds) (1993) *SQUID91, Superconducting Devices and their Applications*, Springer, Berlin, pp. 286–291.

43. Wernsdorfer, W., Hasselbach, K., Benoit, A., Barbara, B., Mailly, D., Tuaillon, J., Perez, J.P., Dupuis, V., Dupin, J.P., Giraud, G., and Perex, A. (1995) *J. Appl. Phys.*, **78**, 7192–7195.

44. Wernsdorfer, W. (2001) *J. Adv. Chem. Phys.*, **118**, 99–190.

45. Wernsdorfer, W. (2009) *Supercond. Sci. Technol.*, **22**, 1–13.

46. (a) Gatteschi, D., Sessoli, R., and Villain, J. (2006) *Molecular Nanomagnets*, Oxford University Press, Oxford; (b) Gatteschi, D., Sessoli, R., Miller, J.S., and Drillon, M. (eds) (2002) *Magnetism: Molecules to Materials III*, Wiley-VCH Verlag GmbH, pp. 63–108.

47. Casimir, H.B.J. and Du Pré, F.K. (1938) *Physica*, V, 507–511.

48. Mc Connell, J. (1980) *Rotational Brownian Motion and Dielectric Theory*, Academic Press, New York.

49. (a) Cole, R.H. (1938) *J. Chem. Phys.*, 6, 385–391; (b) Cole, K.S. and Cole, R.H. (1941) *J. Chem. Phys.*, 9, 341–351.

50. Dekker, C., Arts, A.F.M., de Wijn, H.W., van Duyneveldt, A.J., and Mydosh, J.A. (1989) *Phys. Rev. B*, 40, 11243–11251.

51. Zherlitsyn, S., Hermannsdoerfer, T., Skourski, Y., Sytcheva, A., and Wosnitza, J. (2006) *J. Phys.: Conf. Ser.*, 51, 583–586.

52. Mulay, L.L. and Boudreaux, E. A. (eds) (1976) *Theory and Applications of Molecular Diamagnetism*, John Wiley & Sons, Inc., New York.

53. Schieber, M.M. (1967) *Experimental Magnetochemistry*, North-Holland Publishing Co., Amsterdam.

54. Kuz'ma, Yu.B. and Yarmolyuk, Ya.P. (1971) *Zh. Strukt. Khim.*, 12, 458–461.

55. Nagelschmitz, E.A., Jung, W., Feiten, R., Müller, P., and Lueken, H. (2000) *Z. Anorg. Allg. Chem.*, 627, 523–532.

56. (a) Dronskowski, R., Korczak, K., Lueken, H., and Jung, W. (2002) *Angew. Chem.*, 114, 2638–2642; (2002) *Angew. Chem. Int. Ed.*, 41, 2528–2532; (b) Dronskowski, R. (2005) *Computational Chemistry of Solid State Materials*, Wiley-VCH Verlag GmbH, Weinheim; (c) Samolyuk, G.D., Fokwa, B.P.T., Dronskowski, R., and Miller, G.J. (2007) *Phys. Rev. B*, 76, 094404-1–094404–12; (d) Fokwa, B.P.T., Lueken, H., and Dronskowski, R. (2007) *Chem. Eur. J.*, 13, 6040–6046.

57. (a) Néel, L. (1949) *C. R. Hebd. Seances Acad. Sci.*, 228, 664; (b) Néel, L. (1949) *Ann. Géophys.*, 5, 99–136.

58. Brown, W.F. (1963) *Phys. Rev.*, 130, 1677–1686.

59. Andrä, W. and Nowak, H. (eds) (2007) *Magnetism in Medicine*, 2nd edn, Wiley-VCH Verlag GmbH.

60. Pankhurst, Q.A., Conolly, J., Jones, S.K., and Dobson, J. (2003) *J. Phys. D: Appl. Phys.*, 36, R167–R181.

61. Massover, W.H. (1993) *Micron*, 24, 389–437.

62. Luis, F., del Barco, E., Hernánez, J.M., Remiro, E., Bartolomé, J., and Tejada, J. (1999) *Phys. Rev. B*, 59, 11837–11846.

63. Mulder, C.A.M., van Duynefeldt, A.J., and Mydosh, J.A. (1981) *Phys. Rev. B*, 23, 1384–1396.

64. Maletta, H. and Zinn, W. (1989) in *Handbook on the Physics and Chemistry of the Rare Earths*, vol. 12 (eds K.A. Gschneidner Jr. and L. Eyring), North-Holland Publishing Co., Amsterdam, pp. 213–356.

65. Mydosh, J.A. (1986) *Hyperfine Interact.*, 31, 347–362.

66. (a) Rudermann, M.A. and Kittel, C. (1954) *Phys. Rev.*, 96, 99–102; (b) Kasuya, T. (1956) *Prog. Theor. Phys.*, 16, 45–57; (c) Yoshida, K. (1957) *Phys. Rev.*, 106, 893–898.

67. Mydosh, J.A. (1996) *J. Magn. Magn. Mater.*, 157–158, 606–610.

68. Newman, C.M. and Stein, D.L. (2003) *J. Phys.: Condens. Matter*, 15, R1319–R1364.

69. Jonsson, T., Jonason, K., Jonsson, P., and Nordblad, P. (1999) *Phys. Rev. B*, 59, 8770–8777.

70. Krzakala, F., Ricci-Tersenghi, F., and Zdeborová, L. (2010) *Phys. Rev. Lett.*, 104, 207208-1–207208-4.

71. van der Post, N., Mydosh, J.A., and van Ruitenbeek, J.M. (1996) *Phys. Rev. B*, 53, 15106–15112.

72. Kahn, O. and Martinez, C.J. (1998) *Science*, 279, 44–48.

73. Mironov, V.S. (2006) *Dokl. Phys. Chem.*, 408 (Part 1), 130–136.

74. Coronado, E. and Gatteschi, D. (2006) *J. Mater. Chem.*, 16, 2513–2515.

75. Cirera, J., Ruiz, E., Alvarez, S., Neese, F., and Kortus, J. (2009) *Chem. Eur. J.*, 15, 4078–4087.

76. Sessoli, R., Gatteschi, D., Caneschi, A., and Novak, M.A. (1993) *Nature*, 365, 141–143.

77. Pohjola, T. and Schoeller, H. (2000) *Phys. Rev. B*, 62, 15026–15041.

78. Gatteschi, D., Caneschi, A., Pardi, L., and Sessoli, R. (1994) *Science*, 265, 1054–1058.

79. Gatteschi, D. and Sessoli, R. (2003) *Angew. Chem. Int. Ed.*, 42, 269–297.

80. Kampert, E., Janssen, F.B.J., Boukhvalov, D.W., Russcher, J.C., Smits, J.M.M., de

Gelder, R., de Bruin, B., Christianen, P.C.M., Zeitler, U., Katsnelson, M.I., Maan, J.C., and Rowan, A.E. (2009) *Inorg. Chem.*, **48**, 11903–11908.

81. Rousochatzakis, I., Ajiro, Y., Mitamura, H., Kögerler, P., and Luban, M. (2005) *Phys. Rev. Lett.*, **94**, 147204-1–147204-4.

82. (a) Debye, P. (1929) *Polare Molekeln*, Kap. V, Hirzel, Leipzig. (b) (1945) *Polar Molecules*, Chapter V, Dover Publications, New York.

83. (a) Zachmann, H.G. (1994) *Mathematik für Chemiker*, 5th enlarged edn, Wiley-VCH Verlag GmbH, Weinheim, pp. 515–521; (b) Sokolnikoff, I.S. and Sokolnikoff, E.S. (1941) *Higher Mathematics for Engineers and Physicists*, McGraw-Hill, New York, London.

26
Transmission Electron Microscopy

Hans-Joachim Kleebe, Stefan Lauterbach, and Mathis M. Müller

■ **Method Summary**

Acronyms, Synonyms
- Transmission electron microscopy (TEM)
- Conventional transmission electron microscopy (CTEM)
- Analytical transmission electron microscopy (ATEM)
- Bright field/dark field (BF/DF)
- Convergent beam electron diffraction (CBED)
- High resolution transmission electron microscopy (HRTEM).

Benefits (Information Available)

- Imaging (bright field, dark field, and high resolution)
- Distinction between crystalline and amorphous material
- Imaging of crystal defects (e.g., line defects, planar defect, interfaces)
- Diffraction (selected area diffraction, convergent beam electron diffraction)
- High lateral resolution ($\sim$0.1 nm) in both imaging and analysis.

Limitations (Information Not Available)

- Liquid or gaseous samples cannot be examined
- Single point defects (vacancies, Schottky, Frenkel) cannot be imaged
- Spectroscopy (EDS, EELS) requires additional equipment.

26.1
Introduction

Transmission electron microscopy (TEM) is a widely used characterization technique for solid samples, where a beam of electrons is, in contrast to scanning electron microscopy (SEM), *transmitted* through a very thin foil of the material of interest. SEM also utilizes an electron beam which, however, is *scanned across* a solid, thick sample surface. Therefore, TEM imaging allows the detection of

Methods in Physical Chemistry, First Edition. Edited by Rolf Schäfer and Peter C. Schmidt.
© 2012 Wiley-VCH Verlag GmbH & Co. KGaA. Published 2012 by Wiley-VCH Verlag GmbH & Co. KGaA.

structures and details that are embedded within the material, that is, features that are not necessarily visible on the sample surface. Moreover, due to the short wavelength of the electrons, the resolution of state-of-the-art instruments nowadays reaches $1.0\,\text{Å} = 10^{-10}\,\text{m}$ and even below, which allows imaging on the atomic level. Due to the ultra-high resolution capability and to the possibility of performing chemical analysis with a similar high lateral resolution, this technique has become a powerful characterization tool that is employed not only in nanotechnology but also throughout a wide range of research fields in natural sciences [1–3].

The era of TEM started slowly with the discovery of the electron by Sir J.J. Thomson in 1897 (Nobel Prize in Physics in 1906); however, it still took until 1931, when Max Knoll and Ernst Ruska (Nobel Prize in Physics in 1986) first presented their newly developed TEM. The resolving power of this first instrument was however lower than that of an optical microscope, but by 1933, they crossed this magic line, which led to the development of the first commercial TEM in 1939 by Siemens. Since then, major steps in design and reliable performance of TEM instruments have been achieved, leading to a worldwide application of this exciting and intriguing imaging and analysis method. It should be noted that the commonly used analytical techniques of energy-dispersive X-ray spectroscopy (EDS) and electron energy-loss spectroscopy (EELS) are not covered in this chapter on TEM; for more details, the reader is referred to corresponding textbooks [4–10]; or corresponding chapters in this book (e.g., those on X-ray diffraction, neutron diffraction, electron diffraction at surfaces, and EELS).

26.1.1
Why Electrons?

In 1873, Ernst Karl Abbe published his fundamental work on the resolution limit of optical microscopes. The resolution is defined as the minimum distance of two structural elements, as for example neighboring atoms within a unit cell, which can be imaged as two individual objects: $d_{min} = \lambda/\text{NA}$. Here, NA is the numerical aperture, defined as the sine of the half aperture angle α multiplied by the refractive index n of the medium between the objective and the object. Lambda stands for the wavelength of light and, hence, the following correlation between resolution and wavelength was deduced by Abbe:

$$d_{min} = \frac{\lambda}{n \cdot \sin \alpha} \tag{26.1}$$

It is important to note that this general concept introduced by Abbe for optical microscopy in the late nineteenth century also holds for nowadays electron microscopes; however, with one major variance: the wavelength of electrons (de Broglie wavelength; Prince Louis-Victor Pierre Raymond de Broglie, Nobel Prize in Physics in 1929) is approximately 6 orders of magnitude shorter than the wavelength of light (380–750 nm). Depending on the acceleration voltage utilized in modern instruments, the electron wavelength λ ranges between 0.004 and 0.00087 nm (cf. Equation 26.2):

$$\lambda_{el} = \frac{h}{\left[2m_0eE\left(\dfrac{eE}{2m_0c^2}\right)\right]^{1/2}} \tag{26.2}$$

where h is the Planck's constant, m_0 the rest mass of the electron, e the electron charge, E the potential difference (acceleration voltage measured in volts), and m_0c^2 the rest energy. For a rough estimation of the wavelength expression $\lambda_{el} = 1.22/(E)^{1/2}$ can be used. Although the diffraction angles are smaller for electrons than for light, the noticeably shorter wavelength of electrons results in the higher resolving power of modern electron microscopes. Moreover, this correlation rationalizes why in the late 1980s TEM instruments were manufactured with extreme accelerating voltages, as for example the atomic resolution microscope (ARM) in Stuttgart with an acceleration voltage of 1250 kV giving a resolution limit of 0.1 nm.

A variety of signals are *simultaneously* generated during the *electron beam specimen interaction* such as the generation of Auger and secondary electrons, X-rays (EDS), or cathodoluminescence (CL). The electron scattering can occur in different ways: elastic coherent (electron diffraction at low angles), elastic incoherent (Rutherford scattering at high angles), or inelastic, which is always incoherent (EELS). Due to the electron transparent nature of the TEM specimen, forward scattered electrons can also be collected. In fact, the high accelerating potential causes most of the scattering to occur in a forward direction, with only a rather low fraction of backscattered electrons for typical TEM sample thicknesses.

26.2
Components of a TEM Instrument

In general, a TEM is similar to a light microscope (LM), just turned upside down. There is a light source (TEM = electron gun) at the top (LM = at the bottom). Then there is the sample and an objective lens and finally a viewing screen at the bottom of the TEM (your eye at the top of the LM). The most important differences are the light source and the lenses. While the light for the optical microscope is normal visible light, TEM uses fast electrons. In optical microscopes glass lenses are used, while in electron microscopes electromagnetic lenses are employed to control the electron beam. It should be noted that the upper half of a TEM is very similar to that of a SEM. There is a similar electron gun, an accelerating potential (although this is much higher in a TEM with 100–300 kV) and condenser lenses (typically 2 or 3) to control and collimate the beam onto the sample. In an SEM the final lens before the specimen (the objective) focuses the electron beam onto the sample surface, while in a TEM additional electromagnetic lenses are activated, depending on the specific performance: an intermediate lens (electron diffraction) and projector lenses (high-resolution imaging), see also Figure 26.1.

As depicted in Figure 26.1, a TEM is based on three main parts. From top to bottom there is the gun, the column, including the condenser system, the objective

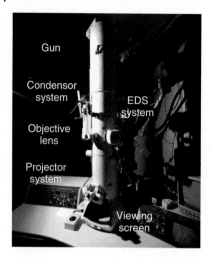

Figure 26.1 Photograph of a modern analytical TEM showing the main components: gun, condenser system, objective lens, projector system, viewing screen, and a CCD camera, in addition to the analytical EDS system.

lens, and the projector system, as well as the camera chamber (or CCD camera). The gun generates the electrons and accelerates them to the nominal voltage. Common emitters for electrons are tungsten or LaB_6 filaments (cathode) for thermionic emission via resistance heating, or a field emission gun (FEG). The beam current density increases strongly from a tungsten cathode to a FEG emitter; the latter can be a so-called cold FEG or a Schottky-type emitter ("warm" FEG). The condenser system is used to condense/focus the beam down to a certain spot size (probe) for specimen illumination. Most TEMs are equipped with two independent condenser lenses labeled C1 and C2. The C1 lens controls the spot size, while C2 is used to focus the beam on the sample for local analysis, with a small spot yielding high lateral resolution, or to form a parallel beam for conventional or high-resolution imaging.

The condenser system is followed by the objective lens, often termed the "heart" of the microscope. The objective lens forms an image from beams that have been scattered as they pass through the specimen. The quality of the electron optics inside the objective lens controls the image quality of the TEM. In general, the electromagnetic lenses obey the same rules as glass lenses. The magnification M is defined by: $M = v/u$; where v is the distance from the center of the lens to the first image plane and u is the distance from the object plane to the center of the lens. To achieve the high magnification, needed to image the atomic nature of solid samples, u has to be very small. The technical solution to reduce u as far as possible is to place the sample inside the objective lens, or to split the lens and put the sample between an upper and a lower polpiece. This technique also ensures that the magnetic field surrounding the sample is homogeneous, resulting in good image quality. The major drawback is the severely restricted maximum size of the specimen holder utilized and the sample size (3 mm in diameter and ~1 mm in depth). But even for such a limited space inside the objective lens, a variety of holders for experimental purposes are available, such as cooling and heating holders, electrical stages (to apply voltage and current), straining stages, double-tilt holders (alignment of specific crystallographic directions; zone axes), or nanoindenters.

Depending on whether an image or a diffraction pattern is formed, the lenses and apertures of the microscope are operated in different ways. As can be seen from Figure 26.2, displaying a schematic drawing of the ray path in a TEM in diffraction (a) and imaging (b) mode. An intermediate lens (system) is mounted below the objective lens. This lens makes the TEM absolutely unique! By changing the magnetic strength of the lens (by simply changing the lens current), one can either form a magnified image of the sample on the viewing screen (observation in real space) or generate an electron diffraction pattern on the screen (observation in reciprocal space). Thus, TEM has the main advantage that it enables the user to gain complementary information from real *and* reciprocal space! The projector system is a multi-lens system and controls the final magnification of the image formed on the viewing screen.

Another option to add scan coils to the TEM column, which allows scanning of the incident beam over a small sample region instead of employing a stationary electron beam. Here, the transmitted signal of the scanned beam is monitored below the sample in the so-called scanning transmission electron microscopy

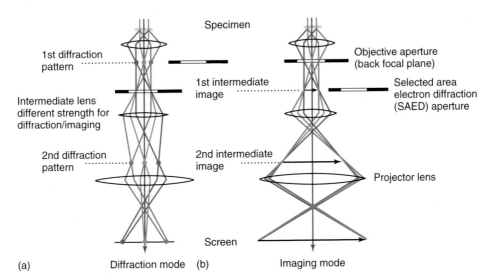

Figure 26.2 Schematic ray diagrams in a TEM column showing the operation of the post-specimen lenses to form either an image or a diffraction pattern on the viewing screen. The back focal and image planes of each lens are also shown.

(STEM) mode. Various detectors are available: bright field (BF) and dark field (DF) detectors record the transmitted and scattered electrons, while a high-angle annular dark-field (HAADF) detector allows the detection of high angle scattering events, which, similar to the back scattered electron (BSE) signal in a SEM, is sensitive to the atomic number of the elements composing the specimen (Z-contrast).

26.2.1
Lens Aberrations

The *theoretical resolution* of a TEM is only restricted by λ, n, and α, the electron wavelength, the refractive index of the viewing medium (here $n = 1$ for vacuum), and the semi-angle of collection (α in mrad) of the magnifying lens, respectively (see Equation 26.1). In reality, however, lens aberrations strongly limit the resolution of the instrument. There are four main lens aberrations which limit the final resolution of the TEM:

1) **Diffraction aberration**, which, according to the Rayleigh theorem, images a point source always as a small disk with radius r_{diff}:

$$r_{\text{diff}} = 0.61 \frac{\lambda}{\alpha} \qquad (26.3)$$

2) **Chromatic aberration, C_c;** electrons that are generated in the gun do not all have the same energy and hence vary in wavelength, they are, similar to the dispersion of visible light, refracted with different angles by an electromagnetic lens. Since the electron beam has to pass through the sample, the specimen–beam interaction causes an even more pronounced variation in electron energy $\Delta E/E$. As a consequence, a point source is imaged as a small disk with radius r_c:

$$r_c = C_c \frac{\Delta E}{E} \alpha \qquad (26.4)$$

3) **Spherical aberration, C_s;** the magnetic field inside the central hole of the lens is stronger at the rim than in the center (optic axis). Therefore, electrons traveling not along the optical axis are deflected more strongly the further they are away from the optic axis, resulting in a small disk with radius r_s:

$$r_s = C_s \alpha^3 \qquad (26.5)$$

4) **Lens astigmatism** occurs due to mechanical lens imperfections. The magnetic field of the lens is not perfectly homogeneous and, as a consequence, there are two different focal lengths for the meridional and sagittal plane. Thus, a point in the image will be projected as a thin line. By combining such a lens with two quadrupole lenses, the lens astigmatism can be corrected.

The three lens aberrations r_{diff}, r_s, and r_c all depend on α. For a very thin sample, C_c can be neglected at first approximation, thus combining Equations 26.3 and 26.5, an optimal α for the best imaging conditions can be calculated. With this value, the experimental resolution r_{min} of the TEM is given by:

$$r_{min} \approx 0.91 \left(C_s \lambda^3\right)^{1/4} \qquad\qquad (26.6)$$

The recent development of *aberration correctors* such as the C_s *corrector* for the objective or condenser lens systems, allowing for the correction of the spherical aberration of the electromagnetic lens, yielded a greatly improved resolving power of modern state-of-the-art TEMs, even down to 0.05 nm. For optical microscopy, the correction of C_s has been available for a long time by inserting one additional lens with a negative C_s value into the ray path, correcting for the positive C_s of the objective lens. This method, however, cannot be realized for the objective or condenser lens system in a TEM, because there is no rotational symmetric single pole electromagnetic lens with a negative C_s value. In 1947, not too long after the invention of the TEM, Prof. Otto Scherzer (TH Darmstadt) started to develop the concept of C_s aberration correction. Later, in the early 1980s, Prof. Harald Rose (TU Darmstadt) set the theoretical background for the construction of a C_s corrector for TEM. Together with Dr. Max Haider he built the first C_s-corrected TEM in Heidelberg in 1998 [11–14]. The C_s corrector is a combination of several multipole electromagnetic lenses. A quadrupole stretches the electron beam into a line crossover that is affected by the divergent field of an octupole, resulting in a spherical corrected exit beam. A second set of multipole lenses is needed to correct for the orthogonal direction. The most recent development in lens aberration correction focuses on the correction of the *chromatic aberration* C_c. With the possibility to correct both C_s and C_c lens aberrations, the resolving power of the TEMs nearly reaches the theoretical resolution limit.

26.3
Specimen Preparation

An elementary requirement for TEM investigations is the electron transparency of the sample. The maximum thickness of the electron transparent areas typically varies between 20 and 100 nm. In particular, when high-resolution imaging or local EELS analysis is anticipated, very thin TEM foils need to be prepared. Therefore, sophisticated specimen preparation techniques are employed worldwide. The preparation of such thin samples can be realized by mechanical and chemical methods, but also by plasma-etching techniques or by electrochemical thinning. The method of choice depends strongly on the material of interest. The intended result is an electron transparent sample, free of any preparation artifacts, still revealing the characteristic micro-, nano-structural features of the original material.

The most common preparation technique is the combination of an automatic grinding (dimpling) procedure, followed by Ar^+-ion milling. Owing to the specific geometry of all TEM sample holders being 3 mm in diameter, the sample is, in a first step, shaped into a 3 mm disk by ultrasonic disk cutting. In a second step, one surface of the disk is ground to remove contamination and polished to create a flat surface; ideally without any scratches. The selection of the abrasive depends on the scientific question to be addressed, since contamination from the abrasive material can occur during mechanical grinding, in particular, when porous materials are prepared. Commercially available abrasives are B_4C, SiC, Al_2O_3, and SiO_2, while the final polish is usually done with diamond-coated lapping films or diamond pastes. In a third step, the sample is ground and polished down from the other side to a final thickness of approximately 10–20 μm. This can be realized either by creating a thin section by manual polishing or a small dimple in the center of the disk, using a so-called *dimple grinder*. As noted earlier, a thickness of 20–100 nm is required for TEM imaging or chemical analysis and, hence, the final step of Ar^+-ion-milling is essential. Thereby, Ar^+-ions are accelerated in an electric field (1–5 kV) toward the sample surface at a small angle of 2–15° where they remove small amounts of material, subsequently creating a perforation in the sample with electron transparent areas near its edges. A schematic of the Ar^+-ion milling procedure is shown in Figure 26.3.

Major drawbacks using this type of TEM-foil preparation are the mechanical stress the sample is subjected to while grinding and the damage introduced by the relatively heavy Ar^+-ions on the sample surface. The latter causes an amorphous thin layer on the surface, deteriorating the image quality, in particular for high resolution TEM. Therefore, a second Ar^+-ion thinning step at a lower acceleration voltage is used to partly remove the amorphous surface layer. Finally, samples that are not conductive need to be lightly carbon-coated in order to avoid charging under the incident electron beam.

The method described is normally used for ceramics and other non-metallic materials, where mechanical stress does not affect the material that much. In contrast, for metallic materials this technique is not applicable since shear stresses during grinding can cause, for example, the formation of dislocations, thus altering the original structural information of the sample. To avoid the introduction of

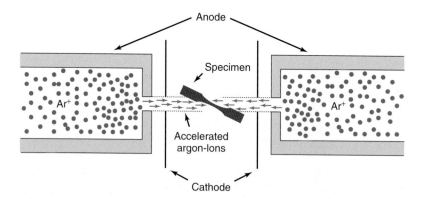

Figure 26.3 Schematic revealing the Ar^+-ion thinning process. The sample rotates while Ar^+-ions bombard the top and bottom side of the sample, continuously removing small amounts of material until a tiny hole is generated in the center and the sample becomes electron transparent at the rim of the perforation.

artifacts, metallic materials are therefore prepared by *electrochemical thinning*. For this purpose, a disk, 3 mm in diameter, is cut out of a thin metal foil and mounted in an electrolytic bath, connecting the sample either to the cathode or anode, depending on the metal prepared. The material around the region of interest should be masked by a varnish to protect and stabilize the TEM sample. The composition of the bath and the voltage used depends on the material and can be looked up in the appropriate literature [15]. Similar to ion-milled samples, a small hole is created in the center of the metal foil, where the edge is thin enough for TEM observations. Figure 26.4 shows three examples of TEM samples prepared by regular Ar$^+$-ion milling and upon electrochemical thinning.

These two types of preparation are very effective in achieving high quality TEM foils for the majority of materials and scientific questions. It should be noted that, for modern TEM, the demands on sample quality have increased as much as the performance of modern instruments has improved, which has led to the development of novel preparation techniques.

With new applications in materials science such as surface coatings, multilayer systems for novel solar cells, and the rapid downscaling in semiconductor fabrication, the preparation of cross-sections and target preparations of defined sample areas becomes more and more important. For cross-sections of multilayer systems or surface coatings, an already known technique, which had fallen into oblivion, was revitalized with high-end equipment – the *tripod polisher*. This ingenious device enables the operator to polish a wedge-shaped sample to such an extent that the edge of the wedge is electron transparent. This is done by means of a special holder allowing both plane-parallel polishing of the surfaces of the sample for defined initial conditions and polishing at a shallow angle of 0.4–0.6° to form the wedge. The process is controlled with an optical microscope by checking for the occurrence of refraction fringes parallel to the edge of the sample, indicating that the wedge is thin enough to be electron transparent. The finished sample is glued to a standard TEM grid for mechanical support and, depending on the material, coated with

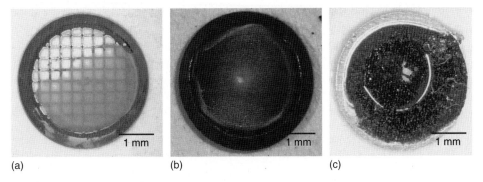

(a) (b) (c)

Figure 26.4 Optical micrographs of thinned TEM samples; (a) an Ar$^+$-ion milled thin section, (b) an Ar$^+$-ion milled dimpled sample, and (c) upon electrochemical thinning.

carbon similar to regularly Ar^+-ion thinned samples. The major advantage of this method is that Ar^+-ion thinning is not required.

While the tripod polisher provides access to cross-sections and samples with large electron transparent areas, it is more or less impossible to predict precisely where these transparent areas will occur and, hence, to select the location of the region of interest within the sample. This drawback in TEM specimen preparation was eliminated with the invention of the *focused ion beam* (FIB). A FIB can be imagined as a normal SEM but instead of electrons, accelerated Ga^+-ions are scanned across the surface. The interaction of these heavy ions with the sample surface sputters off material and allows precise cutting of thin sections. Moreover, sufficient secondary electrons are generated to image the sample surface and the specimen preparation process. As the Ga^+-ion beam is controlled and focused with electromagnetic lenses and deflection coils similar to the electron beam in the SEM, the user has the option to choose precisely the location where sputtering takes place. By controlling the beam via specialized software and hardware, a thin TEM foil can be cut out of the sample from the specific region of interest. Modern FIB machines are equipped with a second column for electrons to control and image the ongoing cutting process; so-called *dual-beam* or *cross-beam FIB*. The Ga^+-ions, similar to the Ar^+-ions, also generate an amorphous surface layer which can be removed by gently milling with low energy Ga^+-ions in a finishing step inside the FIB, externally with Ar^+-ions or internally using a *triple-beam FIB*. The final cross section is mounted on a TEM grid to allow handling, since these foils are strongly limited in size to approximately 5 μm by 20 μm.

26.4
Electron Diffraction

The formation of a *diffraction pattern* on the fluorescent screen is basically related to the intermediate lens current (focal length) which is adjusted in such a way that the plane of the screen is conjugate with the back focal plane of the objective lens. A *selected area electron diffraction* (SAED) aperture can be used to select a specific region of interest from which the pattern is formed. This aperture is inserted in the first image plane, which itself is conjugate with the object plane. The SAED aperture is de-magnified back to the object plane by approximately a factor of 50× so that a region of the sample of less than 1 μm in size can be selected. In this way, electron diffraction allows a direct view of the reciprocal lattice of any given crystalline structure by "simply pressing a button" (cf. also Figure 26.2, which shows a schematic of the ray diagrams for imaging versus diffraction).

The periodic structure of any crystalline material, viewed along a specific zone axis, acts as a diffraction grating, reflecting the characteristic symmetry of the crystal along the specific viewing direction, for example, the threefold symmetry along the [111] direction of a cubic crystal. Moreover, if characteristic extinctions are observed, the presence of glide planes and/or a screw axis can be determined. It is important to note that this diffraction technique also allows unequivocal distinction between crystalline and *amorphous materials*, which otherwise is a rather difficult task, in particular, when only small regions filled with amorphous residue are present in the material. A diffraction pattern from an amorphous phase will be a ring pattern, but the rings will be rather diffuse (Figure 26.5a). Such a diffraction pattern corresponds to the radial distribution function (RDF), which reflects the probability of finding an atom at a certain distance away from neighboring atoms. The diffuseness of such ring patterns is a result of the absence of long-range order in amorphous materials. Diffraction from *polycrystalline materials*, with a large number of randomly oriented small crystallites, can be described by rotating the individual crystals (and thus their reciprocal lattice) about all axes to produce a set of consecutive diffraction cones. When intersected by the *Ewald sphere*, continuous rings of reflections are generated, as shown in Figure 26.5b, while discrete diffraction spots are seen, illuminating a single-crystalline area (Figure 26.5c).

If the beam, however, is focused to a spot on the sample, a *convergent beam electron diffraction* (CBED) pattern is formed in the back focal plane of the objective lens. This is often called μ-*diffraction* or *nano-diffraction*, due to the extremely small volume sampled by the incident electron beam. In this case, the diffraction pattern is composed of disks rather than spots and the detail within the disks is sensitive to the crystal symmetry (Figure 26.6). Moreover, local strains within the crystal which, for example, can arise from intrinsic crystal defects are detectable with this technique.

The crucial difference with the convergence of the incident illumination is the excitation of different diffraction vectors $\mathbf{g}_{(hkl)}$. While, with parallel illumination, typically only diffraction vectors are excited that do not contain a component parallel to the incident electron beam (*zero order Laue zone*, ZOLZ), a higher convergence angle results in numerous directions of incidence (illumination cone) and, as a consequence, the excitation of additional diffraction vectors with their component along the optical axis of the microscope (parallel to the electron beam), so-called *higher order Laue zones* (HOLZ). Therefore, CBED allows for the reconstruction of

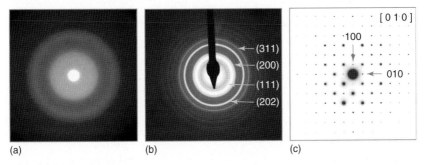

(a)　　　　　　(b)　　　　　　　　(c)

Figure 26.5 SAED patterns of an (a) amorphous material revealing the characteristic halo ring pattern and (b) a diffraction pattern of a polycrystalline sample (nano-sized tetragonal ZrO_2 particles) with consecutive, continuous rings. Note that the distance of these rings from the origin can be used for the determination of interplanar spacings. In (c) a regular SAED image with distinct diffraction spots of a $Sr_2MgGe_2O_7$ (melilithe) single crystal is shown.

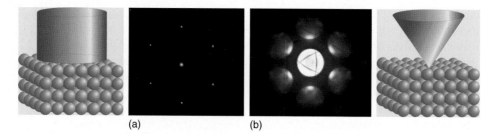

(a)   (b)

Figure 26.6 Comparison of (a) a SAED diffraction pattern and (b) a CBED pattern from a [0001] axis of Al$_2$O$_3$ taken at 200 kV (courtesy Dr. R. Schierholz). Note that (a) shows sixfold symmetry (Friedel's law), while the CBED image (b) reveals a trigonal symmetry within the 000-disk (center), which is consistent with the symmetry of alumina imaged along the [0001] zone axis. The schematic drawings show the corresponding convergence of the electron beam.

the three-dimensional crystal symmetry. Moreover, it allows distinction between, for example, trigonal and hexagonal symmetry (the distinction between α-Al$_2$O$_3$ with space group R-3m and α-SiC (6H polytype) with space group P6$_3$mc), which would not be possible employing conventional SAED patterns, since every SAED diffraction pattern is centrosymmetric (Laue symmetry), giving rise to a sixfold symmetry, even in the case of alumina. If the convergence angle (setting of condenser lens C2) is rather small (~10 mrad) the individual diffraction disks are still separated and such patterns are termed *Kossel-Möllenstedt diagrams*, while at higher angles the disks overlap, which is referred to as *Kossel diagrams*. It is beyond the scope of this introduction to TEM and related imaging techniques to describe CBED in depth; for further reading the reader is referred to corresponding textbooks [16–21].

Since TEM samples are electron transparent in thin regions, electrons that have been forward scattered can be collected. The high accelerating voltage results in a forward scattering with only a small fraction of backscattered electrons. Electrons are either *elastically* or *inelastically* scattered (particle description), but they can also be considered *coherent* or *incoherent* (wave description). Coherent electrons remain in phase after the scattering event, while incoherent electrons are out-of-phase upon scattering. Elastic scattering is usually coherent (thin specimen region) and occurs at low angles in the forward direction (1–10°). At higher angles, however, even elastic scattering becomes incoherent (Rutherford scattering). Inelastic scattering is nearly exclusively incoherent and occurs at very low scattering angles (<1°). In the following, the interaction of the electron with the crystalline matter during the scattering event is addressed in more detail.

The *atomic scattering factor*, $f_{el}(\theta)$, is a measure of the amplitude of an electron wave scattered from an isolated atom. For X-rays, the corresponding amplitude $f_x(\theta)$ can be derived by Equation 26.7:

$$f_x(\theta) = \int_{atom} \rho(r) \exp(-2\pi i\mathbf{K} \cdot \mathbf{r}) \, d\mathbf{r} \qquad (26.7)$$

where θ is the scattering angle, ρ the electron charge density in the crystal, $\mathbf{K}$ is the scattering vector in reciprocal space with $|\mathbf{K}| = 2\sin\theta/\lambda$, and $\mathbf{r}$ is the distance from the electron to the atom core. For electrons, the corresponding equation is (Equation 26.8):

$$f_{el}(\theta) = \frac{m_0 e^2}{2h^2}\left(\frac{\lambda}{\sin\theta}\right)^2 (Z - f_X(\theta)) \tag{26.8}$$

where h is the Planck's constant, m_0 the rest mass, e the electron charge, and Z the atomic number of the respective elements within the unit cell. Since X-rays only interact with the electron cloud, the second term of Equation 26.8 is related to scattering events from the electron cloud, while the first term is due to *Rutherford scattering* from the atomic nucleus. The *atomic scattering factors* for electrons of various elements are shown in Figure 26.7. It should be realized that electrons are scattered by orders of magnitude stronger than X-rays.

If we now consider the scattering event, not only from a single atom but from the entire unit cell of a crystalline material, the atomic contributions of all the atoms within the unit cell can simply be added up; however, taking the position of each atom relative to the origin of the unit cell into account via a phase factor, $2\pi i(\mathbf{K}\cdot\mathbf{r}_i)$, with $\mathbf{K}$ being the diffraction and $\mathbf{r}_i$ the lattice vector of the ith atom. The summation over the entire unit cell gives the so-called *structure factor*, $F(\theta)$, cf. Equation 26.9:

$$F(\theta) = \sum_i f_i(\theta)\exp(-2\pi i\mathbf{K}\cdot\mathbf{r}_i) \tag{26.9}$$

Considering electron diffraction patterns of crystalline matter, one is only interested in the value of $F(\theta)$ for a specific reciprocal lattice point and thus the scattering vector $\mathbf{K}$ can be replaced by $\mathbf{g}$, the *reciprocal lattice vector*. Hence, $F(\theta)$ becomes:

$$F_g = \sum_i f_i(\theta)\exp(-2\pi i\mathbf{g}\cdot\mathbf{r}_i) \tag{26.10}$$

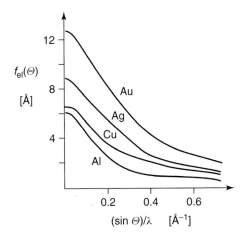

Figure 26.7 Atomic scattering factors of the elements Au, Ag, Cu, and Al as a function of $\sin\Theta/\lambda$.

Note that constructive interference only occurs if $\mathbf{g} \cdot \mathbf{r}_i = N$, with N being an integer. Here, $\mathbf{g}$ is the reciprocal lattice vector which is related to the interplanar spacings d_{hkl} via $d_{hkl} = |1/g|$. This criterion is termed the *Laue criterion* which is equivalent to *Bragg's Law*: $n\lambda = 2d \cdot \sin\theta$. When considering the interaction between the electron beam and matter from the perspective of the electron wave, it follows that the scattering potential P of the crystal (with n unit cells) can be expressed as [21]:

$$P(\mathbf{r}) = \frac{h^2}{2\pi m_0 e V_u} \sum_n F_g \exp(-2\pi i \mathbf{g} \cdot \mathbf{r}_n) \tag{26.11}$$

Here, $\mathbf{r}_n$ stands for the distance between the position of the n-th unit cell and the origin of the crystal. It is important to note that there is a quantity termed *extinction length*, ξ_g, which is closely related to the structure factor, F_g, which is also very important for contrast formation in TEM imaging; see also Figure 26.13; (Pendellösung). ξ_g is given by Equation 26.12:

$$F_g = \frac{\pi V_u \cos\theta}{\lambda \cdot \xi_g} \rightarrow \xi_g = \frac{\pi V_u \cos\theta}{\lambda \cdot F_g} \tag{26.12}$$

Here V_u is the volume of the unit cell and θ is the Bragg angle. The value of the extinction length strongly depends on the material of interest. For example, at 100 kV acceleration voltage, ξ_g for gold, germanium, silicon, and MgO is $\xi_{111(Au)} = 16$ nm, $\xi_{111(Ge)} = 43$ nm, $\xi_{111(Si)} = 60$ nm, and $\xi_{111(MgO)} = 273$ nm, respectively.

For rather thin crystals, the *kinematic theory* holds for the diffraction event considered, that is, only a single scattering event occurs. In this case, the intensity of a reflection, recorded in a diffraction pattern, is proportional to the square of the absolute value of the structure factor: $I_g \propto |F_g|^2$. In general, this approximation is not valid, but most of the contrast variations observed in TEM images can be described qualitatively by the kinematic approximation.

For the region of the sample illuminated by the electron beam, the *amplitude* of a diffracted beam, ϕ_g, is given by the sum over all contributing unit cells (n):

$$\phi_g = \sum_n F_g \exp(-2\pi i \mathbf{K} \cdot \mathbf{r}_n) \tag{26.13}$$

where F_g is the scattering amplitude from the n-th unit cell at the distance $\mathbf{r}_n$ from the origin of the crystal ($\mathbf{r}_n = x_1\mathbf{a} + x_2\mathbf{b} + x_3\mathbf{c}$, with x_1, x_2, and x_3 being integers). Note that, due to translational symmetry, all unit cells of the crystal are considered identical. Moreover, in order to be able to describe diffraction phenomena observed in TEM from crystals with finite dimension, it has to be allowed that the diffraction vector $\mathbf{g}_{(hkl)}$ does not necessarily lie exactly on the *Ewald sphere*, but can be slightly away from it, still giving some intensity. This specific situation is described by the so-called *excitation error*, $\mathbf{s}$, as shown schematically in Figure 26.8.

Therefore, one can define the actual scattering vector $\mathbf{K}$ as $\mathbf{K} = \mathbf{g} + \mathbf{s}$, describing the deviation of the scattering vector $\mathbf{g}$ from the exact position on the Ewald sphere (with a small s-value). Thus, when the crystal is slowly tilted during the experiment, the reciprocal space sweeps through the Ewald sphere from $-\mathbf{s}$ to $+\mathbf{s}$ (or vice versa); $\mathbf{s}$ is negative if $\mathbf{g}$ lies outside the Ewald sphere, as in Figure 26.8, and is

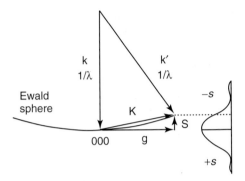

Figure 26.8 Schematic diagram showing the scattering geometry for a thin foil normal to the incident electron beam.
The graph reveals the intensity variation of the diffracted
beam depending on the extinction error ($I = |\varphi_g|^2$).

positive when it is inside. Note that the sign of **s** can be determined experimentally
by employing *Kikuchi lines*, which are generated in a similar way as in the SEM;
cf. *electron backscattered diffraction* (EBSD). Taking the excitation error into account,
and since all terms of $\mathbf{g} \cdot \mathbf{r}_n$ are integers, Equation 26.13 can be rearranged, yielding
for the amplitude of the diffracted beam, given by Equation 26.14:

$$\phi_g = \sum_n F_g \exp(-2\pi i(\mathbf{g} + \mathbf{s}) \cdot \mathbf{r}_n) = \sum_n F_g \exp(-2\pi i(\mathbf{s} \cdot \mathbf{r}_n)) \qquad (26.14)$$

Both **s** and $\mathbf{r}_n$ can be expressed as a lattice vector $\mathbf{r}_n = x\mathbf{a} + y\mathbf{b} + z\mathbf{c}$ and the
excitation error $\mathbf{s} = u\mathbf{a}^* + v\mathbf{b}^* + w\mathbf{c}^*$ (reciprocal space). Under the approximation
that the crystal studied contains a rather large number of unit cells n within
the illuminated volume V_C ($n_x \cdot n_y \cdot n_z$ along the x-, y-, and z-directions and with
$A = n_x \cdot \mathbf{a}$, etc.), the sum in Equation 26.14 can be expressed as a triple integral,
leading to Equation 26.15:

$$\phi_g = \frac{F_g}{V_C} \int_0^A \int_0^B \int_0^C \exp(-2\pi i(ux + vy + wz)) \, dxdydz \qquad (26.15)$$

The integral in Equation 26.15 can be rewritten as shown in Equation 26.16:

$$|\phi_g| = \frac{F_g}{V_C} \cdot \frac{(\sin \pi Au)}{(\pi u)} \cdot \frac{(\sin \pi Bv)}{(\pi v)} \cdot \frac{(\sin \pi Cw)}{(\pi w)} \qquad (26.16)$$

where u, v, w are the components of **s** along the x, y, and z directions. The intensity
distribution is proportional to the square of the absolute amplitude value of the
diffracted beam $I = |\phi_g|^2$, while the resulting shape of the diffraction spot is also
related to a $\sin^2$-function, as shown schematically in Figure 26.9 (see also Equations
26.17 and 26.18).

In TEM, the samples are typically thin foils and can be described as a paral-
lelepiped (A · B · C) with C of thickness t (20–50 nm) and A and B being much

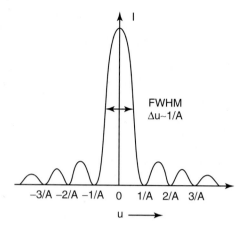

Figure 26.9 Schematic showing the intensity distribution along u for $v = w = 0$. The full width half maximum (FWHM) $\Delta u \sim 1/A$ is indicated (with $A = n_x \cdot a$).

larger. Then, the intensity of the diffracted beam can be expressed for such a thin foil of thickness t and the excitation error s, being perpendicular to the crystal surface and also to $\mathbf{g}$, as given in Equation 26.17 (with $|\mathbf{k}| = 1/\lambda$ being the diameter of the Ewald sphere):

$$I_g(s, t) = \frac{F_g^2}{k^2 V_C^2 \cos^2 \theta} \frac{\sin^2 \pi ts}{(\pi s)^2} \tag{26.17}$$

Equation 26.17 can be re-expressed in terms of the extinction length, ξ_g, as shown in Equation 26.18

$$I_g(s, t) = |\phi_g|^2 = \left(\frac{\pi t}{\xi_g}\right)^2 \frac{\sin^2 \pi ts_{\text{eff}}}{(\pi s_{\text{eff}})^2} \quad \text{with } s_{\text{eff}} = \sqrt{s^2 + \frac{1}{\xi_g^2}} \tag{26.18}$$

For kinematic diffraction or the so-called *two-beam case*, where only one single reflection is very strong, it follows under the assumption $I_0 \approx I_g$ and the total intensity equals $I_{\text{tot}} = I_0 + I_g = 1$ that $I_0 = (1 - I_g)$. Equation 26.18 gives the intensity of the Bragg-diffracted beam and is a very important equation in interpreting contrast variations in TEM images such as *bending contours* (fixed t, changing s_{eff}) and *thickness fringes* (fixed s_{eff}, changing t). It should be emphasized however that the kinematic approach breaks down when thicker sample regions are investigated.

The *dynamical theory of diffraction* considers the wave field in the periodic potential of the crystal and takes into account all *multiple scattering events*. In contrast to kinematic diffraction, which describes the intensity of diffraction peaks (Bragg peaks) in reciprocal space, under the assumption of a single scattering event, dynamical theory includes extinction and interference effects as well as the shape and width of the peaks. Here, *dispersion surfaces* are used to graphically describe dynamical scattering events around reciprocal lattice points [22–24].

The technique of *precession electron diffraction* or simply electron precession was first developed by Vincent and Midgley in Bristol in the early 1990s [25]. In recent years, this technique has gained great attention worldwide, since commercial precession systems have now become available. In electron precession, the electron beam is rocked in a hollow cone above the sample surface and de-rocked again below the specimen. Here, similar to SAED and μ-diffraction, if a parallel illumination is used, the precession pattern is composed of sharp spots while small disks are seen with a convergent beam. If the area of interest is aligned so that the unprecessed beam is parallel to a zone axis, then the precession results in an apparent zone axis pattern which displays the same general symmetry elements as the aligned crystal; however, with this setting, the electron beam is never parallel to the specific zone axis. As a result, dynamical effects and, in particular, *multi-beam interactions* are greatly minimized or even suppressed. Therefore, such electron precession patterns reveal integrated intensities and allow a more quantitative analysis of diffraction intensities and, consequently, are expected to be established as a potential routine for crystal structure determination via electron diffraction, even on the nanometer scale.

26.5
Image Contrast

There are two basic mechanisms concerning the formation of an image in a TEM: (i) for low to intermediate magnifications, approximately up to 150 k, the so-called *amplitude contrast* is effective. At higher magnifications, the *phase contrast*, which will be discussed in more detail in Section 26.6, becomes the dominant image forming mechanism. In conventional TEM, *BF imaging* is typically used to image the overall microstructure of the sample. Here, the incident electron beam is selected by the objective aperture (cf. Figure 26.10a), which is positioned in the back focal plane of the objective lens. The objective aperture is adjustable and can be shifted so that either the direct beam or one diffracted beam is selected for image formation.

As we will see later, any image is, in general, formed with a selected number of electron beams (often just one), using the objective (contrast) aperture. Only those beams that are allowed to pass the aperture will contribute to the image. The conventional *DF imaging* (see Figure 26.10b) uses a beam which is away from the optic axis, thus increasing the contribution of lens aberrations to the image. To avoid such image distortions, one can tilt the incident beam so that now the diffracted beam runs along the optic axis of the microscope, generating an *on-axis DF* image (Figure 26.10c) with a minimized effect of lens aberrations. Figure 26.11 reveals a pair of BF and DF images of plate-like precipitates of the theta-phase (Al_2Cu) embedded in an Al single crystal. Note that the precipitates are aligned parallel to the {100} facets of the Al crystal.

A special imaging mode, which is analogous to the DF mode, is the *weak beam DF*. Here, a special set-up for the diffracted beam contributing to the image is

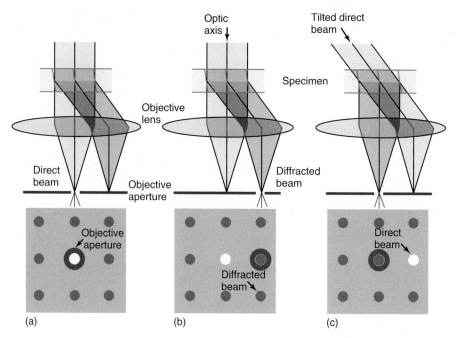

(a)   (b)   (c)

Figure 26.10 Imaging techniques using amplitude contrast by selecting the direct or one diffracted beam; (a) bright field (BF) imaging using the direct beam, (b) dark field (DF) imaging utilizing one diffracted beam away from the optic axis, and (c) on-axis dark field imaging, where the incident beam is tilted to shift the diffracted beam onto the optic axis to minimize the effect of lens aberrations.

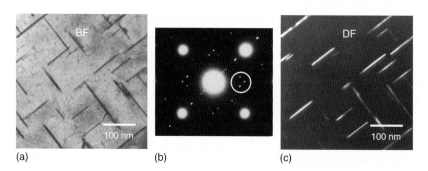

(a)   (b)   (c)

Figure 26.11 TEM images of thin platelet-like precipitations of theta (Al_2Cu) in an Al single crystal; (a) bright-field image and (c) corresponding dark-field image. The central image (b) shows the SAED pattern with the diffraction spots used for dark-field imaging indicated (circle).

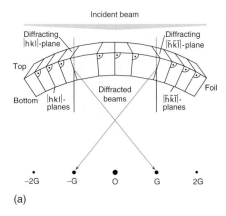

(a) (b)

Figure 26.12 (a) Schematic showing the scattering event from a bend specimen. (b) TEM bright-field image revealing dark contrast lines, so-called bend contours, originating from the buckled nature of the thin TEM sample (Al foil). Here, G represents a reciprocal lattice point (not a reciprocal lattice vector (**g**)).

chosen. This mode is particularly useful for imaging line or planar defects in materials such as dislocations or antiphase boundaries. For a detailed description of this imaging technique the reader is referred to textbooks [2, 3, 22–24].

Commonly observed *contrast variations* in TEM images are due to the fact that the TEM thin foils are not perfectly flat. If over a particular region of the specimen the thickness is reasonably constant, but the specimen is bent or buckled, for example, like a saucer or saddle, then *bend contours* arise, as depicted in Figure 26.12.

Another often observed contrast in TEM micrographs is *thickness fringes*, which are a consequence of TEM specimen preparation via Ar^+-ion thinning or tripod polishing, creating a shallow wedge at the edge of the thin foil. Solving the Equations for the amplitude of the direct and one strongly diffracted beam ($I_0 \approx I_g$) reveals that the intensities of the diffracted and direct beam are coupled and oscillate, depending on the excitation length ξ (Equation 26.12), while the beam travels through the sample in the z-direction. The variation in beam amplitude ($\phi_g =$ diffracted beam, $\phi_o =$ direct beam) is described by the *Howie Whelan equations* (Equation 26.19) also termed *Pendellösung* [22–24].

$$\frac{d\phi_g}{dz} = \frac{\pi i}{\xi_0}\phi_g + \frac{\pi i}{\xi_g}\phi_0 e^{-2\pi i s z} \quad \text{and} \quad \frac{d\phi_0}{dz} = \frac{\pi i}{\xi_0}\phi_0 + \frac{\pi i}{\xi_g}\phi_g e^{2\pi i s z} \qquad (26.19)$$

A more detailed examination of these Equations shows that the amplitude of the beams $I_0(s,t)$ and $I_g(s,t)$ oscillates as a function of thickness t and excitation error s_z. As a consequence, alternating bright and dark lines that run parallel to the edge of the sample are seen in the TEM micrograph, as illustrated in Figure 26.13.

Besides materials for the semiconductor industry (e.g., silicon single crystals) synthetic or natural crystals are far from being perfect. These imperfections are due to intrinsic *crystal defects* such as individual point defects, which cannot be imaged

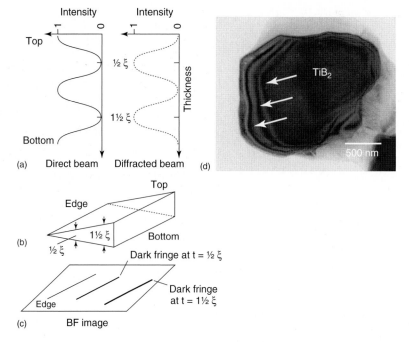

Figure 26.13 Schematic of thickness fringes, (a) for two beam condition ($s_z = 0$), the normalized intensities of the direct and diffracted beam oscillate in a complementary way (here s_z is fixed and t varies), (b) for a wedge specimen, the fringe separation in (c) is determined by the wedge angle and the extinction distance ξ_g. (d) Bright field TEM image showing thickness fringes in a TiB_2 precipitate (marked by arrows).

in a TEM, line defects such as *edge* or *screw dislocations* and planar defects such as *anti-phase domain boundaries* or *stacking faults*. The line and planar defects can be described in a crystallographic way by including a displacement vector **R** and the reciprocal lattice vector **g**, defining the displacement of an atom at any point of the lattice with respect to the original lattice. The Howie–Whelan Equations can then be rewritten as:

$$\frac{d\phi_g}{dz} = \frac{\pi i}{\xi_0}\phi_g + \frac{\pi i}{\xi_g}\phi_0 e^{-2\pi i(s_z + g\cdot R)} \quad \text{and} \quad \frac{d\phi_0}{dz} = \frac{\pi i}{\xi_0}\phi_0 + \frac{\pi i}{\xi_g}\phi_g e^{2\pi i(s_z + g\cdot R)} \quad (26.20)$$

If the displacement vector of a specific defect is known, these equations allow the theoretical prediction of contrast effects originating from local crystal defects.

26.6
High Resolution Imaging

According to the derivations of Buseck *et al.* [26], Spence [27], and Williams and Carter [3], the two-dimensional projection of the spatial information of a sample onto a screen or a CCD camera (imaging process) can mathematically be described by the following image function g(r):

$$g(r) = \int f(r')h(r - r')dr' = f(r) \otimes h(r - r') \tag{26.21}$$

where $f(r) = f(x, y)$ is the *specimen function*, which describes the spatial distribution of information within the specimen, convoluted (multiplied and integrated) with $h(r - r')$, which is the *point spread function* taking into account that, due to lens imperfections, a point in the specimen becomes a disk in the projected image. The specimen function $f(x,y)$ can be expressed as shown in Equation 26.22:

$$f(x, y) = A(x, y) \exp(-i\phi_t(x, y)) \tag{26.22}$$

Here, the term $A(x,y)$ is related to the amplitude of the specimen function, while the second term describes the phase shift of the electron wave passing through the specimen. Operating the TEM in *high resolution* means that the contrast in the image is not created by the exclusion of strongly scattered electrons via the insertion of an objective aperture below the sample (which would cause an amplitude contrast), but by the phase shift caused by the specimen itself. Therefore, the term $A(x,y)$ in Equation 26.22 can be set to 1, which greatly simplifies the specimen function. The phase shift ϕ_t in the second term of Equation 26.22 takes into account the changes in electron phase while the incident beam passes through the projected potential $V_t(x,y)$. Equation 26.22 can be rearranged to:

$$f(x, y) = \exp\left[i\sigma V_t(x, y) - \mu(x, y)\right] \tag{26.23}$$

with $\sigma = \pi/\lambda E$ being the interaction constant. Equation 26.23 includes the model that the specimen can be described as an independent phase object generating the phase shift given in Equation 26.22, which is termed the *phase object approximation*. The right term in Equation 26.23 $\mu(x,y)$ takes the absorption into account while the electron beam passes through the volume V_t. In the case of very thin samples (with $V_t \ll 1$) this approximation can be simplified to the so-called *weak phase object approximation* (WPOA) neglecting the absorption contribution:

$$f(x, y) = 1 - i\sigma V_t(x, y) \tag{26.24}$$

Including this term into Equation 26.21 (with $h(r - r')$ expressed as $h(x, y) = \cos(x, y) + i \cdot \sin(x, y)$) and doing some rearrangement [3], the observed intensity distribution within the projected high-resolution image is given by:

$$I = 1 + 2\sigma V_t(x, y)A\sin(x, y) \tag{26.25}$$

The intensity distribution given in Equation 26.25 is combined with the damping property of the lens system itself expressed by the envelope function $E(u)$ and the influence of the apertures used, accounted for by the aperture function $A(u)$. The information being transferred to the image is then described by the *objective lens transfer function*, with **u** being a reciprocal lattice vector, the spatial frequency for a particular direction. A multiplication and integration in realspace (convolution) (Equation 26.25) gives a simple multiplication of functions in reciprocal space (Equation 26.26):

$$T(u) = E(u)A(u)2\sin\chi(u) \tag{26.26}$$

$\chi(u)$ describes the phase of the specimen and is related to the defocus Δf (of the objective lens), the spherical aberration C_s, the electron wavelength λ, and the spatial frequency u by:

$$\chi = \pi \Delta f \lambda u^2 + \tfrac{1}{2} \pi C_s \lambda^3 u^4 \tag{26.27}$$

Figure 26.14 shows a plot of $\sin \chi(u)$ versus u, which highlights the correlation between the objective lens transfer function and the information obtained from a high-resolution TEM micrograph.

The *zero order pass band*, the area from $u = 0$ to the first crossover of $\sin \chi(u)$ with the zero line (Figure 26.14), is generally used to define the resolution limit d_{min}, which is the value for $1/u$ at $\sin \chi(u) = 0$ in Å. Please note that information is transferred up to much higher spatial frequencies, but the interpretation of these contrasts needs computational support. In non C_s-corrected microscopes, the spherical aberration is constant and the wavelength controlled by the high tension can, as a first approximation, also be considered as constant. The value of $\sin \chi(u)$ is therefore strictly coupled to the defocus Δf of the objective lens (Equation 26.27). Otto Scherzer noticed that the image distortion introduced by the spherical aberration can be somewhat balanced by defocusing the objective lens. The defocus value giving the optimum resolution is named after him:

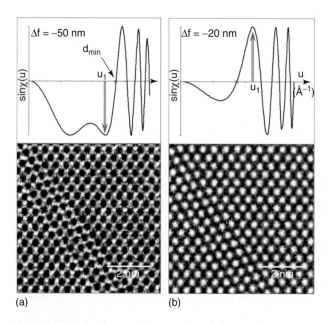

(a)   (b)

Figure 26.14 The plot $\sin \chi(u)$ versus u and the corresponding HRTEM micrographs of a boron suboxide single crystal imaged at different defocus settings: (a) Scherzer defocus ($\Delta f = -50$ nm) and (b) at a defocus value of -20 nm ($E_o = 200$ keV, $C_s = 2.3$ mm).

Scherzer defocus [28]:

$$\Delta f_{\text{Scherzer}} = -1.2(C_s\lambda)^{\frac{1}{2}} \tag{26.28}$$

The corresponding resolution limit at Scherzer defocus is:

$$d_{\min} = 0.66 \left(C_S\lambda^3\right)^{\frac{1}{4}} \tag{26.29}$$

Summarizing, a high-resolution TEM image of a specimen is created by a phase shift of the incident electron beam passing through the observed sample volume, magnified by the objective lens. The resolution is limited by the electron wavelength and the aberrations of the lens system. The thickness of the specimen correlates with the observed volume, which should be as small as possible, in order to minimize dynamic effects and ensure that the weak phase object assumption is valid, thus making HRTEM images interpretable within the zero order pass band.

The lens transfer function, shown in Figure 26.14, reveals maxima and minima that correlate with the contrast transferred to the image plane. If $T(u)$ is negative, a positive phase contrast is seen, meaning that atoms appear dark against a bright background. If $T(u) = 0$, no contrast appears for this spatial frequency, regardless of whether there is an atom or not. If $T(u)$ is positive, negative contrast is observed (atoms bright on a dark background). Due to the oscillation of the lens transfer function, both negative and positive contrast will contribute to the final image. Micrographs recorded under such conditions are difficult to interpret intuitively with respect to the real structure, that is, to the atom positions. Therefore, *image simulation* is required to interpret these HRTEM micrographs (see Figure 26.15).

For this purpose, several parameters under which the HRTEM micrographs were recorded have to be known. To compare the real images with the simulated ones, at least the thickness of the specimen, the exact defocus value and the spherical aberration of the objective lens are required. The most common method for image

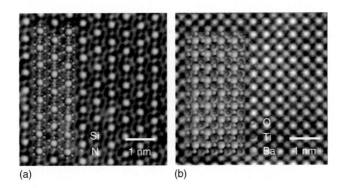

(a) (b)

Figure 26.15 HRTEM micrographs of (a) Si_3N_4 and (b) $BaTiO_3$ with the real atomic positions shown as inset. The simulation reveals that depending on the defocus, positive phase contrast can occur for (a) intrinsic channels in the structure or (b) real atom positions.

simulation is the *multi-slice approach* [29–31]. Here, a virtual specimen is segmented into several slices, whereby the scattering event of the electron beam is calculated for each slice. The electron scattering can be simulated by different methods, such as the reciprocal space or real space approach, the Bloch wave approach or the FFT formalism. In the resulting image, the atom positions are known, since they are the basis of the image simulation, and the simulated contrast can be compared directly with the experimental HRTEM micrograph. Image simulation is a powerful tool to correlate the observed contrast variations in the micrograph to the real structure of the material under investigation.

26.7
Summary

This brief outline on TEM techniques can only provide a "taste" of what is nowadays possible with modern instrumentation. For further reading the reader is referred to the corresponding textbooks given below. However, what should have become transparent is the fact that TEM is a very versatile and unique technique, which allows detailed microstructural characterization on the atomic level. With the most recent developments that enable the correction of lens aberrations (spherical and chromatic), delocalization free imaging, in addition to analysis of single atom columns, has become available; clearly a quantum jump in TEM.

References

1. Goodhew, P.J., Humphreys, J., and Beanland, R. (2000) *Electron Microscopy and Analysis*, 3rd edn, Taylor & Francis.
2. De Graef, M. (2003) *Introduction to Conventional Transmission Electron Microscopy*, Cambridge University Press.
3. Williams, D.B. and Carter, C.B. (2009) *Transmission Electron Microscopy*, 2nd edn, Springer-Verlag.
4. Goldstein, J., Joy, D.C., and Romig, A.D. Jr. (eds) (1986) *Principles of Analytical Electron Microscopy*, Springer-Verlag.
5. Echlin, P., Fiori, C.E., Goldstein, J., Joy, D.C., and Newbury, D.E. (1986) *Advanced Scanning Electron Microscopy and X-Ray Microanalysis*, Plenum Press.
6. Flegler, S.L., Heckman, J.W. Jr., and Klomparens, K.L. (1993) *Scanning and Transmission Electron Microscopy: An Introduction*, Oxford University Press.
7. Egerton, R. (1996) *Electron Energy-Loss Spectroscopy in the Electron Microscope*, Springer-Verlag.
8. Brydson, R. (2001) *Electron Energy Loss Spectroscopy*, Garland Science.
9. Goldstein, J., Newbury, D.E., Joy, D.C., Lyman, C.E., Echlin, P., Lifshin, E., Sawyer, L., and Michael, J.R. (2003) *Scanning Electron Microscopy and X-Ray Microanalysis*, Springer-Verlag.
10. Reed, S.J.B. (2010) *Electron Microprobe Analysis and Scanning Electron Microscopy in Geology*, Cambridge University Press.
11. Scherzer, O. (1947) *Optik*, **2**, 114–132.
12. Rose, H. (1990) *Optik*, **85**, 19–24.
13. Haider, M., Uhlemann, S., Schwan, E., Rose, H., Kabius, B., and Urban, K. (1998) *Nature*, **392**, 768–769.
14. Erni, R. (2010) *Aberration-Corrected Imaging in Transmission Electron Microscopy: An Introduction*, Imperial College Press.
15. (1999) *Metallographic Etching: Techniques for Metallography, Ceramography, Plastography*, American Society for Metals, Metal Park, OH.

16. Edington, J.W. (1975) *Electron Diffraction in the Electron Microscope*, Macmillan.

17. Spence, J.C.H. and Zuo, J.M. (1992) *Electron Microdiffraction*, Springer-Verlag, Berlin.

18. Loretto, M.H. (1994) *Electron Beam Analysis of Materials*, 2nd edn, Chapman & Hall.

19. Fultz, B. and Howe, J.M. (2007) *Transmission Electron Microscopy and Diffractometry of Materials*, Springer.

20. Reimer, L. and Kohl, H. (2008) *Transmission Electron Microscopy: Physics of Image Formation*, Springer-Verlag.

21. Hirsch, P.B. (1999) *Topics in Electron Diffraction and Microscopy of Materials*, IOP Publishing Ltd, Bristol.

22. Hirsch, P.B., Howie, A., Nicholson, R.B., Pashley, D.W., and Whelan, M.J. (1965) *Electron Microscopy of Thin Crystals*, Butterworths, London.

23. Egerton, R. (2005) *Physical Principles of Electron Microscopy: An Introduction to TEM, SEM, and AEM*, Springer-Verlag.

24. Reimer, L. (2009) *Transmission Electron Microscopy: Physics of Image Formation and Microanalysis*, Springer, New York.

25. Vincent, R. and Midgley, P.A. (1994) *Ultramicroscopy*, **53**, 271–282.

26. Buseck, P., Cowley, J., and Eyring, L. (1988) *High-Resolution Transmission Electron Microscopy and Associated Techniques*, Oxford University Press.

27. Spence, J.C.H. (2009) *High-Resolution Electron Microscopy*, Oxford University Press.

28. Scherzer, O. (1949) *J. Appl. Phys.*, **20**, 20–29.

29. Stadelmann, P.A. (1979) *Ultramicroscopy*, **21**, 131–145.

30. Crowley, J.M. and Moodie, A.F. (1957) *Acta Crystallogr.*, **10**, 609–619.

31. Kirkland, E.J. (1998) *Advanced Computing in Electron Microscopy*, Springer, New York.

Appendix

Substances and samples

a

acetamide, threshold ionization and mass spectrometry 74, 75

acetone, strong field fragmentation 44

acetonitrile, CARS 655–657

activated carbon, DRIFTS 470–472

adsorbates
- adsorbates on Ni, TOF-SIMS 556–557
- adsorption on nano-ZrO_2, NMR 142, 144
- disordered layers, LEED-IV 639–640
- CO on Ni(111), LEED 634–636, 638–639
- Cu_2S on TiO_2, XPS 490–491
- In on Ni(111), PAC 343–344
- NO on CeO_2, DRIFTS 464–467
- NO/O_2 on CeO_2, DRIFTS 467–470
- NO + H_2 on Rh(110), PEEM, SPEM, LEED 526–534
- $PdCl_4^{2-}$ adlayer on Au(110), STM 613–614
- octadecanethiol on Au, EQCM 593–594
- oxygen and carbon on CdTe, XPS 484–485
- water adsorption on GaAs(110), UPS 504–505

Al (foil), TEM 815

Al_2O_3, SAED and CBED 807–808

Al_2Cu in Al single crystal, TEM 814

Ag(100), PAC 341–342

Ag (low-index faces), STM 610, 611

Ag deposition on Au(111), STM 615–617

Ag electrode (oxidation/reduction), EQCM and cyclic voltammogram 585–586, 592–593

$Ag_7P_3S_{11}$ (solid ion conductor), NMR 112–115

Al (metal surface), PEEM 540

aminophenol, rotational states and conformer selection 19–21

ammonia isotopomers, REMPI spectra 36–37

ammonia, femtosecond-MIR-spectroscopy 674–677

amorphous B_2O_3, conductivity 225–226

amorphous materials, XRD and SAED 287–288, 807

anthracene, THz time domain spectra 694

aromatic compounds, trace analysis 50, 51

aromatic π-systems (liquid crystalline phases), solid state NMR 139–141

Au(100), STM 611–613

azide anions, femtosecond infrared spectroscopy 650–651

b

backspillover oxygen 535

$BaTiO_3$ (single crystal), HRTEM 819

BeB_2C_2, ELNES 413–415

benzene isotopomers, Raman spectroscopy 426

benzonitrile, Stark effect and photoelectron spectroscopy 9, 10, 22, 23

biomolecules
- bacteriorhodopsin, THz emission 709
- double-stranded DNA, FRET and force spectroscopy 727, 741–742, 748–749
- ferritin, AC magnetometry 783–786
- five-helix bundle protein, vibrational density of states 698
- glycine, MAS NMR 105, 106
- hybridized oligonucleotides, biosensors 752, 753
- membrane-bound protein cluster 263
- membrane proteins, EMR 180
- nickel enzyme (catalyzing methane), EPR 179, 180
- oxygen-evolving complex (OEC), EXAFS 264–266

Methods in Physical Chemistry, First Edition. Edited by Rolf Schäfer and Peter C. Schmidt.
© 2012 Wiley-VCH Verlag GmbH & Co. KGaA. Published 2012 by Wiley-VCH Verlag GmbH & Co. KGaA.

Index

Methods in Physical Chemistry, First Edition. Edited by Rolf Schäfer and Peter C. Schmidt.
© 2012 Wiley-VCH Verlag GmbH & Co. KGaA. Published 2012 by Wiley-VCH Verlag GmbH & Co. KGaA.